LONDON MATHEMATICAL SOCIETY LECTURE NOTE SERIES

The books in the series listed below are available from booksellers, or, in case of difficulty, from Cambridge University Press.

34 Representation theory of Lie groups, M.F. ATIYAH *et al*
36 Homological group theory, C.T.C. WALL (ed)
39 Affine sets and affine groups, D.G. NORTHCOTT
40 Introduction to H_p spaces, P.J. KOOSIS
46 p-adic analysis: a short course on recent work, N. KOBLITZ
49 Finite geometries and designs, P. CAMERON, J.W.P. HIRSCHFELD & D.R. HUGHES (eds)
50 Commutator calculus and groups of homotopy classes, H.J. BAUES
57 Techniques of geometric topology, R.A. FENN
59 Applicable differential geometry, M. CRAMPIN & F.A.E. PIRANI
66 Several complex variables and complex manifolds II, M.J. FIELD
69 Representation theory, I.M. GELFAND *et al*
74 Symmetric designs: an algebraic approach, E.S. LANDER
76 Spectral theory of linear differential operators and comparison algebras, H.O. CORDES
77 Isolated singular points on complete intersections, E.J.N. LOOIJENGA
79 Probability, statistics and analysis, J.F.C. KINGMAN & G.E.H. REUTER (eds)
80 Introduction to the representation theory of compact and locally compact groups, A. ROBERT
81 Skew fields, P.K. DRAXL
82 Surveys in combinatorics, E.K. LLOYD (ed)
83 Homogeneous structures on Riemannian manifolds, F. TRICERRI & L. VANHECKE
86 Topological topics, I.M. JAMES (ed)
87 Surveys in set theory, A.R.D. MATHIAS (ed)
88 FPF ring theory, C. FAITH & S. PAGE
89 An F-space sampler, N.J. KALTON, N.T. PECK & J.W. ROBERTS
90 Polytopes and symmetry, S.A. ROBERTSON
91 Classgroups of group rings, M.J. TAYLOR
92 Representation of rings over skew fields, A.H. SCHOFIELD
93 Aspects of topology, I.M. JAMES & E.H. KRONHEIMER (eds)
94 Representations of general linear groups, G.D. JAMES
95 Low-dimensional topology 1982, R.A. FENN (ed)
96 Diophantine equations over function fields, R.C. MASON
97 Varieties of constructive mathematics, D.S. BRIDGES & F. RICHMAN
98 Localization in Noetherian rings, A.V. JATEGAONKAR
99 Methods of differential geometry in algebraic topology, M. KAROUBI & C. LERUSTE
100 Stopping time techniques for analysts and probabilists, L. EGGHE
101 Groups and geometry, ROGER C. LYNDON
103 Surveys in combinatorics 1985, I. ANDERSON (ed)
104 Elliptic structures on 3-manifolds, C.B. THOMAS
105 A local spectral theory for closed operators, I. ERDELYI & WANG SHENGWANG
106 Syzygies, E.G. EVANS & P. GRIFFITH
107 Compactification of Siegel moduli schemes, C-L. CHAI
108 Some topics in graph theory, H.P. YAP
109 Diophantine analysis, J. LOXTON & A. VAN DER POORTEN (eds)
110 An introduction to surreal numbers, H. GONSHOR
111 Analytical and geometric aspects of hyperbolic space, D.B.A. EPSTEIN (ed)
113 Lectures on the asymptotic theory of ideals, D. REES
114 Lectures on Bochner-Riesz means, K.M. DAVIS & Y-C. CHANG
115 An introduction to independence for analysts, H.G. DALES & W.H. WOODIN
116 Representations of algebras, P.J. WEBB (ed)
117 Homotopy theory, E. REES & J.D.S. JONES (eds)
118 Skew linear groups, M. SHIRVANI & B. WEHRFRITZ
119 Triangulated categories in the representation theory of finite-dimensional algebras, D. HAPPEL
121 Proceedings of *Groups - St Andrews 1985*, E. ROBERTSON & C. CAMPBELL (eds)

122 Non-classical continuum mechanics, R.J. KNOPS & A.A. LACEY (eds)
124 Lie groupoids and Lie algebroids in differential geometry, K. MACKENZIE
125 Commutator theory for congruence modular varieties, R. FREESE & R. MCKENZIE
126 Van der Corput's method of exponential sums, S.W. GRAHAM & G. KOLESNIK
127 New directions in dynamical systems, T.J. BEDFORD & J.W. SWIFT (eds)
128 Descriptive set theory and the structure of sets of uniqueness, A.S. KECHRIS & A. LOUVEAU
129 The subgroup structure of the finite classical groups, P.B. KLEIDMAN & M.W.LIEBECK
130 Model theory and modules, M. PREST
131 Algebraic, extremal & metric combinatorics, M-M. DEZA, P. FRANKL & I.G. ROSENBERG (eds)
132 Whitehead groups of finite groups, ROBERT OLIVER
133 Linear algebraic monoids, MOHAN S. PUTCHA
134 Number theory and dynamical systems, M. DODSON & J. VICKERS (eds)
135 Operator algebras and applications, 1, D. EVANS & M. TAKESAKI (eds)
136 Operator algebras and applications, 2, D. EVANS & M. TAKESAKI (eds)
137 Analysis at Urbana, I, E. BERKSON, T. PECK, & J. UHL (eds)
138 Analysis at Urbana, II, E. BERKSON, T. PECK, & J. UHL (eds)
139 Advances in homotopy theory, S. SALAMON, B. STEER & W. SUTHERLAND (eds)
140 Geometric aspects of Banach spaces, E.M. PEINADOR and A. RODES (eds)
141 Surveys in combinatorics 1989, J. SIEMONS (ed)
142 The geometry of jet bundles, D.J. SAUNDERS
143 The ergodic theory of discrete groups, PETER J. NICHOLLS
144 Introduction to uniform spaces, I.M. JAMES
145 Homological questions in local algebra, JAN R. STROOKER
146 Cohen-Macaulay modules over Cohen-Macaulay rings, Y. YOSHINO
147 Continuous and discrete modules, S.H. MOHAMED & B.J. MÜLLER
148 Helices and vector bundles, A.N. RUDAKOV et al
149 Solitons, nonlinear evolution equations and inverse scattering, M.J. ABLOWITZ & P.A. CLARKSON
150 Geometry of low-dimensional manifolds 1, S. DONALDSON & C.B. THOMAS (eds)
151 Geometry of low-dimensional manifolds 2, S. DONALDSON & C.B. THOMAS (eds)
152 Oligomorphic permutation groups, P. CAMERON
153 L-functions and arithmetic, J. COATES & M.J. TAYLOR (eds)
154 Number theory and cryptography, J. LOXTON (ed)
155 Classification theories of polarized varieties, TAKAO FUJITA
156 Twistors in mathematics and physics, T.N. BAILEY & R.J. BASTON (eds)
157 Analytic pro-*p* groups, J.D. DIXON, M.P.F. DU SAUTOY, A. MANN & D. SEGAL
158 Geometry of Banach spaces, P.F.X. MÜLLER & W. SCHACHERMAYER (eds)
159 Groups St Andrews 1989 volume 1, C.M. CAMPBELL & E.F. ROBERTSON (eds)
160 Groups St Andrews 1989 volume 2, C.M. CAMPBELL & E.F. ROBERTSON (eds)
161 Lectures on block theory, BURKHARD KÜLSHAMMER
162 Harmonic analysis and representation theory for groups acting on homogeneous trees, A. FIGA-TALAMANCA & C. NEBBIA
163 Topics in varieties of group representations, S.M. VOVSI
164 Quasi-symmetric Designs, M.S. SHRIKANDE & S.S. SANE
166 Surveys in combinatorics, 1991, A.D. KEEDWELL (ed)
167 Stochastic analysis, M.T. BARLOW & N.H. BINGHAM (eds)
174 Lectures on mechanics, J.E. MARSDEN
175 Adams memorial symposium on algebraic topology 1, N. RAY & G. WALKER (eds)
176 Adams memorial symposium on algebraic topology 2, N. RAY & G. WALKER (eds)

London Mathematical Society Lecture Note Series. 176

Adams Memorial Symposium on Algebraic Topology: 2

Manchester 1990

Edited by
N. Ray and G. Walker
Department of Mathematics, University of Manchester

CAMBRIDGE
UNIVERSITY PRESS

Published by the Press Syndicate of the University of Cambridge
The Pitt Building, Trumpington Street, Cambridge CB2 1RP
40 West 20th Street, New York, NY 10011-4211, USA
10 Stamford Road, Oakleigh, Victoria 3166, Australia

First published 1992

Library of Congress cataloguing in publication data available

British Library cataloguing in publication data available

ISBN 0 521 42153 5

Transferred to digital printing 2004

Contents of Volume 2

Preface

The international Symposium on algebraic topology which was held in Manchester in July 1990 was originally conceived as a tribute to Frank Adams by mathematicians in many countries who admired and had been influenced by his work and leadership. Preparations for the meeting, including invitations to the principal speakers, were already well advanced at the time of his tragic death in a car accident on 7 January 1989, at the age of 58 and still at the height of his powers.

Those members of the Symposium, and readers of these volumes, who had the good fortune to know Frank as a colleague, teacher and friend will need no introduction here to the qualities of his intellect and personality. Others are referred to Ioan James's article, published as *Biographical Memoirs of Fellows of the Royal Society*, Vol. 36, 1990, pages 3–16, and to the Memorial Address and the Reminiscences written by Peter May and published in *The Mathematical Intelligencer*, Vol. 12, no. 1, 1990, pages 40–44 and 45–48.

We, the editors of these proceedings, were both research students of Frank's during his years at Manchester, As might be imagined, this was a remarkable and unforgettable experience. There was inspiration in plenty, and, on occasion, humble pie to be eaten as well. The latter became palatable as we learned to appreciate that the vigour of Frank's responses was never directed at us as individuals, but rather towards the defence of mathematics. In fact we both discovered that when suitably prompted, Frank was astonishly willing to repeatedly explain arguments that we had bungled. He also provided warm and understanding support, friendship and guidance far beyond his role as research supervisor.

This was an exciting period for Manchester, where Frank's influence was admirably complemented by Michael Barratt, and for algebraic topology in general. When he came to Manchester in 1962, Frank had just developed the K-theory operations he used to solve the problem of vector fields on spheres. In the years that followed he developed his series of papers on $J(X)$, and regularly lectured on subject matter which eventually became his Chicago Lecture Notes volume "Stable homotopy and generalised homology".

Our opening article, by Peter May, describes these and other achievements in more detail, and forms in a sense an introduction to the whole of the book. Although some attempt has been made to group papers according to the themes which May identifies, we cannot pretend that anything very systematic has been attained, or is even desirable. Most of the contributions here are based on talks given at the Symposium, as the reader will see by consulting lists on pages xi–xii. Aside from this, we feel it sufficient to remark that all the articles have been refereed, and that every attempt has been made to attain a mathematical standard worthy of association with the name of J. F. Adams – with what success we must leave the reader to judge.

We also hope that the Symposium itself might be seen as a significant tribute to his philosophy and powers. In keeping with his views on the value of mathematics in transcending political and geographical boundaries, we were fortunate to attract a large number of participants from many countries, including Eastern Europe and the Soviet Union.

In conclusion, we would like to thank the many organisations and individuals who made possible both the Symposium and these volumes.

The bulk of the initial funding was provided by the Science and Engineering Research Council, with substantial additions being made by the London Mathematical Society and the University of Manchester Research Support Fund. Support for important peripherals was given by the NatWest Bank, Trinity College Cambridge, and the University of Manchester Mathematics Department and Vice-Chancellor's Office. We would especially like to thank John Easterling and Mark Shackleton in this context.

During the Symposium our sanity would not have survived intact without the able assistance of all our Manchester students and colleagues in algebraic topology, and most significantly, the fabulous organisational and front-desk skills of the Symposium Secretary, Jackie Minshull. And the high point of the Symposium, an ascent of Tryfan (Frank's favourite Welsh Peak), would have been far less enjoyable without the presence of Manchester guide Bill Heaton.

Mrs Grace Adams and her family were most helpful in providing photographs and other information, and were very supportive of the Symposium despite their bereavement.

The production of these volumes was first conceived by the Cambridge University Press Mathematical Editor David Tranah, and their birth pangs were considerably eased by his laid-back skills. Our referees rose to the task of supplying authoritative reports within what was often a tight deadline. We should also thank those authors who offered a manuscript which we have not had space to publish.

Finally, we both owe a great debt to our respective families, for sustaining us throughout the organisation of the Symposium, and for continuing to support us as its ripples spread downwards through the following months. Therefore, to Sheila Kelbrick and our daughter Suzanne, and to Wendy Walker, thank you.

These volumes are dedicated to Frank's memory.

Nigel Ray Grant Walker

University of Manchester
September 1991

Contents of Volume 1

Programme of one-hour invited lectures

A. K. Bousfield	On K_*-local stable homotopy theory.
G. E. Carlsson	Applications of bounded K-theory.
F. W. Clarke	Cooperations in elliptic homology.
M. C. Crabb	The Adams conjecture and the J map.
E. S. Devinatz	Duality in stable homotopy theory.
W. G. Dwyer	Construction of a new finite loop space.
P. Goerss	Projective and injective Hopf algebras over the Dyer-Lashof algebra.
M. J. Hopkins	p-adic interpolation of stable homotopy groups.
J. R. Hubbuck	Fields of spaces.
S. Jackowski	Maps between classifying spaces revisited.
J. D. S. Jones	Morse theory and classifying spaces.
N. J. Kuhn	A representation theoretic approach to the Steenrod algebra.
J. Lannes	The Segal conjecture from an unstable viewpoint.
M. E. Mahowald	On the tertiary suspension.
J. P. May	The work of J.F. Adams.
M. Mimura	Characteristic classes of exceptional Lie groups.
S. A. Mitchell	Harmonic localization and the Lichtenbaum-Quillen conjecture.
G. Nishida	p-adic Hecke algebra and $\mathrm{Ell}_*(X_\Gamma)$.
S. B. Priddy	The complete stable splitting of BG.
D. C. Ravenel	The telescope conjecture.
C. A. Robinson	Ring spectra and the new cohomology of commutative rings.
Y. Rudjak	Orientability of bundles and fibrations and related topics.
V. Vershinin	The Adams spectral sequence as a method of studying cobordism theories.
C. W. Wilkerson	Lie groups and classifying spaces.

Programme of contributed lectures

A. J. Baker	*MSp* from a chromatic viewpoint.
M. Bendersky	v_1-periodic homotopy groups of Lie groups — II.
C.-F. Bödigheimer	Homology operations for mapping class groups.
B. Botvinnik	Geometric properties of the Adams-Novikov spectral sequence.
D. M. Davis	v_1-periodic homotopy groups of Lie groups.
B. I. Gray	Unstable periodicity.
J. P. C. Greenlees	Completions, dimensionality and local cohomology.
J. Harris	Lannes' T functor on summands of $H^*(B(Z/p)^n)$.
H.-W. Henn	Refining Quillen's description of $H^*(BG; F_p)$.
K. Hess	The Adams-Hilton model for the total space of a fibration.
J. R. Hunton	Detruncating Morava K-theory.
S. Hutt	A homotopy theoretic approach to surgery on Poincaré spaces.
A. Jeanneret	Topological realisation of certain algebras associated to Dickson algebras.
K. Y. Lam	The geometric dimension problem according to J.F. Adams.
J. R. Martino	The dimension of a stable summand of BG.
J. McCleary	Hochschild homology and the cobar construction.
J. McClure	Integral homotopy of $THH(bu)$ — an exercise with the Adams spectral sequence.
N. Minami	The stable splitting of $BSL_3(Z)$.
J. Morava	The most recent bee in Ed Witten's bonnet.
F. Morel	The representability of mod p homology after one suspension.
E. Ossa	Vector bundles over loop spaces of spheres.
M.M. Postnikov	Simplicial sets with internal symmetries.
H. Sadofsky	The Mahowald invariant and periodicity.
R. Schwänzl	Hermitian K-theory of A_∞-rings.
K. Shimomura	On a spectrum whose BP_*-homology is $(BP_*/I_n)[t_1, \ldots, t_k]$.
V. P. Snaith	Adams operations and the determinantal congruence conjecture of M.J. Taylor.
M. C. Tangora	The theorems of Poisson, Euler and Bernoulli on the Adams spectral sequence.
C. B. Thomas	Characteristic classes of modular representations.
R. M. W. Wood	The boundedness conjecture for the action of the Steenrod algebra on polynomials.

Programme of Posters

D. Arlettaz: The Hurewicz homomorphism in dimension 2.

M. Beattie: Proper suspension and stable proper homotopy groups.

T. Bisson: Covering spaces as geometric models of cohomology operations.

D. Blanc Operations on resolutions and the reverse Adams spectral sequence.

J. M. Boardman: Group cohomology and gene splitting.

P. Booth: Cancellation and non-cancellation amongst products of spherical fibrations.

C. Casacuberta and M. Pfenniger: On orthogonal pairs in categories and localization.

S. Edwards: Complex manifolds with c_1 non-generating.

V. Franjou: A short proof of the $\mathcal{U}$-injectivity of H^*RP^∞

V. G. Gorbunov: Symplectic bordism of projective spaces and its application.

T. Hunter: On Steenrod algebra module maps between summands in $H^*((Z/2)^s; F_2)$.

K. Ishiguro Classifying spaces of compact simple Lie groups and p-tori.

N. Iwase: Generalized Whitehead spaces with few cells.

M. Kameko: Generators of $H^*(RP^\infty \times RP^\infty \times RP^\infty)$.

S. Kochman: Lambda algebras for generalized Adams spectral sequences.

I. Leary and N. Yagita: p-group counterexamples to Atiyah's conjecture on filtration of $R_C(G)$.

A. T. Lundell Concise tables of homotopy of classical Lie groups and homogeneous spaces.

G. Moreno: Lower bounds for the Hurewicz map and the Hirzebruch Riemann-Roch formula.

R. Nadiradze Adams spectral sequence and elliptic cohomology.

N. Oda: Localisation of the homotopy set of the axes of pairings.

A. A. Ranicki: Algebraic L-theory assembly.

N. Ray: Tutte algebras of graphs and formal groups.

J. Rutter: The group of homotopy self-equivalence classes of non-simply connected spaces: a method for calculation.

C. R. Stover On the structure of $[\Sigma\Omega\Sigma X, Y]$, described independently of choice of splitting $\Sigma\Omega\Sigma X \longrightarrow \bigvee_{n=1}^{\infty} \Sigma X^{(n)}$.

P. Symonds: A splitting principle for group representations.

Z. Wojtkowiak: On 'admissible' maps and their applications.

K. Xu: Representing self maps.

Participants in the Symposium

Jaume Aguadé *(Barcelona)*
Sadoon Al-Musawe *(Birmingham)*
Dominique Arlettaz *(Lausanne)*
Peter Armstrong *(Edinburgh)*
Tony Bahri *(Rider Coll, New Jersey)*
Andrew Baker *(Manchester)*
Michael Barratt *(Northwestern)*
Malcolm Beattie *(Oxford)*
Martin Bendersky *(CUNY)*
Terence Bisson *(Buffalo)*
David Blanc *(Northwestern)*
Michael Boardman *(Johns Hopkins)*
C.-F. Bodigheimer *(Göttingen)*
Imre Bokor *(Zurich)*
Peter Booth *(Newfoundland)*
Boris Botvinnik *(Khabarovsk)*
Pete Bousfield *(UIC)*
Ronnie Brown *(Bangor)*
Shaun Bullett *(QMWC, London)*
Mike Butler *(Manchester)*
David Carlisle *(Manchester)*
Gunnar Carlsson *(Princeton)*
Carles Casacuberta *(Barcelona)*
Francis Clarke *(Swansea)*
Fred Cohen *(Rochester)*
Michael Crabb *(Aberdeen)*
Don Davis *(Lehigh)*
Ethan Devinatz *(Chicago)*
Albrecht Dold *(Heidelberg)*
Emmanuel Dror-Farjoun *(Jerusalem)*
Bill Dwyer *(Notre Dame)*
Peter Eccles *(Manchester)*
Steven Edwards *(Indiana)*
Michael Eggar *(Edinburgh)*
Sam Evens *(Rutgers)*
Vincent Franjou *(Paris)*
Paul Goerss *(Washington)*
Marek Golasinski *(Toru, Poland)*
Vassily Gorbunov *(Novosibirsk)*
Brayton Gray *(UIC)*
David Green *(Cambridge)*
John Greenlees *(Sheffield)*
J. Gunawardena *(Hewlett-Packard)*
Derek Hacon *(Rio de Janeiro)*
Keith Hardie *(Cape Town)*
John Harris *(Toronto)*
Adam Harrison *(Manchester)*
Philip Heath *(Newfoundland)*
Hans-Werner Henn *(Heidelberg)*
Matthias Hennes *(Bonn)*
Kathryn Hess *(Stockholm)*
Peter Hilton *(SUNY, Binghamton)*
Peter Hoffman *(Waterloo, Canada)*
Mike Hopkins *(MIT)*
John Hubbuck *(Aberdeen)*
Reinhold Hübl *(Regensburg)*
Tom Hunter *(Kentucky)*
John Hunton *(Manchester)*
Steve Hutt *(Edinburgh)*
Kenshi Ishiguro *(Purdue)*
Norio Iwase *(Okayama)*
Stefan Jackowski *(Warsaw)*
Jan Jaworowski *(Indiana)*
Alain Jeanneret *(Neuchâtel)*
David Johnson *(Kentucky)*
Keith Johnson *(Halifax, Nova Scotia)*
John Jones *(Warwick)*
Masaki Kameko *(Johns Hopkins)*
Klaus Heiner Kamps *(Hagen)*
Nondas Kechagias *(Queens, Ont)*
John Klippenstein *(Vancouver)*
Karlheinz Knapp *(Wuppertal)*
Stan Kochman *(York U, Ont)*
Akira Kono *(Kyoto)*

Piotr Krason *(Virginia)*
Nick Kuhn *(Virginia)*
Kee Yuen Lam *(Vancouver)*
Peter Landweber *(Rutgers)*
Jean Lannes *(Paris)*
Ian Leary *(Cambridge)*
Kathryn Lesh *(Brandeis)*
Al Lundell *(Boulder, Col)*
Maria Luisa Sa Magalheas *(Porto)*
Zafer Mahmud *(Kuwait)*
Mark Mahowald *(Northwestern)*
Howard Marcum *(Ohio)*
John Martino *(Yale)*
Tadeusz Marx *(Warsaw)*
Yoshihoru Mataga *(UMDS, Japan)*
Honoré Mavinga *(Wisconsin)*
Peter May *(Chicago)*
John McCleary *(Vassar Coll)*
Jim McClure *(Kentucky)*
Chuck McGibbon *(Wayne State)*
Haynes Miller *(MIT)*
Mamoru Mimura *(Okayama)*
Norihiko Minami *(MSRI)*
Bill Mitchell *(Manchester)*
Steve Mitchell *(Washington)*
Jack Morava *(Johns Hopkins)*
Fabien Morel *(Paris)*
Guillermo Moreno *(Mexico)*
Fix Mothebe *(Manchester)*
Roin Nadiradze *(Tbilisi)*
Goro Nishida *(Kyoto)*
Nobuyuki Oda *(Fukuoka)*
Bob Oliver *(Aarhus)*
Erich Ossa *(Wuppertal)*
Akimou Osse *(Neuchâtel)*
John Palmieri *(MIT)*
Markus Pfenniger *(Zurich)*
Mikhail Postnikov *(Moscow)*
Stewart Priddy *(Northwestern)*
Andrew Ranicki *(Edinburgh)*
Douglas Ravenel *(Rochester)*
Nige Ray *(Manchester)*
Alan Robinson *(Warwick)*
Yuli Rudjak *(Moscow)*
John Rutter *(Liverpool)*
Hal Sadofsky *(MIT)*
Brian Sanderson *(Warwick)*
Pepe Sanjurjo *(Madrid)*
Bill Schmitt *(Memphis, Tenn)*
Roland Schwanzl *(Osnabruck)*
Lionel Schwartz *(Paris)*
Graeme Segal *(Oxford)*
Paul Shick *(John Carroll Univ)*
Don Shimamoto *(Swarthmore Coll)*
Katsumi Shimomura *(Tottori)*
Hubert Shutrick *(Karlstad, Sweden)*
Raphael Sivera *(Valencia)*
Vic Snaith *(McMaster, Ont)*
Richard Steiner *(Glasgow)*
Christopher Stover *(Chicago)*
Chris Stretch *(Ulster)*
Neil Strickland *(Manchester)*
Michael Sunderland *(Oxford)*
Wilson Sutherland *(Oxford)*
Peter Symonds *(McMaster)*
Martin Tangora *(UIC)*
Charles Thomas *(Cambridge)*
Rob Thompson *(Northwestern)*
Japie Vermeulen *(Cape Town)*
Vladimir Vershinin *(Novosibirsk)*
Rainer Vogt *(Osnabruck)*
Grant Walker *(Manchester)*
Andrsej Weber *(Warsaw)*
Clarence Wilkerson *(Purdue)*
Steve Wilson *(Johns Hopkins)*
Zdzislaw Wojtkowiak *(Bonn)*
Reg Wood *(Manchester)*
Lyndon Woodward *(Durham)*
Xu Kai *(Aberdeen)*
Nobuaki Yagita *(Tokyo)*

Addresses of Contributors

J P May	Department of Mathematics University of Chicago 5734 University Avenue Chicago, Illinois 60637, U.S.A.
Kathryn P Hess	Département de Mathematiques Ecole Polytechnique Fédérale de Lausanne CH-1015 Lausanne, Switzerland
J D S Jones	Mathematics Institute University of Warwick Coventry CV4 7AL, U.K.
J McCleary	Department of Mathematics Vassar College Poughkeepsie, New York 12601, U.S.A.
Z Fiedorowicz	Department of Mathematics Ohio State University 231 West 18th Avenue Columbus, Ohio 43210-1174, U.S.A.
R. Schwanzl	Fachbereich Mathematik/Informatik Universität Osnabrück 45 Osnabrück, Postfach 4469 Germany.
R. Vogt	Fachbereich Mathematik/Informatik Universität Osnabrück 45 Osnabrück, Postfach 4469 Germany.

Dominique Arlettaz	Institut de Mathématiques Université de Lausanne CH-1015 Lausanne, Switzerland.
K. Y. Lam	Department of Mathematics University of British Columbia Vancouver, B.C. V6T 1Y4, Canada
D. Randall	Department of Mathematics Loyola University New Orleans, Louisiana 70018, U.S.A.
Mamoru Mimura	Department of Mathematics Faculty of Science, Okayama University Okayama 700, Japan
C. A. McGibbon	Department of Mathematics Wayne State University Detroit, Michigan 48202, U.S.A.
J. M. Møller	Mathematisk Institut Universitetsparken 5 DK-2100 København Ø, Denmark
David Blanc	Department of Mathematics The Hebrew University Givat Ram Campus 91 000 Jerusalem, Israel
Christopher Stover	Department of Mathematics University of Chicago 5734 University Avenue Chicago, Illinois 60637, U.S.A.

Nobuyuki Oda	Department of Applied Mathematics Faculty of Science Fukuoka University 8.19.1 Nanakuma Jonan-ku Fukuoka 814-01, Japan
I. M. James	Mathematical Institute 24-29 St Giles Oxford OX1 3LB.
Ronald Brown	School of Mathematics University of Wales Dean Street, Bangor LL57 1UT, U.K.
Carles Casacuberta	SFB 170, Mathematisches Institut Universität Göttingen 3400 Göttingen, Germany
Georg Peschke	Department of Mathematics University of Alberta Edmonton T6G 2G1, Canada
Markus Pfenniger	School of Mathematics University of Wales Dean Street, Bangor LL57 1UT, U.K.
Peter Hilton	Department of Mathematical Sciences SUNY at Binghamton Binghamton, New York 13901, U.S.A.
Victor P. Snaith	Department of Mathematics McMaster University Hamilton, Ontario L8S 4K1, Canada

M. C. Crabb
Department of Mathematics
University of Aberdeen
The Edward Wright Building
Dunbar Street, Aberdeen AB9 2TY.

J. R. Hubbuck
Department of Mathematics
University of Aberdeen
The Edward Wright Building
Dunbar Street, Aberdeen AB9 2TY.

Kai Xu
present address unknown
Please send c/o J. R. Hubbuck
at University of Aberdeen, see above

Zdzisław Wojtkowiak
Département de Mathématiques
Université de Nice
Parc Valrose, F-06034 Nice, France

Kenshi Ishiguro
SFB 170, Mathematisches Institut
Universität Göttingen
3400 Göttingen, Germany

John Martino
Department of Mathematics
University of Virginia
Charlottesville, Virginia 22903, U.S.A.

Stewart Priddy
Department of Mathematics
Northwestern University
Evanston, Illinois 60208, U.S.A.

Douglas C. Ravenel
Department of Mathematics
University of Rochester
Rochester, New York 14627, U.S.A.

A. K. Bousfield	Department of Mathematics University of Illinois at Chicago Chicago, Illinois 60680, U.S.A.
John R. Hunton	DPMMS University of Cambridge 16 Mill Lane, Cambridge CB2 1SB
N. P. Strickland	Department of Mathematics University of Manchester Manchester M13 9PL, U.K.
Donald M. Davis	Department of Mathematics Lehigh University Bethlehem, Pennsylvania 18015, U.S.A.
Mark Mahowald	Department of Mathematics Northwestern University Evanston, Illinois 60208, U.S.A.
Martin Bendersky	Department of Mathematics CUNY, Hunter College New York 10021, U.S.A.
Dianne Barnes	Department of Mathematics Northwestern University Evanston, Illinois 60208, U.S.A.
David Poduska	Department of Mathematics Case Western Reserve University Cleveland Ohio 44106, U.S.A.

Paul Shick
Department of Mathematics
John Carroll University
Cleveland, Ohio 44118, U.S.A.

Goro Nishida
Research Institute for Mathematical Sciences
Kyoto University
Kitashirakawa Sakyo-ku, Kyoto 606, Japan

Francis Clarke
Department of Mathematics
University College of Swansea
Singleton Park, Swansea SA2 8PP, U.K.

Keith Johnson
Department of Mathematics
Dalhousie University
Halifax, Nova Scotia B3H 3J5, Canada

J.P.C. Greenlees
Department of Mathematics
University of Sheffield
Hicks Building, Hounsfield Road
Sheffield S3 7RH

Martin C. Tangora
Department of Mathematics
University of Illinois at Chicago
Chicago, Illinois 60680, U.S.A.

Alain Jeanneret
Institut de Mathématique
Université de Neuchatel
Chantemerle 20
Neuchatel, CH-2000, Switzerland.

D. P. Carlisle
Department of Computer Science
University of Manchester
Manchester M13 9PL, U.K.

R. M. W. Wood
Department of Mathematics
University of Manchester
Manchester M13 9PL, U.K.

Mohamed Ali Alghamdi,
Department of Mathematics
Faculty of Science
King Abdulaziz University
PO Box 9028, Jeddah 21413, Saudi Arabia

Nicholas J. Kuhn
Department of Mathematics
University of Virginia
Charlottesville, Virginia 22903, U.S.A.

Andrew Baker
Department of Mathematics
University of Glasgow
University Gardens, Glasgow G12 8QW, U.K.

V. G. Gorbunov
Department of Mathematics
University of Manchester
Manchester M13 9PL, U.K.

V. V. Vershinin
Institute of Mathematics
Universitetskii Pr. 4
Novosibirsk, USSR 630090

Jack Morava
Department of Mathematics
Johns Hopkins University
Baltimore, Maryland 21218, U.S.A.

Progress Report on the Telescope Conjecture

Douglas C. Ravenel *
University of Rochester
Rochester, New York 14627

The Telescope Conjecture (made public in a lecture at Northwestern University in 1977) says that the v_n-periodic homotopy of a finite complex of type n has a nice algebraic description. It also gives an explicit description of certain Bousfield localizations. In this paper we outline a proof that it is *false* for $n = 2$ and $p \geq 5$. A more detailed account of this work will appear in [Rav]. In view of this result, there is no longer any reason to think it is true for larger values of n or smaller primes p.

In Section 1 we will give some background surrounding the conjecture. In Section 2 we outline Miller's proof of it for the case $n = 1$ and $p > 2$. This includes a discussion of the localized Adams spectral sequence. In Section 3 we describe the difficulties in generalizing Miller's proof to the case $n = 2$. We end that section by stating a theorem (3.5) about some differentials in the Adams spectral sequence, which we prove in Section 4. This material is new; I stated the theorem in my lecture at the conference, but said nothing about its proof. In Section 5 we construct the parametrized Adams spectral sequence, which gives us a way of interpolating between the Adams spectral sequence and the Adams–Novikov spectral sequence. We need this new machinery to use Theorem 3.5 to disprove the Telescope Conjecture. This argument is sketched in Section 6.

*Partially supported by the National Science Foundation and MSRI

1 Background

Recall that for each prime p there are generalized homology theories $K(n)_*$ (the Morava K–theories) for each integer $n \geq 0$ with the following properties:

(i) $K(0)_*$ is rational homology and $K(1)_*$ is one of $p-1$ isomorphic summands of mod p complex K–theory.

(ii) For $n > 0$, $K(n)_*(\text{pt.}) = \mathbf{Z}/(p)[v_n, v_n^{-1}]$ with $|v_n| = 2p^n - 2$.

(iii) There is a Künneth isomorphism

$$K(n)_*(X \times Y) \cong K(n)_*(X) \otimes K(n)_*(Y).$$

(iv) If X is a finite spectrum with $K(n)_*(X) = 0$, then $K(n-1)_*(X) = 0$.

(v) If the p–localization of X (as above) is not contractible, then

$$K(n)_*(X) \neq 0 \quad \text{for} \quad n \gg 0.$$

The last two properties imply that we can make the following.

Definition 1.1 *A noncontractible finite p–local spectrum X has* **type** n *if n is the smallest integer such that $K(n)_*(X) \neq 0$.*

Definition 1.2 *If X as above has type n then a v_n–***map** *on X is a map*

$$\Sigma^d X \xrightarrow{f} X$$

with $K(n)_*(f)$ an isomorphism and $K(m)_*(f) = 0$ for all $m \neq n$.

The Periodicity Theorem of Hopkins–Smith [HS] says that such a map always exists and is unique in the sense that if g is another such map then some iterate of f is homotopic to some iterate of g. The Telescope Conjecture concerns the telescope $\widehat{X}$, which is defined to be the homotopy direct limit of the system

$$X \xrightarrow{f} \Sigma^{-d} X \xrightarrow{f} \Sigma^{-2d} X \xrightarrow{f} \cdots$$

The Periodicity Theorem tells us that this is independent of the choice of the v_n–map f.

The motivation for studying $\widehat{X}$ is that the associated Adams–Novikov spectral sequence has nice properties. We will illustrate with some simple examples. Suppose

$$BP_*(X) = BP_*/I_n = BP_*/(p, v_1, \cdots v_{n-1}).$$

This happens when X is the Toda complex $V(n-1)$. These are known to exist for small n and large p. Then

$$BP_*(\widehat{X}) = v_n^{-1}BP_*/I_n.$$

The E_2-term of the associated Adams–Novikov spectral sequence is

$$E_2^{s,t} = \mathrm{Ext}^{s,t}_{BP_*(BP)}(BP_*, v_n^{-1}BP_*/I_n),$$

which can be computed directly. For more details, see 5.1.14 and Chapter 6 of [Rav86]. It is a free module over $K(n)_*$. In particular when $n=2$ and $p \geq 5$ (in which case the spectrum $V(1)$ is known to exist) it has total (for all values of s) rank 12 and vanishes for $s > 4$. This means that the Adams–Novikov spectral sequence collapses and there are no extension problems.

The computability of this Ext group was one of the original motivations for studying v_n–periodic homotopy theory.

However, we do not know that this Adams–Novikov spectral sequence converges to $\pi_*(\widehat{X})$. It is known [Rav87] to converge to $\pi_*(L_nX)$, where L_nX denotes the Bousfield localization of X with respect to $E(n)$–theory. (When X is a finite spectrum of type n, this is the same as the localization with respect to $K(n)$–theory.) Since $\widehat{X}$ is $K(n)_*$–equivalent to X, there are maps

$$X \xrightarrow{i} \widehat{X} \xrightarrow{\lambda} L_nX.$$

The Telescope Conjecture says that λ is an equivalence, or equivalently that the Adams–Novikov spectral sequence converges to $\pi_*(\widehat{X})$. This statement is trivial for $n=0$, known to be true for $n=1$ ([Mil81] and [Mah82]). The object of this paper is to sketch a counterexample for $n=2$ and $p \geq 5$.

2 Miller's proof for $n = 1$ and $p > 2$

It is more or less a formality to reduce the Telescope Conjecture for a given value of n and p to proving it for one particular p–local finite spectrum of type n. We will outline Miller's proof for the mod p Moore spectrum $V(0)$. In that case the v_1–map

$$\Sigma^{2p-2}V(0) \xrightarrow{\alpha} V(0) \tag{2.1}$$

is the map discovered long ago by Adams in [Ada66]. There is a map

$$S^{2p-3} \longrightarrow S^0 \longrightarrow V(0)$$

which corresponds to an element in the Adams–Novikov spectral sequence called $h_{1,0}$. The Telescope Conjecture says that

$$\pi_*(\widehat{V(0)}) = K(1)_* \otimes E(h_{1,0}) \tag{2.2}$$

where $E(\cdot)$ denotes an exterior algebra.

Miller studies this problem by looking at the classical Adams spectral sequence for $\pi_*(V(0))$. In its E_2-term there is an element

$$v_1 \in E_2^{1,2p-1}$$

that corresponds to the Adams map α. One can formally invert this element and get a localized Adams spectral sequence converging to $\pi_*(\widehat{V(0)})$. (This convergence is not obvious, and is proved in [Mil81].)

We will describe the construction of this localized Adams spectral sequence. Recall that the classical Adams spectral sequence for the homotopy of spectrum X is constructed as follows. One has an *Adams resolution for* X, which is a diagram of the form

$$\begin{array}{ccccccc} X = X_0 & \longleftarrow & X_1 & \longleftarrow & X_2 & \longleftarrow & \cdots \\ {\scriptstyle f_0}\big\downarrow & & {\scriptstyle f_1}\big\downarrow & & {\scriptstyle f_2}\big\downarrow & & \\ K_0 & & K_1 & & K_2 & & \end{array}$$

with the following properties.

(i) Each K_s is a wedge of suspensions of mod p Eilenberg–Mac Lane spectra.

(ii) Each map f_s induces a monomorphism in mod p homology.

(iii) X_{s+1} is the fibre of f_s.

The *canonical Adams resolution* for X is obtained by setting

$$K_s = X_s \wedge H/(p).$$

A map $g : X \to Y$ induces a map of Adams resolutions, i.e., a collection of maps $g_s : X_s \to Y_s$ with suitable properties. The map g has *Adams filtration* $\geq t$ if it lifts to a map $g' : X \to Y_t$. In this case it is automatic that g_s lifts to Y_{s+t}.

Now consider the example at hand, namely $X = V(0)$. The map α has Adams filtration 1, so we have maps

$$V(0) = X_0 \xrightarrow{\alpha'_0} \Sigma^{-q} X_1 \xrightarrow{\alpha'_1} \Sigma^{-2q} X_2 \xrightarrow{\alpha'_2} \cdots,$$

where $q = 2p - 2$. We define $\widehat{X}_s$ to be the limit of

$$X_s \xrightarrow{\alpha'_s} \Sigma^{-q} X_{s+1} \xrightarrow{\alpha'_{s+1}} \Sigma^{-2q} X_{s+2} \xrightarrow{\alpha'_{s+2}} \cdots,$$

and $\widehat{K}_s$ to be the cofibre of the map $\widehat{X}_{s+1} \to \widehat{X}_s$, or equivalently the limit of

$$K_s \longrightarrow \Sigma^{-q} K_{s+1} \longrightarrow \Sigma^{-2q} K_{s+2} \longrightarrow \cdots,$$

Like K_s, it is a bouquet of mod p Eilenberg–Mac Lane spectra. These spectra are defined for *all integers* s, not just for $s \geq 0$ as in the classical case.

Thus we get a *localized Adams resolution*, i.e., a diagram

$$\begin{array}{ccccccccc} \cdots & \longleftarrow & \widehat{X}_s & \longleftarrow & \widehat{X}_{s+1} & \longleftarrow & \widehat{X}_{s+2} & \longleftarrow & \cdots \\ & & {\scriptstyle \hat{f}_s}\downarrow & & {\scriptstyle \hat{f}_{s+1}}\downarrow & & {\scriptstyle \hat{f}_{s+2}}\downarrow & & \\ & & \widehat{K}_s & & \widehat{K}_{s+1} & & \widehat{K}_{s+2} & & \end{array} \tag{2.3}$$

and a spectral sequence converging to the homotopy of the telescope $\widehat{V(0)}$, which is the limit of

$$\widehat{X}_0 \longrightarrow \widehat{X}_{-1} \longrightarrow \widehat{X}_{-2} \longrightarrow \cdots$$

To prove the spectral sequence converges, one must show that the inverse limit of the $\widehat{X}_s$ is contractible.

Unlike the classical Adams spectral sequence, which is confined to the first quadrant, the localized Adams spectral sequence is a full plane spectral sequence with $E_1^{s,t}$ conceivably nontrivial for all integers s and t. However, it can be shown that the E_2-term has a vanishing line of slope $1/q$, namely

$$E_2^{s,t} = 0 \quad \text{for} \quad s > \frac{t-s+1}{q}.$$

Fortunately the E_2-term of the localized Adams spectral sequence is far simpler than that of the usual Adams spectral sequence. In order to describe it we need to recall some facts about the Steenrod algebra A. Its dual is

$$A_* = E(\tau_0, \tau_1, \cdots) \otimes P(\xi_1, \xi_2, \cdots)$$

where $P(\cdot)$ denotes a polynomial algebra over $\mathbf{Z}/(p)$. We will denote these two factors by Q_* and P_* respectively.

We will use the homological (as opposed to cohomological) formulation of the Adams spectral sequence for $\pi_*(X)$, so the E_2-term is

$$\operatorname{Ext}_{A_*}(\mathbf{Z}/(p), H_*(X)) \tag{2.4}$$

where $H_*(X)$ (the mod p homology of X) is regarded as a comodule over A_*.

There is an extension of Hopf algebras

$$P_* \longrightarrow A_* \longrightarrow Q_*$$

which leads to a Cartan–Eilenberg spectral sequence converging to (2.4) with

$$E_2 = \mathrm{Ext}_{P_*}(\mathbf{Z}/(p), \mathrm{Ext}_{Q_*}(\mathbf{Z}/(p), H_*(X))).$$

The inner Ext group is easy to compute since Q_* is dual to an exterior algebra. For $X = V(0)$ it is

$$P(v_1, v_2, \cdots) \quad \text{with} \quad v_n \in \mathrm{Ext}^{1,2p^n-1}.$$

(The elements v_n correspond so closely to the generators of $\pi_*(BP)$ that we see no point in making a notational distinction between them.)

For odd primes the Cartan–Eilenberg spectral sequence collapses. (See [Rav86, 4.4.3]. It is stated there only for $X = S^0$, but the proof given will work for any X.) It follows that

$$\mathrm{Ext}^s_{A_*}(\mathbf{Z}/(p), H_*(X)) \cong \bigoplus_{i+j=s} \mathrm{Ext}^i_{P_*}(\mathbf{Z}/(p), \mathrm{Ext}^j_{Q_*}(\mathbf{Z}/(p), H_*(X))). \quad (2.5)$$

We can pass to the telescope $\widehat{V(0)}$ by inverting v_1. Then we have the following very convenient change–of–rings isomorphism.

$$\begin{aligned} \mathrm{Ext}_{P_*}(\mathbf{Z}/(p), v_1^{-1}P(v_1, v_2, \cdots)) &\cong \mathrm{Ext}_{B(1)_*}(\mathbf{Z}/(p), K(1)_*) \quad (2.6)\\ &\cong K(1)_* \otimes \mathrm{Ext}_{B(1)_*}(\mathbf{Z}/(p), \mathbf{Z}/(p)) \end{aligned}$$

where $K(1)_*$ as usual denotes the ring $v_1^{-1}P(v_1)$ and

$$B(1)_* = P(\xi_1, \xi_2, \cdots)/(\xi_i^p).$$

This Hopf algebra has a cocommutative coproduct, so its Ext group is easy to compute and we have

$$\mathrm{Ext}_{B(1)_*}(\mathbf{Z}/(p), \mathbf{Z}/(p)) \cong E(h_{1,0}, h_{2,0}, \cdots) \otimes P(b_{1,0}, b_{2,0}, \cdots)$$

where

$$\begin{aligned} h_{i,0} &\in \mathrm{Ext}^{1,2p^i-2}\\ b_{i,0} &\in \mathrm{Ext}^{2,2p^{i+1}-2p}. \end{aligned}$$

This should be compared with the localized form of the Adams–Novikov spectral sequence, in which the E_2–term is

$$\mathrm{Ext}_{BP_*(BP)}(BP_*, v_1^{-1}BP_*/(p)).$$

One can get a spectral sequence converging to this called the algebraic Novikov spectral sequence by filtering BP_* by powers of the ideal

$$I = (p, v_1, v_2, \cdots). \tag{2.7}$$

The E_2–term of this spectral sequence is a regraded form of (2.6). We denote the rth differential in this spectral sequence by δ_r. These can all be computed by algebraic methods coming from BP–theory. In this case we have

$$\delta_2(h_{i+1,0}) = v_1 b_{i,0} \quad \text{for} \quad i > 0.$$

Miller uses this to deduce that there are similar differentials in the localized Adams spectral sequence, namely

$$d_2(h_{i+1,0}) = v_1 b_{i,0}.$$

This gives

$$E_3 = E_\infty = K(1)_* \otimes E(h_{1,0}),$$

which proves the Telescope Conjecture for $n = 1$ and $p > 2$.

3 Difficulties for $n = 2$

One can mimic Miller's argument for $n = 2$ and $p \geq 5$. In that case one has the spectrum

$$V(1) = S^0 \cup_p e^1 \cup_{\alpha_1} e^{2p-1} \cup_p e^{2p},$$

which is the cofibre of the Adams map α of (2.1). There is a v_2–map

$$\Sigma^{2p^2-2} V(1) \xrightarrow{\beta} V(1)$$

constructed by Larry Smith [Smi71] and H. Toda [Tod71]. The Adams E_2–term is

$$\mathrm{Ext}_{P_*}(\mathbf{Z}/(p), P(v_2, v_3, \cdots)).$$

We can use the map β to localize this Adams spectral sequence in the same way as Miller localized the one for $V(0)$. The resulting E_2–term is

$$K(2)_* \otimes \mathrm{Ext}_{B(2)_*}(\mathbf{Z}/(p), \mathbf{Z}/(p))$$

where

$$B(2)_* = P(\xi_1, \xi_2, \cdots)/(\xi_i^{p^2}).$$

This does not have a cocommutative coproduct, so its Ext group is not as easy to compute as (2.6), but it is still manageable. It is a subquotient of the cohomology of the cochain complex

$$C^{*,*} = E(h_{1,0}, h_{2,0}, \cdots; h_{1,1}, h_{2,1}, \cdots) \otimes P(b_{1,0}, b_{2,0}, \cdots; b_{1,1}, b_{2,1}, \cdots)$$

where

$$\begin{aligned} h_{i,j} &\in C^{1,2p^j(p^i-1)} \\ b_{i,j} &\in C^{2,2p^{j+1}(p^i-1)} \end{aligned}$$

and the coboundary ∂ is given by

$$\begin{aligned} \partial(h_{i,0}) &= \pm h_{1,0}h_{i-1,1} \\ \partial(h_{i,1}) &= 0 \\ \partial(b_{i,0}) &= \pm h_{1,1}b_{i-1,1} \\ \partial(b_{i,1}) &= 0. \end{aligned} \tag{3.1}$$

There is also an algebraic Novikov spectral sequence with the following differentials.

$$\delta_2(h_{i,0}) = \pm v_2 b_{i-2,1} \tag{3.2}$$

$$\delta_{1+p^{i-1}}(h_{i,1}) = \pm v_2^{p^{i-1}} b_{i-2,0} \quad \text{for } i \geq 3 \tag{3.3}$$

The reader may object to (3.2) on the grounds that $h_{i,0}$ is not a cocycle in $C^{*,*}$, and he would be correct. It would be more accurate to say that the algebraic Novikov spectral sequence has differentials formally implied by (3.2), such as

$$\begin{aligned} \delta_2(h_{1,0}h_{i,0}) &= \pm v_2 h_{1,0} b_{i-2,1} \quad \text{and} \\ \delta_2(h_{i-1,1}h_{i,0}) &= \pm v_2 h_{i-1,1} b_{i-2,1}. \end{aligned}$$

In any case these differentials kill the elements $b_{i,j}$ and $h_{i+2,j}$ for all $i > 0$, and the E_2-term of the Adams–Novikov spectral sequence is the cohomology of

$$K(2)_* \otimes E(h_{1,0}, h_{1,1}, h_{2,0}, h_{2,1})$$

with the coboundary given by (3.1), namely

$$\partial(h_{2,0}) = \pm h_{1,0}h_{1,1}.$$

This is a $K(2)_*$-module of rank 12 with basis

$$E(h_{2,1}) \otimes \{1,\ h_{1,0},\ h_{1,1},\ h_{1,0}h_{2,0},\ h_{1,1}h_{2,0},\ h_{1,0}h_{1,1}h_{2,0}\}. \tag{3.4}$$

This is the value of $\pi_*(\widehat{V(1)})$ predicted by the Telescope Conjecture.

The difficulty is that while Miller's methods allow us to translate the algebraic differentials implied by (3.2) into differentials in the localized Adams spectral sequence, they do *not* enable us to do so for those of (3.3). The latter would give us d_r's for arbitrarily large r, and such differentials could be interfered with by other shorter differentials not related to the algebraic Novikov spectral sequence.

The following result says that such interfering differentials *do* occur in the localized Adams spectral sequence.

Theorem 3.5 *In the localized Adams spectral sequence for $\widehat{V(1)}$ for $p \geq 5$,*

$$d_{2p}(h_{i,1}) = \pm v_2 b_{i-1,0}^p \quad \textit{for } i \geq 2$$

modulo nilpotent elements.

The proof of this will be sketched below in Section 4. For $i = 2$ this can be deduced from the Toda differential [Rav86, 4.4.22] by direct calculation.

This result shows that the E_{2p}-term of the localized Adams spectral sequence is a subquotient of

$$E(h_{1,0}, h_{1,1}, h_{2,0}) \otimes P(b_{1,0}, b_{2,0}, \cdots)/(b_{i,0})^p.$$

Even though this is infinite dimensional, it is too small in the sense that it appears to have only two elements with Novikov filtration one (namely $h_{1,0}$ and $h_{1,1}$), while there are three such elements in (3.4).

4 Computing the differentials $d_{2p}(h_{i,1})$

The purpose of this section is to prove Theorem 3.5. The following is rationale for these differentials, which will be made more rigorous and precise below. In the appropriate form of the cobar complex, we have

$$-d(h_{i+2,0}) \equiv h_{1,0}h_{i+1,1} + v_2 b_{i,1} \tag{4.1}$$

modulo terms with higher Adams filtration. It follows that the target of this differential must be a permanent cycle in the localized Adams spectral sequence. Now suppose we knew that

$$d_{2p-1}(b_{i,1}) = h_{1,0} b_{i,0}^p. \tag{4.2}$$

Then combining this with (4.1) would determine the differential on $h_{i+1,1}$, giving Theorem 3.5 up to suitable indeterminacy.

The Toda differential

For $i = 1$, (4.2) is the Toda differential, first established in [Tod67]. The following is a reformulation of Toda's proof. The generator of $\pi_{2p-2}(BU)$ is represented by a map which extends (via the loop space structure of BU) to a map

$$\Omega S^{2p-1} \longrightarrow BU,$$

which can be composed with the map

$$\Sigma\Omega^2 S^{2p-1} \longrightarrow \Omega S^{2p-1}$$

(adjoint to the identity map) to give a vector bundle over $\Sigma\Omega^2 S^{2p-1}$. Its Thom spectrum is the cofibre of a stable map

$$\Omega^2 S^{2p-1} \xrightarrow{f} S^0.$$

Now $\Omega^2 S^{2p-1}$ splits stably into an infinite wedge of finite spectra B_i for $i > 0$ described explicitly by Snaith in [Sna74]. Localization at p makes B_i contractible except when $i \equiv 0$ or $1 \bmod p$, and makes B_{pj+1} equivalent to $\Sigma^{2p-3} B_{pj}$ for $j > 0$. The best way to see this is to look at mod p homology. We have

$$H_*(\Omega^2 S^{2p-1}; \mathbf{Z}/(p)) = E(x_0, x_1, \cdots) \otimes P(y_1, y_2, \cdots)$$

where the dimensions of x_j and y_j are $2p^j(p-1) - 1$ and $2p^j(p-1) - 2$ respectively.

In order to describe the Snaith splitting homologically, it is convenient to assign a *weight* to each monomial. We do this by defining the weight of both x_j and y_j to be p^j. This leads to a direct sum decomposition of the homology corresponding to the Snaith splitting of the suspension spectrum, i.e., $H_*(B_i)$ is spanned by the monomials of weight i.

Now observe that the only generator whose weight is not divisible by p is x_0, which is an exterior generator. It follows that multiplication by x_0 gives an isomorphism from the subspace spanned by monomials with weight divisible by p to the that spanned by the ones with weight congruent to 1 mod p. This isomorphism can be realized by a p-local equivalence $\Sigma^{2p-3} B_{pi} \to B_{pi+1}$. Moreover, every monomial has weight congruent to 0 or 1 mod p.

Also note that the first monomial of weight pi is y_1^i, which has dimension $2i(p^2 - p - 1)$. It follows that

$$(B_{pi})_{(p)} = \Sigma^{2i(p^2-p-1)} D_i$$

for some (-1)-connected finite spectrum D_i.

Thus the Snaith splitting (after localizing at p) has the form

$$\Omega^2 S_+^{2p-1} \simeq (S^0 \vee S^{2p-3}) \wedge \bigvee_{i \geq 0} \Sigma^{2i(p^2-p-1)} D_i.$$

In particular the resulting map

$$S^{2p-3} \longrightarrow S^0$$

is α_1, the generator of the $(2p-3)$–stem corresponding to the element $h_{1,0}$ in the Adams spectral sequence.

D_1 is the mod p Moore spectrum and the map

$$\Sigma^{2(p^2-p-1)} D_1 \xrightarrow{f_1} S^0$$

is β_1 on the bottom cell, i.e., the generator of the $2(p^2-p-1)$–stem, which corresponds to $b_{1,0}$ in the Adams spectral sequence. In general, the bottom cell of D_i is mapped in by β_1^i.

D_p is a 4–cell complex of the form

$$D_p = S^0 \cup_p e^1 \cup_{\alpha_1} e^{2p-2} \cup_p e^{2p-1},$$

where the third cell is attached to the bottom cell by α_1.

The restriction of f_p to the bottom cell is β_1^p. The fact that this extends over the third cell means that $\alpha_1\beta_1^p = 0$ in $\pi_*(S^0)$. This means that $h_{1,0}b_{1,0}^p$ must be the target of a differential in the Adams spectral sequence for the sphere spectrum.

If one computes the Adams E_2–term through the relevant range of dimensions, one finds that the only possible source for this differential is $b_{1,1}$. However, one can also deduce this by studying the map f_p more closely. Let

$$S^0 = Y_0 \longleftarrow Y_1 \longleftarrow Y_2 \longleftarrow \cdots$$

be an Adams resolution for S^0. Then routine calculations show that the restrictions of f_p to various skeleta of D_p lift to various Y_i. Let the relevant suspensions of these skeleta be denoted for brevity by

$$S^{2p(p^2-p-1)} = D_p^{(0)} \longrightarrow D_p^{(1)} \longrightarrow D_p^{(2)} \longrightarrow D_p^{(3)} = \Sigma^{2p(p^2-p-1)} D_p.$$

Then we have liftings

$$\begin{array}{ccccccc} D_p^{(0)} & \longrightarrow & D_p^{(1)} & \longrightarrow & D_p^{(2)} & \longrightarrow & D_p^{(3)} \\ \downarrow & & \downarrow & & \downarrow & & \downarrow \\ Y_{2p} & \longrightarrow & Y_{2p-1} & \longrightarrow & Y_2 & \longrightarrow & Y_1 \end{array}$$

In each case the corresponding map to Y_s/Y_{s+1} (the generalized Eilenberg–Mac Lane spectrum whose homotopy is $E_1^{s,*}$) factors through the top cell of the finite complex. The four resulting elements in the Adams E_1–term are $b_{1,0}^p$, $b_{1,0}^{p-1}h_{1,1}$, $b_{1,1}$ and $h_{1,2}$.

This, along with the fact that the third cell of D_p is attached to the first by α_1, gives the Toda differential

$$d_{2p-1}(b_{1,1}) = h_{1,0}b_{1,0}^p.$$

Generalizing the Toda differential to $i > 1$

One might hope to generalize Toda's proof of (4.2) for $i = 1$ to larger values of i by constructing a map

$$\Omega^2 S^{2p^i-1} \xrightarrow{f} S^0$$

with suitable properties. In particular, the bottom cell, S^{2p^i-3} would have to represent $h_{i,0}$. However, this is impossible since the latter is not a permanent cycle.

We can get around this difficulty by replacing S^0 by the spectrum $T(i-1)$, which is a connective p–local ring spectrum characterized by

$$BP_*(T(i-1)) = BP_*[t_1, t_2, \cdots t_{i-1}].$$

In particular, $T(0) = S^0$. To construct these spectra for $i > 0$, recall that $\Omega SU \simeq BU$ by Bott periodicity, so for each $n > 0$ we have a map

$$\Omega SU(n) \longrightarrow BU$$

which induces a stable vector bundle over $\Omega SU(n)$. We denote the resulting Thom spectrum by $X(n)$. After localizing at p, each of these admits a splitting generalizing the Brown–Peterson splitting of $MU_{(p)}$, which is the case $n = \infty$. This is proved in [Rav86, 6.5.1]. The resulting minimal summand is $T(i)$ for $p^i \le n \le p^{i+1} - 1$.

Now consider the commutative diagram of spaces

$$\begin{array}{ccccc}
\Omega^2 SU(p^i) & \longrightarrow & \text{pt.} & \longrightarrow & \Sigma\Omega^2 SU(p^i) \\
\downarrow & & \downarrow & & \downarrow \\
\Omega^2 S^{2p^i-1} & \longrightarrow & \Omega SU(p^i-1) & \longrightarrow & \Omega SU(p^i)
\end{array}$$

where the top row is a cofibre sequence and the bottom row is a fibre sequence, and the left most map is induced by the usual projection of $SU(p^i)$ onto S^{2p^i-1}. There is a natural stable vector bundle over every space in sight, and Thomification leads to a diagram of p–local spectra

$$\begin{array}{ccccc}
\Omega^2 SU(p^i) & \longrightarrow & S^0 & \longrightarrow & T \\
\downarrow & & \downarrow & & \downarrow \\
\Omega^2 S^{2p^i-1} & \longrightarrow & X(p^i-1) & \longrightarrow & X(p^i) \\
\| & & \downarrow & & \downarrow \\
\Omega^2 S^{2p^i-1} & \xrightarrow{f} & T(i-1) & \longrightarrow & T(i)
\end{array}$$

where T is the Thom spectrum of the bundle over $\Sigma\Omega^2 SU(p^i)$

This gives us the map f we are looking for. We could use it to prove a statement similar to (4.2) in the Adams spectral sequence for $T(i-1)$. However, for $i > 1$ the element $h_{1,0}$ is trivial in this setting, and $b_{i,1}$ is actually a permanent cycle. (The latter can be seen by observing that the first element of Novikov filtration $2p+1$ is $h_{i,0}b_{i,0}^p$, whose dimension exceeds that of $b_{i,1}$ when $i > 1$.)

Fortunately, all is not lost. $T(i-1)$ is a split ring spectrum, and $T(i-1) \wedge T(i-1)$ is a wedge of suspensions of $T(i-1)$, indexed by the monomials in the t_j's for $j < i$. Thus for each such monomial we get a map from $T(i-1)$ to an appropriate suspension of itself. These maps induce cohomology operations in $T(i-1)$-theory. The Quillen operations in BP-theory are constructed in the same way.

We are interested in the map

$$T(i-1) \xrightarrow{r_1} \Sigma^{2p-2} T(i-1)$$

corresponding to the monomial t_1. This is analogous to the first Steenrod reduced power operation $\mathcal{P}^1$. The induced map in homotopy, which we also denote by r_1, lowers degree by $2p-2$.

Using arguments similar to Toda's one can use the map f to prove the following.

Theorem 4.3 *For $i > 1$,*

$$r_1(b_{i,1}) = b_{i,0}^p$$

in $\pi_*(T(i-1))$.

Completing the proof of Theorem 3.5

Theorem 4.3 can be used to prove an analog of Theorem 3.5 in the localized Adams spectral sequence for $V(1) \wedge T(i-1)$. This determines $d_{2p}(h_{i+1,1})$ in the localized Adams spectral sequence for $V(1)$ itself modulo the kernel of the map from $\widehat{V(1)}$ to $\widehat{V(1)} \wedge T(i-1)$. This is good enough because disproving the Telescope Conjecture requires only that $d_{2p}(h_{i+1,1})$ be an element which is not nilpotent.

Let

$$\cdots \longleftarrow \widehat{X}_{-1} \longleftarrow \widehat{X}_0 \longleftarrow \widehat{X}_1 \longleftarrow \widehat{X}_2 \longleftarrow \cdots \qquad (4.4)$$

be a localized Adams resolution (as defined in (2.3)) for $V(1) \wedge T(i-1)$. Then we have

$$h_{i+1,0},\ h_{i,1} \in \pi_*(\widehat{X}_1/\widehat{X}_2).$$

The resolution of (4.4) can be obtained by smashing $T(i-1)$ with a similar resolution for $V(1)$. It follows that for $i > 1$, the map r_1 is defined on the entire resolution and (4.1) implies that

$$r_1(h_{i+1,0}) = -h_{i,1}.$$

Differentials on these elements in the localized Adams spectral sequence correspond to their images under the map δ induced by

$$\widehat{X}_1/\widehat{X}_2 \longrightarrow \Sigma\widehat{X}_2.$$

One can also deduce from (4.1) that

$$\delta(h_{i+1,0}) = -v_2 b_{i,1}.$$

The following diagram commutes

$$\begin{array}{ccc} \hat{X}_1/\hat{X}_2 & \xrightarrow{\delta} & \Sigma\hat{X}_2 \\ {\scriptstyle r_1}\downarrow & & \downarrow{\scriptstyle r_1} \\ \Sigma^{q+1}\hat{X}_1/\hat{X}_2 & \xrightarrow{\delta} & \Sigma^{q+1}\hat{X}_2 \end{array}$$

so we have

$$\begin{aligned} \delta(h_{i,1}) &= -\delta(r_1(h_{i+1,0})) \\ &= -r_1(\delta(h_{i+1,0})) \\ &= r_1(v_2 b_{i,1}) \\ &= v_2 r_1(b_{i,1}) \\ &= v_2 b_{i,0}^p, \end{aligned}$$

which is the desired result.

5 A parametrized Adams spectral sequence

In this section we will describe a variant of the Adams spectral sequence that we need to disprove the Telescope Conjecture.

Again we need to recall how the Adams spectral sequence is set up. Let F denote the mod p Eilenberg–Mac Lane spectrum H/p, let $\overline{F}$ denote the fibre of the map

$$\overline{F} \longrightarrow S^0 \longrightarrow F$$

and let $\overline{F}^{(s)}$ denote the sth smash power of $\overline{F}$.

Then for any spectrum X we have a tower

$$X \longleftarrow X \wedge \overline{F} \longleftarrow X \wedge \overline{F}^{(2)} \longleftarrow \cdots.$$

This gives us a collection of cofibre sequences

$$X \wedge \overline{F}^{(s+1)} \longrightarrow X \wedge \overline{F}^{(s)} \longrightarrow X \wedge \overline{F}^{(s)} \wedge F,$$

which in turn give long exact sequences of homotopy groups. These form an exact couple which gives the Adams spectral sequence. If X is a connective p–torsion spectrum, then the spectral sequence converges to $\pi_*(X)$, and for any X the E_2–term is

$$E_2^{s,t} = \operatorname{Ext}_{A_*}^{s,t}(\mathbf{Z}/(p), H_*(X)).$$

The Adams–Novikov spectral sequence is constructed in the same way, replacing $\overline{F}$ by $\overline{E}$, the fibre of the map

$$\overline{E} \longrightarrow S^0 \longrightarrow BP.$$

Let

$$X_{i,j} = X \wedge \overline{E}^{(i)} \wedge \overline{F}^{(j)}$$

so we have a diagram

$$\begin{array}{ccccccc}
X & \longleftarrow & X_{1,0} & \longleftarrow & X_{2,0} & \longleftarrow & \cdots \\
\uparrow & & \uparrow & & \uparrow & & \\
X_{0,1} & \longleftarrow & X_{1,1} & \longleftarrow & X_{2,1} & \longleftarrow & \cdots \\
\uparrow & & \uparrow & & \uparrow & & \\
X_{0,2} & \longleftarrow & X_{1,2} & \longleftarrow & X_{2,2} & \longleftarrow & \cdots \\
\uparrow & & \uparrow & & \uparrow & & \\
\vdots & & \vdots & & \vdots & &
\end{array} \tag{5.1}$$

The left edge of this diagram is the tower giving the classical Adams spectral sequence, while the top edge gives the Adams–Novikov spectral sequence.

Question 5.2 *Is there an algebraic structure that exploits the diagram (5.1) the way a spectral sequence exploits a tower?*

Such a structure would be very useful. For example, the computations of [Rav86, 4.4] indicate that the 2–component of the stable homotopy groups of spheres can be computed through a respectable range of dimensions simply by comparing the E_2–terms of the Adams spectral sequence and the Adams–Novikov spectral sequence. The structure of the two spectral sequences each imply the existence of nontrivial differentials in the other. It would be nice to have a more systematic way of doing this.

We will construct some more spectral sequences associated with (5.1), but we do not think this is the definitive answer to 5.2. The situation is still like the parable of the blind men and the elephant. (When I first used this metaphor in a lecture, I actually saw an elephant on the Rochester campus the next day.)

We can assume that all maps in (5.1) are inclusions (see [Rav84, 3.1] for a proof), so it makes sense to speak of unions and intersections of the various $X_{i,j}$ as subspectra of X.

Now fix a pair of relatively prime, nonnegative integers m and n, and define

$$W_s = \bigcup_{mi+nj\geq s} X_{i,j}.$$

This gives a tower

$$X = W_0 \longleftarrow W_1 \longleftarrow W_2 \longleftarrow \cdots \tag{5.3}$$

from which we can derive a generalization of the Adams spectral sequence. It is the classical Adams spectral sequence when $m = 0$ and $n = 1$, and the Adams–Novikov spectral sequence when $m = 1$ and $n = 0$.

The fact that the map

$$\overline{E} \longrightarrow S^0$$

factors through $\overline{F}$ can easily be seen to imply the following.

Proposition 5.4 *The spectral sequence thus described is the classical Adams spectral sequence whenever $n \geq m$.*

In view of this result, the case $n > m$ is superfluous and we may as well assume that $m \geq n$. Let ϵ denote the rational number n/m. Then we have $0 \leq \epsilon \leq 1$ and the extreme values of ϵ give the Adams–Novikov spectral sequence and the classical Adams spectral sequence respectively.

Definition 5.5 *For a rational number $\epsilon = n/m$ (with m and n relatively prime) between 0 and 1, the* **Adams spectral sequence parametrized by** ϵ *is the homotopy spectral sequence based on the exact couple associated with the tower*

$$X = W_0 \longleftarrow W_{1/m} \longleftarrow W_{2/m} \longleftarrow \cdots$$

with

$$W_s = \bigcup_{i+\epsilon j\geq s} X \wedge \overline{E}^{(i)} \wedge \overline{F}^{(j)}$$

where the union is over nonnegative integers i and j and it is understood that $\overline{E}^{(0)} = \overline{F}^{(0)} = S^0$. (This notation is not the same as in (5.3); W_s here is W_{ms} there.)

Notice that there are inclusion maps

$$X \wedge \overline{E}^{\lceil s \rceil} \longrightarrow W_s \longrightarrow X \wedge \overline{F}^{\lceil s \rceil}$$

(where $\lceil s \rceil$ denotes the smallest integer $\geq s$) which induce maps

Adams–Novikov spectral sequence
↓
parametrized Adams spectral sequence
↓
classical Adams spectral sequence

once suitable indexing conventions have been adopted. The composite is the usual reduction map. Thus the parametrized Adams spectral sequences interpolate between the Adams–Novikov spectral sequence and the classical Adams spectral sequence.

We have adopted a convention that allows the filtration grading s to be any nonnegative multiple of $1/m$; the same will be true of the differential index r. (The reader who is uncomfortable with these fractional indices is free to replace the rational numbers r, s and t with the integers mr, ms and $t + (m-1)s$ throughout the discussion.) With this understanding, we have the usual

$$E_r^{s,t} \xrightarrow{d_r} E_r^{s+r,t+r-1}. \tag{5.6}$$

The difference $t - s$ is still an integer, i.e., $E_r^{s,t}$ vanishes when $t - s$ is not an integer.

The usual E_2-term is replaced by the $E_{1+\epsilon}$-term, at least when ϵ is a reciprocal integer. Recall (2.5) that in the classical (i.e., $\epsilon = 1$) case we have

$$E_2^{s,t} \cong \bigoplus_{i+j=s} \mathrm{Ext}_{P_*}^{i,t-j}(\mathbf{Z}/(p), \mathrm{Ext}_{Q_*}^{j}(\mathbf{Z}/(p), H_*(X))).$$

Theorem 5.7 *In the parametrized Adams spectral sequence with $\epsilon = 1/m$ (and $m > 1$ if $p = 2$), if X is such that $E_2 = E_m$ in the classical Adams spectral sequence for $X \wedge BP$,*

$$E_{1+\epsilon}^{s,t} = \bigoplus_{i+j\epsilon=s} \mathrm{Ext}_{P_*}^{i,t-\epsilon j}(\mathbf{Z}/(p), \mathrm{Ext}_{Q_*}^{j}(\mathbf{Z}/(p), H_*(X))).$$

In other words, the element v_n, which has filtration 0 in the Adams–Novikov spectral sequence and filtration 1 in the classical Adams spectral sequence, has filtration ϵ in the parametrized Adams spectral sequence.

The hypothesis on X is satisfied when X is S^0, $V(0)$, $V(1)$, or any spectrum for which the Adams spectral sequence for $X \wedge BP$ collapses. In this case there is an isomorphism

$$\mathrm{Ext}^j_{Q_*}(\mathbf{Z}/(p), H_*(X)) \cong I^j BP_*(X)/I^{j+1}BP_*(X)$$

of P_*–comodules, where I is as in (2.7). It follows that the $E_{1+\epsilon}$–term of the parametrized Adams spectral sequence is isomorphic (up to reindexing) to the E_1–term of the algebraic Novikov spectral sequence. Then we have

Theorem 5.8 *Let $\epsilon = 1/m$ and suppose X is such that the classical Adams spectral sequence for $BP \wedge X$ collapses from E_2. In the parametrized Adams spectral sequence for X, let $x \in E_r^{s,t}$ be represented by an element $\tilde{x} \in E_{1+\epsilon}^{s,t}$ which corresponds to an element $\hat{x}$ in the algebraic Novikov spectral sequence which is* not *a permanent cycle. Then for m sufficiently large (depending on x), $d_r(x)$ in the parametrized Adams spectral sequence corresponds to $\delta_{mr}(\hat{x})$ in the algebraic Novikov spectral sequence.*

In other words, in many cases the differential on x can be computed by BP–theoretic methods when ϵ is sufficiently small.

A similar statement can be made about permanent cycles in the Adams–Novikov spectral sequence.

Theorem 5.9 *In the parametrized Adams spectral sequence, let ϵ and X be as in 5.8. Let $x \in E_r^{s,t}$ be represented by an element $\tilde{x} \in E_{1+\epsilon}^{s,t}$ which correpsonds to a permanent cycle in both the algebraic Novikov spectral sequence and the Adams–Novikov spectral sequence. Then for m sufficiently large (depending on the dimension $t-s$), x is a permanent cycle in the parametrized Adams spectral sequence.*

6 Disproving the Telescope Conjecture

Now we can outline our disproof of the Telescope Conjecture. As noted above (3.4), the predicted value of $\pi_*(\widehat{V(1)})$ is

$$K(2)_* \otimes E(h_{2,1}) \otimes \{1,\ h_{1,0},\ h_{1,1},\ h_{1,0}h_{2,0},\ h_{1,1}h_{2,0},\ h_{1,0}h_{1,1}h_{2,0}\}.$$

This can be shown to imply that

$$\pi_*(\widehat{V(1)} \wedge T(2)) \cong K(2)_* \otimes P(v_3, v_4) \otimes E(h_{3,0},\ h_{3,1},\ h_{4,0},\ h_{4,1}).$$

We will disprove the Telescope Conjecture by showing that $h_{4,1}$ is *not* in $\pi_*(\widehat{V(1)} \wedge T(2))$. If it were, then for some $N \gg 0$, $v_2^N h_{4,1}$ would be a permanent cycle in the Adams–Novikov spectral sequence for $\pi_*(V(1) \wedge T(2))$.

Using 5.9, this means that a similar element would be a permanent cycle in the parametrized Adams spectral sequence for sufficiently large m. It follows that $h_{4,1}$ would be a permanent cycle in the localized parametrized Adams spectral sequence for all sufficiently small $\epsilon > 0$.

Here is what we know about the localized parametrized Adams spectral sequence for the spectrum $V(1) \wedge T(2)$ for $\epsilon = 1/m$. We have

$$\begin{aligned} E_{1+\epsilon} &= K(2)_* \otimes P(v_3, v_4) \otimes E(h_{3,0}, h_{4,0}, \cdots; h_{3,1}, h_{4,1}, \cdots) \\ &\quad \otimes P(b_{3,0}, b_{4,0}, \cdots; b_{3,1}, b_{4,1}, \cdots). \end{aligned}$$

and

$$d_{1+\epsilon}(h_{i+2,0}) = \pm v_2 b_{i,1} \qquad \text{for } i \geq 3,$$

which gives

$$\begin{aligned} E_{1+2\epsilon} &= K(2)_* \otimes P(v_3, v_4) \otimes E(h_{3,0}, h_{4,0}; h_{3,1}, h_{4,1}, \cdots) \\ &\quad \otimes P(b_{3,0}, b_{4,0}, \cdots). \end{aligned}$$

Now we combine 3.5 and (3.3) to get the heuristic formula

$$d(h_{i,1}) = \pm v_2 b_{i-1,0}^p \pm v_2^{p^{i-1}} b_{i-2,0}, \tag{6.1}$$

where the second term is understood to be trivial for $i = 4$ and both terms are trivial for $i = 3$. The second term has lower filtration when

$$\epsilon < \frac{2p-2}{p^{i-1}-1},$$

i.e., for small values of i, while the first term has lower filtration for $i \gg 0$.

Suppose for example that $\epsilon = 1/p^3$. Then (6.1) gives

$$\begin{aligned} d(h_{4,1}) &= \pm v_2 b_{3,0}^p \\ d(h_{5,1}) &= \pm v_2 b_{4,0}^p \pm v_2^{p^4} b_{3,0} \end{aligned}$$

Hence

$$d_{1+p}(h_{5,1}) = \pm v_2^{p^4} b_{3,0}$$

and $b_{3,0}^p$ is dead in E_{2p-1}, so the expected differential on $h_{4,1}$ is trivial. However (6.1) also gives

$$\begin{aligned} d(h_{5,1}(d(h_{5,1})^{p-1})) &= (d(h_{5,1}))^p \\ &= \pm v_2^p b_{4,0}^{p^2} \pm v_2^{p^5} b_{3,0}^p, \end{aligned}$$

from which we get

$$d(h_{4,1} \pm v_2^{1-p^5} h_{5,1}(d(h_{5,1})^{p-1})) = \pm v_2^{1+p-p^5} b_{4,0}^{p^2},$$

which gives the differential

$$d_{2p^2-1+(1+p-p^5)\epsilon}(h_{4,1}) = \pm v_2^{1+p-p^5} b_{4,0}^{p^2}.$$

We need to be sure that this differential is nontrivial, i.e., that $b_{4,0}^{p^2}$ has not been killed earlier by another differential. For this value of ϵ, the first term of (6.1) is the dominant one for all $i \geq 6$. It follows that

$$\begin{aligned} E_{2p-1+2\epsilon} &= K(2)_* \otimes P(v_3, v_4) \otimes E(h_{3,0}, h_{4,0}, h_{3,1}, h_{4,1}) \\ &\quad \otimes P(b_{4,0}, b_{5,0}, \cdots)/(b_{5,0}^p, b_{6,0}^p, \cdots). \end{aligned}$$

The first three exterior generators can be shown to be permanent cycles by studying the Adams–Novikov spectral sequence for $V(1) \wedge T(2)$ in low dimensions. The elements $b_{i,0}$ for $i \geq 5$ can be ignored here because they have (after being multiplied by a suitable negative power of v_2 to get them in roughly the same dimension as $h_{4,1}$) lower filtration than $h_{4,1}$.

It follows that the indicated differential on $h_{4,1}$ is nontrivial as claimed for $\epsilon = 1/p^3$.

Similar computations can be made for smaller positive values of ϵ. For example when $\epsilon = p^{-j}$ for $j \geq 2$, we get

$$d_r(h_{4,1}) = \pm v_2^e b_{j+1,0}^{p^{j-1}}$$

for suitable values of r and e. It follows that $h_{4,1}$ is not a permanent cycle in the localized parametrized Adams spectral sequence for any positive value of ϵ, and the Telescope Conjecture for n = 2 and $p \geq 5$ is false.

References

[Ada66] J. F. Adams. On the groups J(X), IV. *Topology*, 5:21–71, 1966.

[HS] M. J. Hopkins and J. H. Smith. Nilpotence and stable homotopy theory II. To appear.

[Mah82] M. E. Mahowald. The image of J in the EHP sequence. *Annals of Mathematics*, 116:65–112, 1982.

[Mil81] H. R. Miller. On relations between Adams spectral sequences, with an application to the stable homotopy of a Moore space. *Journal of Pure and Applied Algebra*, 20:287–312, 1981.

[Rav] D. C. Ravenel. A counterexample to the Telescope Conjecture. To appear.

[Rav84] D. C. Ravenel. The Segal conjecture for cyclic groups and its consequences. *American Journal of Mathematics*, 106:415–446, 1984.

[Rav86] D. C. Ravenel. *Complex Cobordism and Stable Homotopy Groups of Spheres.* Academic Press, New York, 1986.

[Rav87] D. C. Ravenel. The geometric realization of the chromatic resolution. In W. Browder, editor, *Algebraic topology and algebraic K-theory*, pages 168–179, 1987.

[Smi71] L. Smith. On realizing complex cobordism modules, IV, Applications to the stable homotopy groups of spheres. *American Journal of Mathematics*, 99:418–436, 1971.

[Sna74] V. Snaith. Stable decomposition of $\Omega^n\Sigma^n X$. *Journal of the London Mathematical Society*, 7:577–583, 1974.

[Tod67] H. Toda. An important relation in the homotopy groups of spheres. *Proceedings of the Japan Academy*, 43:893–942, 1967.

[Tod71] H. Toda. On spectra realizing exterior parts of the Steenrod algebra. *Topology*, 10:53–65, 1971.

UNIVERSITY OF ROCHESTER, ROCHESTER, NEW YORK 14627, USA

On K_*-local stable homotopy theory

A.K. Bousfield*

1 Introduction

Many of the hidden features of stable homotopy theory can be conveniently exposed by localizing the stable homotopy category with respect to appropriate homology theories. In this note we discuss the algebraic structure of K_*-local stable homotopy theory, where K_* is the usual complex K-homology theory. We work integrally, not just at a prime. Our main goal is to give a homotopy classification of the K_*-local spectra, using a *united K-homology theory* which combines the complex, real, and self-conjugate theories. As an interesting byproduct of this work, we also obtain a Kunneth theorem for the united K-homology of a smash product of spectra or a product of spaces. Full technical details will appear in [9] and elsewhere.

We work in the stable homotopy category $\mathcal{S}$ of CW-spectra and begin by recalling the general theory of homological localizations of spectra for a homology theory E_* determined by a spectrum E (see [2], [7], [11]). To expose the part of stable homotopy theory seen by E_*, we may use the category of fractions $\mathcal{S}[\mathcal{E}^{-1}]$ obtained from $\mathcal{S}$ by adjoining formal inverses to the E_*-equivalences. However, to be more concrete and to avoid set theoretic difficulties, we call a spectrum Y *E_*-local* if each E_*-equivalence $V \longrightarrow W$ of spectra induces an isomorphism $[W,Y]_* \cong [V,Y]_*$, and we call a map $X \longrightarrow X_E$ of spectra an *E_*-localization* if it is an E_*-equivalence with X_E E_*-local. By [7]:

1.1 Theorem. *For each $E, X \in \mathcal{S}$, there exists an E_*-localization $X \longrightarrow X_E$.*

This gives the terminal example of an E_*-equivalence out of X and is automatically natural in X. Moreover, the full subcategory $\mathcal{S}_E \subset \mathcal{S}$ of E_*-local spectra is equivalent to the category of fractions $\mathcal{S}[\mathcal{E}^{-1}]$, and $\mathcal{S}_E$ thus captures the part of stable homotopy theory seen by E_*.

---

*Partially supported by the National Science Foundation

When E and X are both connective, then $X \longrightarrow X_E$ is just one of the usual arithmetic localizations or completions at a set of primes. However, the most interesting E_*-localizations at present are those given by the various BP-related periodic homology theories studied so successfully by Ravenel, Hopkins, Devinatz, Smith, and others.

We now turn to the most basic of these nonconnective periodic localizations, namely, the K_*-localization. This is the same as the KO_*-localization since K_* and KO_* (as well as K^* and KO^*) have the same equivalences in $\mathcal{S}$. We believe that the K_*-localization was first studied somewhat speculatively by Adams-Baird and then more substantively by Ravenel. Our version in [7] of their key result is:

1.2 Theorem. *For a spectrum X, there is a natural equivalence $X_K \simeq S^0_K \wedge X$ where S^0_K is the "nonconnective* ImJ *spectrum" determined at a prime p by cofiber sequences*

$$\mathcal{J}_{(p)} \longrightarrow KO_{(p)} \xrightarrow{\psi^r - 1} KO_{(p)}$$

$$(S^0_K)_{(p)} \longrightarrow \mathcal{J}_{(p)} \longrightarrow \Sigma^{-1}HQ$$

with $r = 3$ for $p = 2$ and with r an integer generating the group of units in Z/p^2 for p odd.

The K_*-localization can also be characterized using the K_*-equivalences

$$A : \Sigma^q MZ/p \longrightarrow MZ/p$$

of Adams [1] for Moore spectra at a prime p, where $q = 2p - 2$ for p odd and $q = 8$ for $p = 2$. As shown in [7], work of M. Mahowald and H. Miller implies:

1.3 Theorem. *A spectrum Y is K_*-local if and only if $A : \pi_*(Y; Z/p) \cong \pi_{*+q}(Y; Z/p)$ for each prime p.*

1.4 Theorem. *A map $f : X \longrightarrow Y$ of spectra is a K_*-equivalence if and only if $f_* : \pi_*X \otimes Q \cong \pi_*Y \otimes Q$ and $f_* : \pi_*(X; Z/p)[A^{-1}] \cong \pi_*(Y; Z/p)[A^{-1}]$ for each prime p.*

Thus the K_*-localization makes all mod-p homotopy groups *A-periodic* (or *v_1-periodic* in BP-parlance) with

$$\pi_*(X_K; Z/p) \cong \pi_*(X; Z/p)[A^{-1}].$$

The preceding results give a fairly good homotopy theoretic understanding of K_*-local spectra, but they don't go very far toward classifying these spectra algebraically. For that purpose, we first deal with KO-module spectra, which are important examples of K_*-local spectra and are already very diverse.

2 The classification of KO-module spectra

Let KOS denote the *homotopy category of KO-module spectra* in the elementary sense. Thus an object $G \in KOS$ is just a spectrum G together with an action $KO \wedge G \longrightarrow G$ which is associative and unitary in the stable homotopy category $\mathcal{S}$. Note that a spectrum X determines an extended KO-module spectrum $KO \wedge X$ with

$$\mathrm{Hom}_{KOS}(KO \wedge X, G)_* \cong [X, G]_*.$$

Following terminology of Yosimura [12], we say that spectra X and Y are *quasi KO_*-equivalent* when $KO \wedge X \cong KO \wedge Y$ in KOS. This holds, for instance, when $X_K \simeq Y_K$ or $\Sigma^8 X \simeq Y$. For a KO-module spectrum G or $KO \wedge X$, the most obvious invariant is just the $\pi_* KO$-module $\pi_* G$ or $KO_* X$. However, this $\pi_* KO$-module will often be badly behaved with infinite homological dimension. To overcome this difficulty, we start by introducing an abelian category $\mathcal{CRT}$ of *united K-theoretic* modules, with much better homological algebraic properties than the $\pi_* KO$-modules.

Let

$$\mathcal{B} = \{K, KO, KT\} \subset KOS$$

denote the graded additive full subcategory of KOS given by the KO-module spectra

$$\begin{aligned} K &\cong KO \wedge C(\eta) = KO \wedge (S^0 \cup_\eta e^2) \\ KO &\cong KO \wedge S^0 \\ KT &\cong KO \wedge C(\eta^2) = KO \wedge (S^0 \cup_{\eta^2} e^2) \end{aligned}$$

of complex, real, and self-conjugate K-theory (see [4] or [6]). The spectrum KT, sometimes denoted by KC or KSC, is actually a commutative ring spectrum of period 4 with

$$\pi_i KT \cong \begin{cases} Z & \text{for } i \equiv 0, 3 \bmod 4 \\ Z/2 & \text{for } i \equiv 1 \bmod 4 \\ 0 & \text{for } i \equiv 2 \bmod 4. \end{cases}$$

We regard $\mathcal{B}$ as our coefficient ring, replacing the usual coefficient ring

$$\pi_* KO = \{KO\} \subset KOS.$$

Of course, $\mathcal{B}$ is an additive category with *three* objects, while an ordinary ring is an additive category with *one* object, but the elementary theory and terminology of noncommutative rings extends immediately to such rings with several objects (see e.g. [10]). Now let $\mathcal{CRT}$ denote the abelian category of $\mathcal{B}$-modules, i.e., the category of additive functors from $\mathcal{B}$ to graded abelian groups. Thus a module $M \in \mathcal{CRT}$ consists of graded abelian groups $\{M^C, M^R, M^T\}$ equipped with the various $\mathcal{B}$-operations such as: the complexification $c : M^R_* \longrightarrow M^C_*$, the realification $r : M^C_* \longrightarrow M^R_*$, the Hopf action $\eta : M^R_* \longrightarrow M^R_{*+1}$, the Bott periodicity $B : M^C_* \cong M^C_{*+2}$, the conjugation $\psi^{-1} : M^C_* \cong M^C_*$, etc.

We can now introduce the *united K-homotopy*

$$\pi^{CRT}_* G = \{\pi_*(G \wedge C(\eta)),\ \pi_* G,\ \pi_*(G \wedge C(\eta^2))\} \in \mathcal{CRT}$$

of a KO-module spectrum G and the *united K-homology*

$$K^{CRT}_* X = \{K_* X, KO_* X, KT_* X\} \in \mathcal{CRT}$$

of a spectrum X. Our original interest in the functor K^{CRT}_* was aroused by a purely formal observation:

2.1 Proposition. *Each of the modules $K^{CRT}_*(DC(\eta))$, $K^{CRT}_*(DS^0)$, and $K^{CRT}_*(DC(\eta^2))$ is free on one generator in $\mathcal{CRT}$, where DF denotes the Spanier-Whitehead dual of F.*

Thus, by using K^{CRT}_* in place of KO_*, we create additional projective homologies, and we can hope (rightly) to achieve a general reduction in homological dimensions. Of course, D could be omitted above, because it merely desuspends, but the stated result may be generalized to other united homology theories. For a commutative ring spectrum E and set of finite CW-spectra $\{F_\alpha\}_\alpha$, let $\mathcal{F} = \{E \wedge F_\alpha\}_\alpha$ be the associated full subcategory of E-module spectra, and consider the homology $\mathcal{F}$-modules $E^{\mathcal{F}}_*(X) = \{(E \wedge F_\alpha)_* X\}_\alpha$. Then each $E^{\mathcal{F}}_*(DF_\alpha)$ is a free $\mathcal{F}$-module on one generator.

Returning to united K-theory, recall that Anderson [4] introduced three basic exact sequences involving the complex, real, and self-conjugate K-homology of a space or spectrum. For a module $M = \{M^C, M^R, M^T\} \in \mathcal{CRT}$, there are corresponding Anderson sequences

$$\cdots \longrightarrow M^R_* \xrightarrow{\eta} M^R_{*+1} \xrightarrow{c} M^C_{*+1} \xrightarrow{rB^{-1}} M^R_{*-1} \xrightarrow{\eta} \cdots$$

$$\cdots \longrightarrow M^R_* \xrightarrow{\eta^2} M^R_{*+2} \longrightarrow M^T_{*+2} \longrightarrow M^R_{*-1} \xrightarrow{\eta^2} \cdots$$

$$\cdots \longrightarrow M^C_* \xrightarrow{1-\psi^{-1}} M^C_* \longrightarrow M^T_{*-1} \longrightarrow M^C_{*-1} \xrightarrow{1-\psi^{-1}} \cdots.$$

Successive maps will always compose to zero, and we call M *Anderson exact* (or *CRT-acyclic*) when these sequences are exact.

2.2 Example. For a KO-module spectrum G and spectrum X, $\pi_*^{CRT}G$ and $K_*^{CRT}X$ are Anderson exact.

In [9], we prove:

2.3 Theorem. *For a module $M \in \mathcal{CRT}$, the following are equivalent:*

(i) M is Anderson exact with M_*^C free abelian;

(ii) M is projective in $\mathcal{CRT}$;

(iii) M is free in $\mathcal{CRT}$.

This shows, for instance, that $K_*^{CRT}X$ is free in $\mathcal{CRT}$ whenever K_*X is free abelian, and thus:

2.4 Corollary. *If X is a spectrum with K_*X free abelian, then $K_*^{CRT}X \cong K_*^{CRT}W$ for some wedge W of suspensions of $C(\eta)$, S^0, and $C(\eta^2)$.*

By [9], there is a similar characterization of injectives in $\mathcal{CRT}$, and thus:

2.5 Theorem. *For a module $M \in \mathcal{CRT}$, the following are equivalent:*

(i) *M is Anderson exact;*

(ii) *M has projective dimension ≤ 1 in $\mathcal{CRT}$;*

(iii) *M has injective dimension ≤ 1 in $\mathcal{CRT}$.*

2.6 Corollary. *For a spectrum X and KO-module spectrum G, $K_*^{CRT}X$ and $\pi_*^{CRT}G$ have projective and injective dimensions ≤ 1 in $\mathcal{CRT}$.*

As in [2], or [9], this leads to a universal coefficient theorem.

2.7 Theorem. *For a spectrum X and KO-module spectrum G, there is a natural short exact sequence*

$$\mathrm{Ext}_{\mathcal{CRT}}(K_{*-1}^{CRT}X, \pi_*^{CRT}G) \rightarrowtail [X, G] \twoheadrightarrow \mathrm{Hom}_{\mathcal{CRT}}(K_*^{CRT}X, \pi_*^{CRT}G).$$

This sequence need not be splittable, since it gives a nontrivial extension of $Z/2$ by $Z/2$ when $X = \Sigma MZ/2$ and $G = KO$. Using the case $G = KO \wedge Y$, we obtain:

2.8 Corollary. *Two spectra X and Y are quasi KO-equivalent if and only if $K_*^{CRT}X \cong K_*^{CRT}Y$ in $\mathcal{CRT}$.*

By Theorem 3.4 and Lemma 3.5 below, we have:

2.9 Theorem. *For each Anderson exact module $M \in \mathcal{CRT}$, there exists a spectrum X with $K_*^{CRT}X \cong M$ in $\mathcal{CRT}$.*

Since $\pi_*^{CRT}(KO \wedge X) \cong K_*^{CRT}X$, the above results combine to give the promised classification of KO-module spectra.

2.10 Theorem. *There are one-to-one correspondences between the following classes:*

(i) *the quasi KO-equivalence classes of spectra in $\mathcal{S}$;*

(ii) *the isomorphism classes of KO-module spectra in $KO\mathcal{S}$;*

(iii) *the isomorphism classes of Anderson exact modules in $\mathcal{CRT}$.*

The required correspondences are induced by the functors sending $X \in \mathcal{S}$ to $KO \wedge X \in KO\mathcal{S}$ and sending $G \in KO\mathcal{S}$ to $\pi_*^{CRT}G \in \mathcal{CRT}$. The classical decompositions $K \simeq KO \wedge C(\eta)$ and $KT \simeq KO \wedge C(\eta^2)$ are now seen to be quite typical.

2.11 Corollary. *Each KO-module spectrum G is expressible as $G \simeq KO \wedge X$ for some spectrum X.*

We shall discuss elsewhere the algebraic problem, raised by Theorem 2.10, of classifying the Anderson exact modules in $\mathcal{CRT}$. Each such module M can be built by first choosing M_0^C and M_1^C as arbitrary abelian groups with involution ψ^{-1}. These determine M^C and there is a procedure (involving free presentations) for finding all prolongations of M^C to Anderson exact modules $M \in \mathcal{CRT}$. Such prolongations always exist, and they are unique when the alternating sequence

$$\cdots \longrightarrow M_*^C \xrightarrow{1-\psi^{-1}} M_*^C \xrightarrow{1+\psi^{-1}} M_*^C \xrightarrow{1-\psi^{-1}} M_*^C \xrightarrow{1+\psi^{-1}} \cdots$$

is exact.

As an interesting byproduct of our work on united K-theory, we obtain a Kunneth theorem.

2.12 Theorem. *For spectra X and Y, there is a natural short exact sequence*

$$K_*^{CRT}X \otimes_{\mathcal{CRT}} K_*^{CRT}Y \rightarrowtail K_*^{CRT}(X \wedge Y) \twoheadrightarrow \mathrm{Tor}^{\mathcal{CRT}}(K_{*-1}^{CRT}X, K_*^{CRT}Y).$$

This Kunneth exact sequence need not be splittable, since it expresses $KO_1(MZ/2 \wedge MZ/2)$ as a nontrivial extension of $Z/2$ by $Z/2$. There is a similar Kunneth exact sequence for the united K-homology of a product of spaces or the united K-cohomology of a product of finite CW-complexes. This generalizes Atiyah's Kunneth theorem [5] to include real and self-conjugate K-theory as well as complex K-theory. The required tensor product $M \otimes_{\mathcal{CRT}} N \in \mathcal{CRT}$ of $M, N \in \mathcal{CRT}$ has

$$(M \otimes_{\mathcal{CRT}} N)^C \cong M^C \otimes_{\pi_* K} N^C,$$

but has more complicated real and self-conjugate components which depend on all of M and N. We now give a somewhat abstract definition of $M \otimes_{\mathcal{CRT}} N$ and will provide a more constructive definition elsewhere. We first enlarge our coefficient category $\mathcal{B} = \{K, KO, KT\} \subset \mathcal{KOS}$ to the full subcategory $\overline{\mathcal{B}} \subset \mathcal{KOS}$ given by all finite wedges of K, KO, and KT. By additivity, the $\overline{\mathcal{B}}$-modules are equivalent to the $\mathcal{B}$-modules. Moreover, $\overline{\mathcal{B}}$ is a symmetric monoidal additive category with identity object $KO \in \overline{\mathcal{B}}$ and with multiplication functor $\mu : \overline{\mathcal{B}} \otimes \overline{\mathcal{B}} \longrightarrow \overline{\mathcal{B}}$ given by

$$\mu(KO \wedge W_1, KO \wedge W_2) = KO \wedge W_1 \wedge W_2,$$

where the tensor product category $\overline{\mathcal{B}} \otimes \overline{\mathcal{B}}$ is formed using the cartisian product of object classes and the tensor product of Hom groups. In fact, $\overline{\mathcal{B}}$ is also equipped with a duality functor $(KO \wedge W)^\# \cong KO \wedge DW$ and with function objects $F(G, H) \cong \mu(G^\#, H)$. The multiplication functor $\mu : \overline{\mathcal{B}} \otimes \overline{\mathcal{B}} \longrightarrow \overline{\mathcal{B}}$ determines adjoint restriction and extension functors

$$L : (\overline{\mathcal{B}} \otimes \overline{\mathcal{B}}\text{-modules}) \rightleftarrows (\overline{\mathcal{B}}\text{-modules}) : R.$$

For $M, N \in \mathcal{CRT}$ we can now define $M \otimes_{\mathcal{CRT}} N$ as the object of $\mathcal{CRT}$ corresponding to $L(M \overline{\otimes} N)$, where $M \overline{\otimes} N$ is the $\overline{\mathcal{B}} \otimes \overline{\mathcal{B}}$-module given by the obvious external tensor product of M and N.

3 The classification of K_*-local spectra

To algebraically classify the K_*-local spectra, we must strengthen the united K-homology

$$K_*^{CRT} X = \{K_* X, KO_* X, KT_* X\} \in \mathcal{CRT}$$

by taking account of the stable Adams operations

$$\psi^j : (K_* X)[1/j] \cong (K_* X)[1/j]$$
$$\psi^j : (KO_* X)[1/j] \cong (KO_* X)[1/j]$$
$$\psi^j : (KT_* X)[1/j] \cong (KT_* X)[1/j]$$

for integers $j \neq 0$. For this purpose, we first introduce an abelian category $\mathcal{A}$ of *abelian groups with stable Adams operations*. An object $B \in \mathcal{A}$ is an abelian group with automorphisms $\psi^j : B[1/j] \cong B[1/j]$, for integers $j \neq 0$, satisfying:

(i) $\psi^1 = 1$ and $\psi^{jk} = \psi^j\psi^k$ in $B[1/jk]$;

(ii) $B \otimes Q = \bigoplus_{m \in Z} W_m(B \otimes Q)$ where $W_m(B \otimes Q)$ denotes the set of all $x \in B \otimes Q$ with $\psi^j x = j^m x$ whenever $j \neq 0$;

(iii) each $y \in B$ lies in some finitely generated subgroup $N \subset B$ such that $\psi^j : N[1/j] \cong N[1/j]$ whenever $j \neq 0$ and such that, for each prime p and integer $t \geq 1$, the induced operations $\psi^j : N/p^t N \cong N/p^t N$ with $p \nmid j$ are periodic in j of some period p^s depending on p^t.

To codify the structure of K_*^{CRT} with its stable Adams operations, we introduce an abelian category $\mathcal{ACRT}$ consisting of objects $M = \{M^C, M^R, M^T\} \in \mathcal{CRT}$ with $M_n^C, M_n^R, M_n^T \in \mathcal{A}$ for each n where the operations ψ^j commute appropriately with the KO-module operations. Now $K_*^{CRT}X$ belongs to $\mathcal{ACRT}$ for each spectrum X. By using $\mathcal{ACRT}$ in place of $\mathcal{CRT}$, we sacrifice all nonzero projectives but retain enough injectives and suffer only a slight loss of injective dimension. As shown in [9]:

3.1 Theorem. *A module $M \in \mathcal{ACRT}$ is Anderson exact if and only if M has injective dimension ≤ 2 in $\mathcal{ACRT}$.*

3.2 Corollary. *For a spectrum X, $K_*^{CRT}X$ has injective dimension ≤ 2 in $\mathcal{ACRT}$.*

This is an integral version of the Adams-Baird theorem [3]. That theorem, which was quite remarkable in its time, asserts that

$$\operatorname{Ext}^{s,t}_{K_{(p)*}K_{(p)}}(K_{(p)*}X, K_{(p)*}Y) = 0$$

for $s \geq 3$ when p is an odd prime and X and Y are finite CW-spectra. More generally, by [8, 7.10 and 10.5], all $K_{(p)*}K_{(p)}$-comodules have injective dimension ≤ 2 when p is an odd prime. However, the corresponding statements are false for $p = 2$, even when $KO_{(2)}$ is used in place of $K_{(2)}$. In our search for a 2-local or integral version of the Adams-Baird theorem, we turned to united K-theory after observing Proposition 2.1 above. Our present integral theorem contains the previous p-local Adams-Baird theorems because the category of $K_{(p)*}K_{(p)}$-comodules is equivalent to the full subcategory of p-local Anderson exact modules in $\mathcal{ACRT}$ for an odd prime p by [8, 10.5] and [9, 7.13].

The above theorem opens the way to an Adams spectral sequence with only three nontrivial rows. By [9]:

3.3 Theorem. *For spectra X and Y, there is a K_*^{CRT}-Adams spectral sequence $\{E_r^{s,t}(X,Y)\}$ converging strongly to $[X_K, Y_K]_{t-s}$ with*

$$E_2^{s,t}(X,Y) \cong \mathrm{Ext}^{s,t}_{\mathcal{ACRT}}(K_*^{CRT}X, K_*^{CRT}Y)$$

where $E_2^{s,t}(X,Y) = 0$ for $s \geq 3$ and $E_3^{s,t}(X,Y) = E_\infty^{s,t}(X,Y)$.

The K_*^{CRT}-Adams spectral sequence agrees with its K-theoretic rivals when they are "well behaved." In particular, it reduces to the KO_*-Adams spectral sequence when $X = S^0$ or, more generally, when KO_*X is a free π_*KO-module, and reduces to the K_*-Adams spectral sequence when the generator $\eta \in \pi_1 KO \cong Z/2$ annihilates KO_*X or KO_*Y.

We can now proceed to classify K_*-local spectra. Each spectrum X determines an Anderson exact module $K_*^{CRT}X \in \mathcal{ACRT}$, and conversely:

3.4 Theorem. *For each Anderson exact module $M \in \mathcal{ACRT}$, there exists a K_*-local spectrum X with $K_*^{CRT}X \cong M$ in $\mathcal{ACRT}$.*

This follows by [9, 10.1] where the required X is constructed by realizing a short injective resolution of M. There is a corresponding result in $\mathcal{CRT}$ which we have called Theorem 2.9. It is now deduced using:

3.5 Lemma. *Each Anderson exact module in $\mathcal{CRT}$ can be given a module structure in $\mathcal{ACRT}$.*

The proof of this lemma is somewhat long and will be presented elsewhere.

For an Anderson exact module $M \in \mathcal{ACRT}$ a *realization* consists of a K_*-local spectrum X together with an isomorphism $\alpha : K_*^{CRT}X \cong M$ in $\mathcal{ACRT}$. Two such realization $\alpha : K_*^{CRT}X \cong M$ and $\beta : K_*^{CRT}Y \cong M$ are called *strictly equivalent* if there exists an equivalence $f : X \simeq Y$ with $\beta f_* = \alpha$. The obstruction

$$D(\alpha, \beta) = \beta d_2(\beta^{-1}\alpha)\alpha^{-1} \in \mathrm{Ext}^{2,1}_{\mathcal{ACRT}}(M, M)$$

to this strict equivalence is obtained by applying the K_*^{CRT}-Adams differential d_2 to the "identity" homomorphism $K_*^{CRT}X \longrightarrow K_*^{CRT}Y$. Let $\mathcal{R}(M)$ denote the collection of all strict equivalence classes of realizations of M. Since $\mathcal{R}(M)$ is nonempty by Theorem 3.4, we may choose a "base" realization $\alpha : K_*^{CRT}X \cong M$. By [9, 10.3]:

3.6 Theorem. *For an Anderson exact module $M \in \mathcal{ACRT}$, there is a bijection*

$$D(\alpha, -) : \mathcal{R}(M) \cong \mathrm{Ext}^{2,1}_{\mathcal{ACRT}}(M, M).$$

This completes our algebraic classification of K_*-local spectra. It is not entirely satisfactory because *strict equivalence* may be finer than *homotopy equivalence*. The set $\mathcal{R}(M)$ of *homotopy types* of realizations of M can be expressed as an orbit set

$$\overline{\mathcal{R}}(M) = \mathcal{R}(M)/\mathrm{Aut}(M)$$

using the composition action of $\mathrm{Aut}(M)$ on $\mathcal{R}(M)$. However, at present, we are unable to construct the corresponding composition action of $\mathrm{Aut}(M)$ on $\mathrm{Ext}^{2,1}_{\mathcal{ACRT}}(M,M)$ in a purely algebraic way, except under certain special conditions. For instance, when the generator $\eta \in \pi_1 KO \cong Z/2$ annihilates M^R, then there is a canonical choice of a "base" realization α, and the associated composition action of $\mathrm{Aut}(M)$ on $\mathrm{Ext}^{2,1}_{\mathcal{ACRT}}(M,M)$ is given by conjugation. Also, when M^C is trivial in all odd (or all even) dimensions, then $\mathrm{Ext}^{2,1}_{\mathcal{ACRT}}(M,M) = 0$, and there is a *unique* K_*-local homotopy type realizing M.

Our classification theorems can easily be applied to finite CW-spectra, since by [9, 11.1]:

3.7 Theorem. *A K_*-local spectrum X is the K_*-localization of a finite CW-spectrum if and only if K_*X is of finite type.*

Thus, for instance, Theorem 3.4 shows that each Anderson exact module $M \in \mathcal{ACRT}$ of finite type can be realized as $M \cong K_*^{CRT}F$ for some finite CW-spectrum F.

A long-term goal of this work is to totally algebraicize K_*-local stable homotopy theory, just as Quillen and Sullivan did for rational homotopy theory. Although that goal is still out of reach, we are approaching a reasonable algebraic understanding of K_*-local stable homotopy theory.

References

[1] J.F. Adams, On the groups $J(X)$-IV, *Topology* **5** (1966), 21–71.

[2] J.F. Adams, Stable Homotopy and Generalized Homology, University of Chicago Press, 1974.

[3] J.F. Adams, Operations of the nth kind in K-theory and what we don't know about RP^∞, *London Mathematical Society Lecture Notes*, **Vol. 11**, 1974, 1–10.

[4] D.W. Anderson, Thesis, Berkeley, 1964.

[5] M.F. ATIYAH, Vector bundles and the Kunneth formula, *Topology* **1** (1962), 245–248.

[6] M.F. ATIYAH, K-theory and reality, *Quart. J. Math. Oxford* **17** (1966), 367–386.

[7] A.K. BOUSFIELD, The localization of spectra with respect to homology, *Topology* **18** (1979), 257–281. (Correction in *Comm. Math. Helv.* **58** (1983), 599–600.)

[8] A.K. BOUSFIELD, On the homotopy theory of K-local spectra at an odd prime, *Amer. J. Math.* **107** (1985), 895–932.

[9] A.K. BOUSFIELD, A classification of K-local spectra, *J. Pure Appl. Algebra.* **66** (1990), 121–163.

[10] B. MITCHELL, Rings with several objects, *Adv. Math.* **8** (1972), 1–161.

[11] D.C. RAVENEL, Localization with respect to certain periodic homology theories, *Amer. J. Math.* **106** (1984), 351–414.

[12] Z. YOSIMURA, The quasi KO-homology types of the real projective spaces, *Springer Lecture Notes in Mathematics,* **Vol. 1418**, (1990), 156–174.

DETRUNCATING MORAVA K-THEORY

John Robert Hunton*

Dedicated to the memory of Frank Adams.

§1 Introduction

In [4] we considered the problem of computing the Morava K-theory of extended power constructions $K(n)^*(D_p(X))$ for various spaces X. A solution was given for those X with our property of possessing a *unitary-like embedding* (ULE). Recall that we said, for an odd prime p and some particular p-primary $K(n)$-theory, a space Y was *unitary-like* if the ring $K(n)^*(Y)$ had no nilpotent elements (for the prime 2 we imposed the additional condition that $K(n)^{\mathrm{odd}}(Y) = 0$). Then the space X was said to have a ULE if there existed some map

$$e: X \longrightarrow Y$$

with Y unitary-like and $K(n)^*(e)$ an epimorphism. One might ask which spaces have ULE's. In [4] we gave constructions for ULE's sufficient to allow us to solve the questions considered there. In this paper we address the following finer problem.

Problem. *Given a space X with $K(n)^{\mathrm{odd}}(X) = 0$, does there exist a space Y and a map $e: X \longrightarrow Y$, epimorphic in $K(n)^*(-)$ and with $K(n)^*(Y)$ a formal power series algebra over $K(n)_*$?*

Clearly all such maps e are also ULE's for X. Furthermore, at least for odd primes, the condition $K(n)^{\mathrm{odd}}(X) = 0$ is strictly necessary given the graded commutativity of the multiplication in $K(n)^*(-)$. We show below that for odd primes and finite spaces X this problem always has a positive solution: the $K(n)^{\mathrm{odd}}(X) = 0$ condition is thus equivalent to geometrically viewing $K(n)^*(X)$ as a truncation of a multiplicatively free object. We also discuss the problem for infinite spaces and hope to return to the matters raised there in a future article.

The idea of the construction is as follows. In order to solve the problem we need a good supply of spaces, candidates for Y, whose $K(n)$-cohomologies are formal power series algebras, and we also need lifting theorems for the creation of the required maps e. Our first potential candidates

---

* Supported by an S.E.R.C. postdoctoral fellowship.

for Y are products of the even indexed spaces in the Ω-spectrum for $E(n)$; we prove the lifting result that the map

$$E(n)^*(X) \longrightarrow K(n)^*(X)$$

is an epimorphism for finite complexes X which satisfy $K(n)^{\text{odd}}(X) = 0$. In fact this follows from showing that $K(n)^*(X)$ can be lifted to $\widehat{E(n)}^*(X)$ if $K(n)^{\text{odd}}(X) = 0$, a result that holds for infinite X as well. The $\widehat{E(n)}$ here are the I_n-adically complete spectra of [2] and [3], and so this gives further evidence of the fundamental role of these spectra in the study of Morava K-theory. However, it turns out that the algebras $K(n)^*(\mathbf{E(n)}_r)$ are rather too big for convenience, and we derive our actual solution by restricting to certain spaces in the Ω-spectrum for $BP\langle n\rangle$.

The paper is set out as follows. We study the spaces $\mathbf{E(n)}_r$ and $\mathbf{BP\langle n\rangle}_r$ in §2 and in the process describe the Hopf ring for $E(n)$. The lifting result mentioned above (Theorem 11) and the solution of our problem (Theorem 12) are in §3 with some concluding remarks made in §4. The prime p is to be taken as odd throughout.

The work in this paper represents part of the contents of the author's Ph.D. thesis [5]. We wish to thank Andrew Baker for introducing us to the $\widehat{E(n)}$ spectra and their considerable potential, and for his encouragement throughout, and wish also to thank the referee who kindly took the time to point out many improvements to the original manuscript. Finally, the author would very much like to record his gratitude to his research supervisor, the late Professor Frank Adams, who saw the beginning of this work.

§2 The Hopf ring for E(n)

In this section we prove the results on the Morava K-theory of the spaces we want to use in our construction. In fact we go further than is required for this end and compute (Theorem 5) the Hopf ring $H_*(\mathbf{E(n)}_*; \mathsf{F}_p)$ as well. This is calculated via a result (Corollary 7) about certain of the spaces in the Ω-spectrum for $BP\langle n\rangle$ – this latter is all that is actually required to solve the problem stated in the introduction. The calculations we give here follow a different, and more succinct, approach to that of our original ones in [5]. We intend to present at a future date yet another approach to the $H_*(\mathbf{E(n)}_*; \mathsf{F}_p)$ Hopf ring in joint work with M. J. Hopkins. This will compute the Hopf ring directly from the Landweber exactness property of $E(n)$-theory.

All singular homology groups used will be with mod p coefficients (p odd) and from now on we shall suppress this from the notation. All homology is unreduced. We assume a certain familiarity with the basic notions and terminology of Hopf rings; the reader will find **[6]** a suitable general reference. However, we mention below some of the principal facts that we shall need. In particular, for a spectrum E we use the notation $\mathbf{E}_r$ for the rth space in the Ω-spectrum for E, *i.e.*, $\mathbf{E}_r$ is the space representing the functor $E^r(-)$. The connected component of the basepoint in this space is denoted by $\mathbf{E}'_r$.

Recall that, for sequences of non-negative integers $I = (i_0, i_1, \ldots)$ and $J = (j_0, j_1, \ldots)$ with only finitely many of the i_k and j_l non-zero, we have elements in the Hopf ring for the BP spectrum

$$[v^I] \circ b^J = [p^{i_0} v_1^{i_1} \cdots] \circ b_{(0)}^{\circ j_0} \circ b_{(1)}^{\circ j_1} \circ \cdots.$$

Here, $v \in \pi_{-r}(BP)$, regarded as a map from a point into $\mathbf{BP}_r$, gives rise to the element $[v] \in H_0(\mathbf{BP}_r)$ as the image under v_* of a selected generator of $H_0(\text{point})$. The element $b_{(s)}$ is defined as b_{p^s}, where $b_t \in H_{2t}(\mathbf{BP}_2)$ is the image of β_t, the standard $2t$ dimensional generator of $H_*(CP^\infty)$, under the map given by the complex orientation for BP,

$$\omega_{BP} : CP^\infty \longrightarrow \mathbf{BP}_2.$$

Similarly, for sequences I with $i_k = 0$ for $k > n$, we have elements $[v^I] \circ b^J$ in $H_*(\mathbf{BP\langle n\rangle}_*)$, and, likewise, corresponding elements in the Hopf ring for $E(n)$.

The homology coproduct, ψ say, on all these elements is generated via the Hopf ring formulæ

$$\begin{aligned} \psi([v]) &= [v] \otimes [v], \\ \psi(b_t) &= \sum_{j=0}^{t} b_j \otimes b_{t-j}, \\ \psi(x \circ y) &= \psi(x) \circ \psi(y). \end{aligned} \tag{1}$$

For $E = BP$, $BP\langle n\rangle$ or $E(n)$, denote by $H_*^R(\mathbf{E}_*)$ the free $\mathsf{F}_p[E^*]$ Hopf ring on generators $b_t \in H_{2t}(\mathbf{E}_2)$ and the suspension element $e_1 \in H_1(\mathbf{E}_1)$, modulo the formal group law relations ([6], (3.8)) and the identity $e_1 \circ e_1 = b_1$. Then we have isomorphisms

$$\begin{aligned} H_*^R(\mathbf{E}_*) &\cong H_*^R(\mathbf{BP}_*) \otimes_{\mathsf{F}_p[BP^*]} \mathsf{F}_p[E^*] \\ &\cong H_*(\mathbf{BP}_*) \otimes_{\mathsf{F}_p[BP^*]} \mathsf{F}_p[E^*] \end{aligned} \quad (2)$$

where the first is by construction and the second by [6]. There is a natural Hopf ring map

$$\tau_E : H_*^R(\mathbf{E}_*) \longrightarrow H_*(\mathbf{E}_*).$$

We recall further results from [6]. There it is shown that the Hopf algebras $H_*(\mathbf{BP}'_r)$ are bipolynomial for r even and exterior for r odd. Moreover, for each r, a certain set of the $[v^I] \circ b^J \in H_*(\mathbf{BP}_*)$ are defined, projecting to a basis of the module of indecomposables $QH_*(\mathbf{BP}_{2r})$. Let us name the indexing sets of these basis elements, registered as certain pairs of sequences (I, J), by $\mathcal{A}_r$. Thus

$$QH_*(\mathbf{BP}'_{2r}) \cong \mathsf{F}_p\{[v^I] \circ b^J | (I,J) \in \mathcal{A}_r\}. \quad (3)$$

Then [6] also allows us to describe the indecomposables in the homology of the odd degree spaces by

$$QH_*(\mathbf{BP}_{2r+1}) \cong \mathsf{F}_p\{[v^I] \circ b^J \circ e_1 | (I,J) \in \mathcal{A}_r\}. \quad (4)$$

Theorem 5. *The map $\tau_{E(n)}$ is an isomorphism. Moreover, the algebra $H_*(\mathbf{E(n)}'_r)$ is exterior for r odd and is polynomial for r even.*

Proof. We begin by proving the second part of the statement. First recall Wilson's splitting theorem:

Theorem 6, (W. S. Wilson, [8], (5.4)). *For $r < 2(p^n + \cdots + p + 1)$ there is an equivalence of H-spaces*

$$\mathbf{BP}_r \simeq \mathbf{BP\langle n\rangle}_r \times \prod_{j>n} \mathbf{BP\langle j\rangle}_{r+2(p^j-1)}.$$

□

We deduce:

Corollary 7. *For r sufficiently small the Hopf algebra $H_*(\mathbf{BP\langle n\rangle}'_r)$ is bipolynomial if r is even and exterior if r is odd. In particular, for r both even and small, $K(n)^*(\mathbf{BP\langle n\rangle}_r)$ is a formal power series algebra.*

Proof. Theorem 6 gives a factorisation of the identity map

$$\mathbf{BP}\langle\mathbf{n}\rangle'_r \xrightarrow{i} \mathbf{BP}'_r \xrightarrow{\pi} \mathbf{BP}\langle\mathbf{n}\rangle'_r$$

for r sufficiently small. This gives a split monomorphism, i_*, in homology, and so $H_*(\mathbf{BP}\langle\mathbf{n}\rangle'_r)$ is polynomial for r even as a subalgebra of the polynomial algebra $H_*(\mathbf{BP}'_r)$. Likewise, π^* in cohomology shows it in fact to be bipolynomial. For r odd, the composite shows that i_* is a monomorphism into an exterior algebra which also induces a monomorphism on indecomposables. Hence $H_*(\mathbf{BP}\langle\mathbf{n}\rangle'_r)$ is itself exterior.

The $K(n)$ cohomology result now follows from a simple application of the Atiyah-Hirzebruch spectral sequence. The sequence collapses as it is all in even dimensions and there are no multiplicative extension problems. □

The element $v_n \in \pi_{2(p^n-1)}(BP\langle n\rangle)$ gives rise to maps

$$\mathbf{BP}\langle\mathbf{n}\rangle_r \xrightarrow{v_n} \mathbf{BP}\langle\mathbf{n}\rangle_{r-2(p^n-1)} \tag{8}$$

which in homology represent $\circ$-multiplying by $[v_n]$. The spaces in the Ω-spectrum for $E(n)$ are given by the direct limits of iterations of these maps:

$$\mathbf{E}(\mathbf{n})_r = \operatorname{colim}\{\mathbf{BP}\langle\mathbf{n}\rangle_r \xrightarrow{v_n} \mathbf{BP}\langle\mathbf{n}\rangle_{r-2(p^n-1)} \xrightarrow{v_n} \cdots\}.$$

The formula for the distribution of $\circ$ over $*$-products, ([**6**], (1.12)(c)(*vi*)),

$$a \circ (b * c) = \Sigma(-1)^{\deg a'' \deg b}(a' \circ b) * (a'' \circ c),$$

where $\psi(a) = \Sigma a' \otimes a''$, shows that the map (8) gives rise in homology to a map of algebras. This follows from the coproduct formula (1) for $[v_n]$. Thus $H_*(\mathbf{E}(\mathbf{n})'_r) = \operatorname{colim}_j\{H_*(\mathbf{BP}\langle\mathbf{n}\rangle'_{r-2j(p^n-1)})\}$ is a direct limit of polynomial (exterior) algebras, when r is even (odd) and hence is itself polynomial (exterior).

Remark 9. The above also allows us to compute the cohomology ring $H^*(\mathbf{E}(\mathbf{n})'_{2r})$. Tor of an exterior algebra is a divided power algebra and so the dual Rothenberg-Steenrod spectral sequence

$$\{\operatorname{Tor}^{H_*(\mathbf{E}(\mathbf{n})_{2r-1})}(\mathsf{F}_p,\mathsf{F}_p)\}^* \Longrightarrow H^*(\mathbf{E}(\mathbf{n})'_{2r})$$

collapses as it is entirely in even dimensions. However, unlike in the case of the MU Hopf ring, the algebra $H_*(\mathbf{E}(\mathbf{n})'_{2r-1})$ is not of finite type, and so we are unable to conclude that the E_∞-term of this sequence, and hence $H^*(\mathbf{E(n)}'_{2r})$, is polynomial – it is in fact far larger.

It remains to show that $\tau_{E(n)}$ is an isomorphism. This will follow by showing that

$$\tau_{BP\langle n\rangle}: H^R_*(\mathbf{BP\langle n\rangle}_r) \longrightarrow H_*(\mathbf{BP\langle n\rangle}_r)$$

is an isomorphism for sufficiently small r.

Consider the diagram

$$\begin{array}{ccc} H_*(\mathbf{BP}_r) & \xrightarrow{q_*} & H_*(\mathbf{BP\langle n\rangle}_r) \\ \| & & \uparrow \tau_{BP\langle n\rangle} \\ H^R_*(\mathbf{BP}_r) & \xrightarrow{\rho} & H^R_*(\mathbf{BP\langle n\rangle}_r). \end{array}$$

The homomorphism ρ is epimorphic by (2). We recall:

Proposition (10), (K. Sinkinson, [7], (2.4)). *For* $r < p^n + \cdots + p + 1$,

$$QH_*(\mathbf{BP\langle n\rangle}'_{2r}) \cong \mathsf{F}_p\{[v^I] \circ b^J | (I, J) \in \mathcal{A}_r \text{ and } i_q = 0 \text{ for } q > n\}. \quad \square$$

As $H_*(\mathbf{BP\langle n\rangle}'_{2r})$ is a polynomial algebra for r sufficiently small, a simple application of the usual (homology) Rothenberg-Steenrod spectral sequence shows that for such r, $H_*(\mathbf{BP\langle n\rangle}_{2r+1})$ is exterior and

$$QH_*(\mathbf{BP\langle n\rangle}_{2r+1}) \cong \mathsf{F}_p\{[v^I] \circ b^J \circ e_1 | (I, J) \in \mathcal{A}_r \text{ and } i_q = 0 \text{ for } q > n\}.$$

Thus q_* and $\tau_{BP\langle n\rangle}$ are epimorphic for r sufficiently small. If we let Qq_* and $Q\rho$ denote the induced maps on indecomposables, then (3) and (4) show that

$$\ker Qq_* \subset \ker Q\rho$$

since ρ acts by tensoring with the coefficients and hence certainly kills any element $[v^I] \circ b^J$ or $[v^I] \circ b^J \circ e_1$ with some i_q, $q > n$, non-zero. The commutative diagram above gives the opposite inclusion, and so these kernals are equal. Then $\ker q_* = \ker \rho$ since $H_*(\mathbf{BP\langle n\rangle}'_r)$ is a free algebra. Hence $\tau_{BP\langle n\rangle}$ is isomorphic for all r sufficiently small. The result for $\tau_{E(n)}$ now follows by passing to direct limits. $\square$

§3 Lifting Results

In this section we complete our work by proving the following two results.

Theorem 11. *If X is a finite complex with $K(n)^{\text{odd}}(X) = 0$, then the map $E(n)^*(X) \longrightarrow K(n)^*(X)$ is an epimorphism.*

Theorem 12. *For each finite complex X satisfying $K(n)^{\text{odd}}(X) = 0$ we can construct a space Y, whose $K(n)$-cohomology is a formal power series algebra over $K(n)_*$, and a map $e: X \longrightarrow Y$ giving an epimorphism in $K(n)^*(-)$.*

Proof of Theorem 11. We must recall the $\widehat{E(n)}$ spectra of [2] and [3]. It is shown there that there is a tower of spectra

$$\cdots \longrightarrow E(n)/I_n^{k+1} \longrightarrow E(n)/I_n^k \longrightarrow \cdots \longrightarrow E(n)/I_n = K(n)$$

where the homotopy of the spectrum $E(n)/I_n^k$ is $E(n)_*/I_n^k$ and I_n is the usual ideal $(p, v_1, \ldots, v_{n-1})$ in $E(n)_*$. Each $E(n)/I_n^k$ is a module spectrum over the ring spectrum $\widehat{E(n)} = \text{holim}_{\leftarrow} E(n)/I_n^k$.

For X a finite complex we have

$$\widehat{E(n)}^*(X) = \lim_{\leftarrow k} E(n)^*(X)/I_n^k \cdot E(n)^*(X).$$

In particular, the natural map $\widehat{E(n)}^*(X) \longrightarrow K(n)^*(X)$ factors through $E(n)^*(X)/I_n \cdot E(n)^*(X)$. Thus for X finite, $\widehat{E(n)}^*(X) \longrightarrow K(n)^*(X)$ is an epimorphism if and only if the map $E(n)^*(X) \longrightarrow K(n)^*(X)$ is.

Now the map $E(n)/I_n^{k+1} \longrightarrow E(n)/I_n^k$ has fibre $\prod_\nu \Sigma^{|\nu|} K(n)$, where ν runs over all $p^{r_0} v_1^{r_1} \cdots v_{n-1}^{r_{n-1}}$ with $r_0 + r_1 + \cdots + r_{n-1} = k$. Thus, as $|\nu|$ is always even, we see that if $K(n)^{\text{odd}}(X) = 0$ then $(E(n)/I_n^k)^{\text{odd}}(X) = 0$ and the homomorphisms $(E(n)/I_n^{k+1})^*(X) \longrightarrow (E(n)/I_n^k)^*(X)$ are epimorphisms for all k. As $\widehat{E(n)}^*(X)$ for a finite complex X is also given by $\lim_{\leftarrow k} (E(n)/I_n^k)^*(X)$, we have:

Lemma 13. *If $K(n)^{\text{odd}}(X) = 0$ then the map $\widehat{E(n)}^*(X) \longrightarrow K(n)^*(X)$ is epimorphic.* □

This completes the proof of Theorem 11. □

Proof of Theorem 12. It suffices to prove the result for X connected. Up to a multiple by a power of v_n, an element $x \in K(n)^{2r}(X)$ can be considered as a map $X \longrightarrow \mathbf{K(n)}'_{2r}$. Theorem 11 allows x to be lifted to a map $\dot{x}: X \longrightarrow \mathbf{E(n)}'_{2r}$. As $\mathbf{E(n)}'_r = \text{colim}_j \{\mathbf{BP}\langle\mathbf{n}\rangle'_{r-2j(p^n-1)}\}$ there is also a lift $\ddot{x}: X \longrightarrow \mathbf{BP}\langle\mathbf{n}\rangle'_{2r-2j(p^n-1)}$, for every j sufficiently large. We can arrange by Corollary 7, perhaps by picking a j even larger, for $K(n)^*(\mathbf{BP}\langle\mathbf{n}\rangle'_{2r-2j(p^n-1)})$ to be a formal power series algebra. Let x run over a set of $K(n)_*$ generators of $K(n)^*(X)$ (or, alternatively, just a set of multiplicative generators), and take Y as the product of all the corresponding spaces $\mathbf{BP}\langle\mathbf{n}\rangle'_{2r-2j(p^n-1)}$. Then the required map $e: X \longrightarrow Y$ is given by the product of all the resulting lifts $\ddot{x}$. □

Remark 14. The same argument as in the above proof shows that we could form a map $e': X \longrightarrow Y'$ from the first lifts $\dot{x}$, the space Y' now being composed from even graded spaces in the Ω-spectrum for $E(n)$. This would not be a satisfactory solution to our original problem as the algebras $K(n)^*(\mathbf{E}(\mathbf{n})'_{2r})$ fail to be polynomial, as noted in Remark 9. However, these algebras are free of nilpotent elements, and so such a map e' would qualify as a ULE in the sense of [4].

Remark 15. Lemma 13 shows that we could drop the requirement of X being a finite complex in the preceeding arguments, or at least reduce it to include spaces X with $K(n)^*(X)$ finitely generated over $K(n)_*$ as a ring, if we had suitably good structure results for the spaces in the Ω-spectrum for $\widehat{E(n)}$. We hope to return to this point in a future paper.

§4 Concluding remarks

We finish with a couple of observations. Firstly, the results of §2 on the Hopf ring $H_*(\mathbf{E}(\mathbf{n})_*)$ allow us to construct the following spectral sequences:

Corollary 16. *There are spectral sequences*

$$\mathrm{Tor}_{**}^{E(n)_*}(E(n)_*(X), K(n)_*) \Longrightarrow K(n)_*(X),$$
$$\mathrm{Ext}_{E(n)_*}^{**}(E(n)_*(X), K(n)^*) \Longrightarrow K(n)^*(X),$$

and similarly, with $K(n)$ replaced by any other suitable $E(n)$-module spectrum.

Proof. These are of course the universal coefficient theorems of [1], lecture I. It suffices to demonstrate that the spectrum $E(n)$ can be written as the direct limit of finite spectra E_α which satisfy the hypotheses of Assumption 20 of [1] (page 28). The demonstrated homological structure of the spaces $\mathbf{E}(\mathbf{n})_r$ show that this can indeed be done, similarly to the way Adams argues for the KU spectrum on page 30 (*vi*) of [1]. □

Secondly, returning to our motivating problem, we note the following analogous result.

Theorem 17. *Let X be a space satisfying $H^{\mathrm{odd}}(X; \mathbf{Z}_{(p)}) = 0$ and with $H^*(X; \mathbf{Z}_{(p)})$ finitely generated over $\mathbf{Z}_{(p)}$ as a ring. Then there is a space Y with $H^*(Y; \mathbf{Z}_{(p)})$ a polynomial algebra and a map $e: X \longrightarrow Y$ giving an epimorphism in $H^*(-; \mathbf{Z}_{(p)})$.*

Proof. The $H^{\mathrm{odd}}(X; \mathbf{Z}_{(p)}) = 0$ condition means that the Atiyah-Hirzebruch spectral sequence

$$H^*(X; BP_*) \Longrightarrow BP^*(X)$$

collapses and the edge homomorphism $BP^*(X) \longrightarrow H^*(X; \mathbf{Z}_{(p)})$ is epimorphic. We can now argue as in Remark 14 and the proof of Theorem 12 using the result from [6] that $H^*(\mathbf{BP}'_{2r}; \mathbf{Z}_{(p)})$ is a polynomial algebra. Note that the finiteness condition on $H^*(X; \mathbf{Z}_{(p)})$ is introduced purely to avoid having to form the map e out of an infinite product of lifts $X \longrightarrow \mathbf{BP}_{2r}$; still more general spaces can be dealt with in particular circumstances. □

References

[1] J. F. Adams, Lectures on generalised cohomology, in Lecture Notes in Mathematics, vol. **99** (1969), Springer–Verlag, Berlin–Heidelberg–New York.

[2] A. Baker, A_∞ structures on some spectra related to Morava K-theories, to appear, *Quart. J. Maths.*

[3] A. Baker and U. Würgler, Liftings of formal groups and the Artinian completion of $v_n^{-1}BP$, *Math. Proc. Camb. Phil. Soc.* **106** (1989), 511-530.

[4] J. R. Hunton, The Morava K-theory of wreath products, *Math. Proc. Camb. Phil. Soc.* **107** (1990), 309-318.

[5] J. R. Hunton, *On Morava's Extraordinary K-Theories*, Ph.D. thesis, Cambridge University, 1989.

[6] D. C. Ravenel and W. S. Wilson, The Hopf ring for complex cobordism, *Journal of Pure and Applied Algebra* **9** (1977), 241-280.

[7] K. Sinkinson, The cohomology of certain spaces associated with the Brown-Peterson spectrum, *Duke Math. J.* **43** (1976), 605-622.

[8] W. S. Wilson, The Ω-spectrum for Brown-Peterson cohomology part II, *Amer. J. Math.* **97** (1975), 101-123.

Manchester University and Trinity College, Cambridge.

On the p-adic Interpolation of Stable Homotopy Groups

N. P. Strickland *

September 27, 1991

1 Philosophy

This paper is based on a talk given by M.J. Hopkins at the Adams Memorial Symposium, amplified by further discussions with him. While it contains a number of theorems, it is primarily intended to describe a certain philosophy for approaching and describing phenomena in stable homotopy theory. The story begins with Frank Adams' ideas about im$J$ and K-theory self-maps and leads via BP theory to Ravenel's chromatic tower. This tower of localization functors with respect to combinations of Morava K-theories can be seen as breaking the p-local stable category into simple layers. The work of Jack Morava establishes a close connection between these layers and the cohomology of modules over certain p-adic Lie groups.

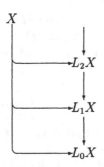

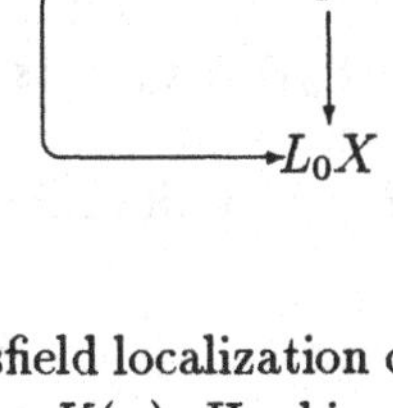

The n'th term L_nX is the Bousfield localization of X with respect to $E(n)$, or equivalently $v_n^{-1}BP$ or $K(0)\vee \ldots K(n)$. Hopkins and Ravenel have proved the

*I am grateful for the support of the Sims Fund.

"Chromatic Convergence Theorem", that for X finite, the tower $\{\pi_* L_n X\}_{n\geq 0}$ is pro-isomorphic to the constant group $\pi_* X$. (This is a stronger statement than that $\pi_* X \simeq \lim \pi_* L_n X$).

- L_0 was understood by Serre – rational stable homotopy is essentially equivalent to rational graded linear algebra.
- $L_1 S^0$ was understood by Adams and others – this is essentially the image of J and is connected with number theory via the Bernoulli numbers.
- The only other global information available about such localizations is the determination of $L_2 M_p$ for $p \geq 5$ by Katsumi Shimomura.

We need a systematic language with new concepts to reflect the patterns seen in such computations. Our whole program is motivated by an attempt to understand the results of Shimomura, and it owes much to conversations with Mark Mahowald, Doug Ravenel and Ethan Devinatz.

To get a "low resolution" picture of the chromatic tower we would like to investigate the difference between successive layers. One way to do this would be to take the fibre of the natural map $L_n X \longrightarrow L_{n-1} X$. A different and more fruitful procedure is to note that L_n is localization with respect to $K(0) \vee \ldots K(n)$ and so look at $L_{K(n)}$.

Morava has defined homology theories $\mathcal{K}_n(X)$ with coefficients:

$$\mathcal{K}_n(S^0) \simeq W\mathsf{F}_{p^n}[\![u_1 \ldots u_{n-1}]\!][u^{\pm 1}] \quad |u_i| = 0 \quad |u| = 2$$

The Morava stabilizer group S_n acts naturally in this theory. (We shall take S_n to mean the group of all automorphisms of the height n formal group law over F_{p^n}, not merely the strict ones). Morava also set up an Adams type spectral sequence:

$$\mathrm{H}^*(S_n; \mathcal{K}_n(X)) \Longrightarrow \pi_* L_{K(n)} X \otimes W\mathsf{F}_{p^n}$$

The Galois group $G = Gal(\mathsf{F}_{p^n}) \simeq C_n$ acts on the spectral sequence and the Galois invariant part converges to $\pi_* L_{K(n)} X$ itself. For $X = S^0$, the spectral sequence collapses for almost all primes.

For $\pi_* L_{K(1)} S^0$ we have the following diagram:

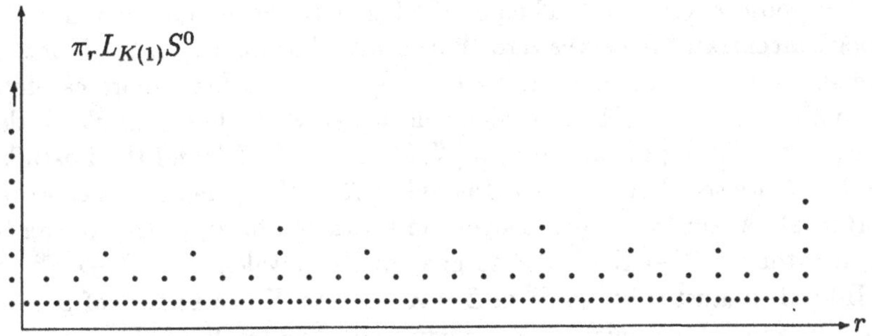

with $\pi_{kq-1} L_{K(1)} S^0 \simeq \mathbb{Z}_p/(pk)$ and $\pi_0 L_{K(1)} S^0 \simeq \mathbb{Z}_p$.

We would hope that $H^*(S_n; \mathcal{K}_n S^0)$ might behave similarly, with the term in dimension r depending on the p-adic shape of r.

2 Picard groups

We now discuss the idea of Picard groups, which begin to allow us to organize these phenomena. The functor $\Sigma^n : X \mapsto S^n \wedge X$ is an automorphism of the category of spectra, which preserves cofibrations and infinite wedges. If T is another such automorphism, then applying Brown representability to $\pi_* TX$ one finds a spectrum S_T with $TX \simeq S_T \wedge X$, and

$$S_{T^{-1}} \wedge S_T \simeq S^0$$

This motivates the following definition:

Definition 2.1 *A spectrum Z is invertible iff*

$$Z \wedge W \simeq S^0$$

for some spectrum W. Pic is the group of isomorphism classes of invertible spectra, with multiplication given by the smash product. Given an isomorphism class $\lambda \in Pic$ we write S^λ for a representative spectrum.

The idea of defining a Picard group of invertible objects of a category was first used in algebraic topology by Adams and Priddy, in their paper on the uniqueness of BSO. Their group was built from modules rather than spectra.

We have become used to the idea that in order to understand a spectrum X fully, we need to understand $\Sigma^n X$ for all n at the same time. The philosophy here is to extend this to all invertible spectra. However, the following result shows that in the full stable category, we get nothing new:

Theorem 2.2 *Pic* $\simeq \mathbb{Z}$ *generated by* S^1

Proof Suppose $Z \wedge W \simeq S^0$. Then by the Künneth theorem, clearly $H\mathbb{Q}_* Z \simeq \mathbb{Q}$, wlog concentrated in degree zero. We need to show that $Z \simeq S^0$. Using the Universal Coefficient theorem and Künneth theorem a little more carefully, we find that $H_* Z \simeq H_* W \simeq \mathbb{Z}$ again in degree zero. One can check that $Z \simeq F(W, S^0)$. Using the adjunction $[X, Z] \simeq [X \wedge W, S^0]$ and the Postnikov tower for S^0 we see that $f : X \to Z$ is null if X is H-acyclic. In other words, Z is H-local. Again by induction over the Postnikov tower for S^0 we can lift the generator $u : Z \to H$ of $H^0 Z$ to give an H-equivalence $v : Z \to S^0$. As Z is H-local, v must split, so $S^0 \simeq Z \vee X$ say. As X is a retract of S^0, it is connective and it has vanishing homology so it is contractible. Thus $Z \simeq S^0$. □

The situation changes, however, if we look at the $K(n)$-local category. Notice that this is not closed under smash products, but that it is again a symmetric monoidal category with the operation $(X, Y) \mapsto L_{K(n)}(X \wedge Y)$ and unit $L_{K(n)} S^0$.

Definition 2.3 *A* $K(n)$*-local spectrum* Z *is invertible iff*

$$L_{K(n)}(Z \wedge W) \simeq L_{K(n)} S^0$$

for some $K(n)$*-local spectrum* W*.* Pic_n *is the group of such spectra.*

We expect the functor $\mathcal{K}_n$ to reflect much of the structure of the $K(n)$-local category and thus to be a powerful invariant for the study of Pic_n. We at least have the following result:

Theorem 2.4 *The following are equivalent:*

1. $L_{K(n)} Z$ *is invertible in the* $K(n)$*-local category.*
2. $\dim_{K(n)_*} K(n)_* Z = 1$
3. $\mathcal{K}_n(Z)$ *is free of rank one over* $\mathcal{K}_n(S^0)$

Proof The hard part is 3 ⇒ 1. Set $W = F(Z, L_{K(n)} S^0)$ (which is already $K(n)$-local). We claim that the evaluation map

$$L_{K(n)}(Z \wedge W) \to L_{K(n)} S^0$$

is an equivalence. Choose a finite spectrum U acyclic for $K(n-1)$ but not $K(n)$. It suffices to show that

$$U \wedge Z \wedge W \to U \wedge L_{K(n)} S^0$$

is a $K(n)$- equivalence. To do this, we consider the category $\mathcal{C}$ of spectra Y such that

$$Z \wedge F(Z, L_{K(n)}Y) \to L_{K(n)}Y$$

is an equivalence. $\mathcal{C}$ is closed under cofibrations and retracts. Using $K(n)^*Z \simeq K(n)^*$ we can show that $K(n) \in \mathcal{C}$. It is a theorem of Hopkins and Ravenel that $L_{K(n)}U$ has a finite $K(n)$-Postnikov tower and thus lies in $\mathcal{C}$. It follows that U itself is in $\mathcal{C}$. As U is finite, $F(Z, L_{K(n)}U) \simeq U \wedge W$. The result follows.

□

Notation:
We shall write $\mathcal{O}$ for the degree zero part $W\mathbb{F}_{p^n}[\![u_1 ... u_{n-1}]\!]$ of $\mathcal{K}_n S^0$.
Pic_n^0 will denote the even dimensional part of Pic_n. It is a subgroup of index two, and $L_{K(n)}S^1$ is a canonical representative of the non-zero coset.

Suppose $Z \in \mathrm{Pic}_n^0$, so that $\mathcal{K}_n Z$ has a free generator e in dimension 0. Thus for $g \in S_n$ there is a unique $\varepsilon_g \in \mathcal{O}^\times$ such that $ge = \varepsilon_g e$. The map $g \mapsto \varepsilon_g$ is a crossed homomorphism, and it varies by a coboundary if we choose a different zero-dimensional generator, so we obtain a well-defined element of $\mathrm{H}^1(S_n; \mathcal{O}^\times)$ which in fact lies in the Galois invariant subgroup. This gives an exact sequence:

$$1 \longrightarrow \kappa_n \longrightarrow \mathrm{Pic}_n^0 \longrightarrow \mathrm{H}^1(S_n; \mathcal{O}^\times)^G$$

κ_n is simply defined as the kernel. Suppose $Z \in \kappa_n$. Then the E_2-term of the $\mathcal{K}_n$- based Adams spectral sequence coincides with that for the sphere, ie $\mathrm{H}^*(S_n; \mathcal{K}_n(S^0))$, although the differentials may differ. If the unit class in H^0 is a permanent cycle, we have a $K(n)$-equivalence $S^0 \to Z$. Hopkins and Ravenel have shown that the spectral sequence has a horizontal vanishing line at E^n for large n. These ideas give a (rather poor) upper bound on κ_n. In particular, it is an inverse limit of finite p-groups, and for fixed n it vanishes when $p \gg 0$.

Conjecture: *κ_n is always a finite p-group.*

This would follow from the nilpotence machinery if we could prove that $\mathrm{H}^*(S_n; \mathcal{K}_n(S^0))$ was always finitely generated over $W\mathbb{F}_{p^n}$.

One might hope for an analog of the Artin reciprocity law which would tell us that the behaviour at small primes is completely determined by the behaviour elsewhere.

A few cases of κ have been determined:

$$\kappa_1 = \begin{cases} 0 & p \text{ odd} \\ \mathbf{Z}/(2) & p = 2 \text{ (due to Adams)} \end{cases}$$

$$\kappa_2 = \begin{cases} 0 & p \geq 5 \\ ? & p = 3 \\ \neq 0 & p = 2 \end{cases}$$

Tentative calculations of Don Davis and Mark Mahowald indicate that κ_2 for $p = 2$ contains a subgroup $\mathbf{Z}/(4) \times \mathbf{Z}/(2)$.

For $n = 1$, $S_1 = \mathbf{Z}_p^\times$ and $\mathcal{O}^\times = \mathbf{Z}_p^\times$ (for p odd) so

$$\mathrm{H}^1(S_1; \mathcal{O}^\times)^G = \hom(\mathbf{Z}_p^\times, \mathbf{Z}_p^\times) = \mathbf{Z}_p \times \mathbf{Z}/(p-1) = \mathrm{Pic}_1^0$$

If $Z \in \mathrm{Pic}_1^0$ corresponds to $\vartheta \in \mathrm{end}(\mathbf{Z}_p^\times)$ then the Adams operation ψ^k acts on K^*Z in dimension zero as multiplication by $\vartheta(k)$. Conversely, if k is a topological generator of $\mathbf{Z}_p^\times$, then Z is the fibre of

$$\psi^k - \vartheta(k) : K_p^\wedge \longrightarrow K_p^\wedge$$

For $n = 1$ and $p = 2$,

$$\mathrm{Pic}_1 = \mathbf{Z}/(4) \times \mathbf{Z}/(2) \times \mathbf{Z}_2$$

We next recall that S_n is a p-adic Lie group. Such groups have been extensively studied and their behaviour is in many ways similar to that of real Lie groups - they have associated Lie algebras, open subgroups of finite cohomological dimension , and maximal tori (although these are not always conjugate). In the case of S_n, the stabilizer T of $(u_1, \ldots u_{n-1})$ is a maximal torus and is isomorphic to $W\mathbf{F}_{p^n}^\times$. We have natural maps:

$$\mathrm{Pic}_n^0 \longrightarrow \mathrm{H}^1(S_n; \mathcal{O}^\times)^G \longrightarrow \mathrm{H}^1(T; W\mathsf{F}_{p^n}^\times)^G \simeq \mathrm{end}(W\mathsf{F}_{p^n}^\times)^G \simeq \mathbf{Z}/(p^n-1) \times \mathbf{Z}_p^n$$

In the case $n = 2, p \geq 5$ the whole composite is an isomorphism. Unfortunately, we can only prove this using the full force of Shimomura's calculations in BP - we would like information to flow in the opposite direction, as it were. For $n > 2$ the composite is not an isomorphism, but the image of Pic_n^0 is the same as the image of the "algebraic Picard group" $\mathrm{H}^1(S_n; \mathcal{O}^\times)^G$. This image consists of the endomorphisms

$$\lambda \mapsto \lambda^a |\lambda|^b$$

of $W\mathsf{F}_{p^n}^\times$, for $a, b \in \mathbf{Z}_p$.

3 Applications

3.1 The generating hypothesis

Freyd conjectured that the spheres generate the finite stable category. More explicitly:
Let $f : X \to Y$ be a map of finite spectra. Then:

$$\pi_* f = 0 \Rightarrow f = 0 \qquad \text{(GH)}$$

Many interesting consequences are known to follow from this. The program described below for attacking it is due to Devinatz. It follows from Chromatic Convergence that GH is equivalent to the conjunction for all n and p of the following:

$$\pi_* f = 0 \Rightarrow L_n f = 0 \qquad (\mathrm{GH}_n)$$

We also consider the subproblem GHS_n in which Y is restricted to be a sphere. GH_0 is easy. Devinatz has proved GHS_1 at odd primes.

By the Brown representability theorem and the injectivity of $\mathbf{Q}/\mathbf{Z}_{(p)}$ there is an essentially unique spectrum I_n such that

$$[X, I_n] \simeq \hom_{\mathbf{Z}}(\pi_0 L_n X, \mathbf{Q}/\mathbf{Z}_{(p)})$$

and natural maps $I_n \longrightarrow I_{n+1}$. (In fact Hopkins and Ravenel have shown that $L_n X = L_n S^0 \wedge X$ which implies that I_n is the Brown-Comenetz dual of $L_n S^0$.) We can try to write I_n as a colimit of finite spectra in a controlled way, say

$$I_n = \lim_{\substack{\rightarrow \\ \alpha}} Y_\alpha$$

It turns out that this provides a universal class of examples of GHS_n. It can also be shown that $\dim_{K(n)_*} K(n)_*(I_n/I_{n-1}) = 1$, so I_n/I_{n-1} defines an element of Pic_n. There is a formula due to Hopkins and B.H. Gross which identifies this element modulo κ_n. This enables one to write I_n as an appropriate kind of colimit. It is hoped that this kind of information may help to settle the conjecture.

3.2 Interpolation of homotopy groups

Given an element λ of Pic_n represented by a spectrum S^λ, we can define:

$$\pi_\lambda X = [S^\lambda, X]$$

for $K(n)$-local spectra X. This idea of indexing objects over p-adic rings etc. was first introduced by Serre, in the theory of modular forms. There is a method due to Iwasawa and Mazur which explains the sense in which the groups π_λ interpolate the usual groups π_n. To explain this, we first define the dual Picard group Pic_n^* which is (roughly speaking) $\hom(\mathrm{Pic}_n^0, W\mathbf{F}_{p^n}^\times)$ and is a profinite group. Next let Λ be the profinite group ring:

$$\Lambda = W\mathbf{F}_{p^n}[\![\mathrm{Pic}_n^*]\!] = \lim_{\substack{\leftarrow \\ U}} W\mathbf{F}_{p^n}[\mathrm{Pic}_n^*/U]$$

where U runs over open subgroups of finite index in Pic_n^*.

We would like to have a single Λ-module πX corresponding to the collection of groups $\{\pi_\lambda X\}_{\lambda \in \mathrm{Pic}_n}$. An element $\lambda \in \mathrm{Pic}_n$ gives an evaluation map $\mathrm{Pic}_n^* \to W\mathbf{F}_{p^n}^\times$ and in turn a ring homomorphism $\Lambda \to W\mathbf{F}_{p^n}$ which we will again denote by λ. We can take the tensor product of the Λ-module π and $W\mathbf{F}_{p^n}$ regarded as a Λ-algebra via λ, and we should get:

$$\pi_\lambda X \simeq \pi \otimes_{\Lambda,\lambda} W\mathbf{F}_{p^n}$$

For $n = 1$ at odd primes we have $\mathrm{Pic}_1^0 \simeq \mathrm{end}(\mathbf{Z}_p^\times)$ and so the map

$$\mathbf{Z}_p^\times \longrightarrow \mathrm{Pic}_1^* \simeq \mathrm{hom}(\mathrm{end}(\mathbf{Z}_p^\times), \mathbf{Z}_p^\times) \quad a \mapsto (\vartheta \mapsto \vartheta(a))$$

is an equivalence. Using the natural splitting $\mathbf{Z}_p^\times \simeq \mathbf{F}_p^\times \times U_1$ where $U_1 = 1 + p\mathbf{Z}_p \simeq \mathbf{Z}_p$ and idempotents corresponding to the characters of $\mathbf{F}_p^\times \subseteq \mathbf{Z}_p^\times$ we can split Λ as

$$\Lambda = \prod_0^{p-2} \mathbf{Z}_p[\![U_1]\!] \simeq \prod_0^{p-2} \mathbf{Z}_p[\![T]\!]$$

T corresponds to $k-1$ for a generator k of U_1 (there is no natural choice of k or T). The even dimensional homotopy of $L_{K(1)}S^1$ is essentially $\mathrm{Im}J$, and corresponds to the Λ-module $\pi = \mathbf{Z}_p$ with Pic_n^* acting trivially. This module is cyclic, with annihilator

$$(T) \times \prod_1^{p-2} \mathbf{Z}_p[\![T]\!] \subseteq \prod_0^{p-2} \mathbf{Z}_p[\![T]\!]$$

For $\vartheta \in \mathrm{Pic}_1^0$ with $\vartheta(k) = k^r$ we have $\pi \otimes_{\Lambda,\vartheta} \mathbf{Z}_p \simeq \mathbf{Z}_p[\![T]\!] \simeq \mathbf{Z}_p/(k^r - 1)$. Here $k^r - 1$ is a unit unless $p - 1 \mid r$ and $k^{(p-1)t} - 1 = (\text{unit}) \times pt$ so

$$\pi_{(p-1)t} S^1 \simeq \mathbf{Z}_p/pt$$

The formal Zariski spectrum of Λ is the disjoint union of $p-1$ copies of the affine line over $\mathbf{Z}_p$. Many of the qualitative features of this answer can be predicted from the fact that π is a cyclic module whose support is the origin in the zero'th copy.

Stable homotopy groups are usually only modules over the stable homotopy ring $\pi_* S^0$. The usual methods of commutative algebra are almost useless for this ring, as all positive dimensional elements are nilpotent. In contrast, Λ is very amenable to such an approach. We expect the behaviour of $\pi L_{K(n)} S^0$ to be largely controlled by the geometry of its support variety, which is in turn connected with the action of S_n on a certain moduli space. This space is equipped with a canonical vector bundle and a connection on it, and so has a rich geometric structure.

This situation is supposed to be analogous to one in algebraic geometry, where one has a variety X and analyzes the relationship between the number of points of X defined over various finite extension fields, the geometry of X, and its Weil ζ-function.

References

[1] J.F. Adams and S.B. Priddy. On the uniqueness of BSO. *Math. Proc. Camb. Phil. Soc.*, 80:475–509, 1978.

[2] M. Hazewinkel. *Formal Groups and Applications.* Academic Press, 1978.

[3] M. Lazard. *Groupes analytiques p-adiques.* Publ. Math. IHES, 1965.

[4] D.C. Ravenel. *Complex Cobordism and Stable Homotopy Groups of Spheres.* Academic Press, 1986.

N. P. Strickland
Department of Mathematics
University of Manchester
Oxford Road
Manchester M13 9PL

Some remarks on v_1-periodic homotopy groups

Donald M. Davis
Lehigh University
Bethlehem, PA 18015

Mark Mahowald
Northwestern University
Evanston, IL 60208 *

This paper is dedicated to the memory of Frank Adams, whose insights laid the foundation for this work.

1 A definition and some examples

In this paper we give a new, canonical, definition of v_1-homotopy groups, show that it is compatible with the original definition of [9], and discuss several examples related to this definition. The final form of the definition is due to Erich Ossa.

In [9], a definition was proposed for the v_1-periodic homotopy groups $v_1^{-1}\pi_*(X)$ of a space X for which $\Omega^L X \xrightarrow{p^e} \Omega^L X$ is null-homotopic for some integers e and L. The prime p is implicit in the $v_1^{-1}\pi_*(-)$ functor; there is a different $v_1^{-1}\pi_*(-)$ for each prime p. Spaces X satisfying the above hypothesis are sometimes said to have an H-space exponent at p; they include spheres and compact Lie groups. ([18], [19], [28])

The definition in [9] involved choices of a null-homotopy, the integer p^e, and an Adams map, and the argument to show that it was independent of these choices was at best strained. Indeed, it relied upon showing that one could choose the Adams map so that there was a split short exact sequence

$$0 \to v_1^{-1}\pi_i(X) \to v_1^{-1}\pi_i(X;\mathbf{Z}/p^e) \to v_1^{-1}\pi_{i-1}(X) \to 0 \qquad (1.1)$$

*Both authors were partially supported by the National Science Foundation. The first author was also supported by the S.E.R.C., and the second author by New College. The authors also thank the Mathematical Institute at Oxford, where this work was performed.

where the middle group, whose definition we recall below, does not involve a null-homotopy, and is independent of the choice of the Adams map. A byproduct of the work of this paper will be that the definition of [9] was also independent of the choice of the exponent p^e. Since the middle group of (1.1) is canonical, the groups $v_1^{-1}\pi_*(X)$ of [9] could have been considered to be canonical to the extent that they could be deduced from the middle group of (1.1). This can be done as long as there is no positive integer f such that $\mathbf{Z}/p^f$ appears as a direct summand in $v_1^{-1}\pi_i(X;\mathbf{Z}/p^e)$ for all integers i, and this has been the case in all applications to date.

The definition itself was not involved in the computations in [9], [2], and [3]. All that was required there was (1.1), so that one could use the calculation of the groups $v_1^{-1}\pi_*(S^{2n+1};\mathbf{Z}/p^e)$ in [21], [26], and [10] to obtain $v_1^{-1}\pi_*(S^{2n+1})$, and the usual sort of exactness property for fibrations.

In this note, we propose a new and canonical definition of $v_1^{-1}\pi_*(X)$, which involves no choice of null-homotopy or p^e, is independent of choice of Adams maps, and satisfies the two requisite properties, namely (1.1) and the exact sequence for fibrations. Moreover, this new definition applies to all spaces X, not just those with H-space exponents, although (1.1) is only claimed for spaces with H-space exponents. Using the proof that our new definition satisfies (1.1), we then show directly that it agrees with the old definition.

We now recall the definition of $v_1^{-1}\pi_*(X;\mathbf{Z}/p^e)$. Let $M^n(k)$ denote the Moore space $S^{n-1}\cup_k e^n$. The mod k homotopy group $\pi_n(X;\mathbf{Z}/k)$ is defined to be the set of homotopy classes $[M^n(k),X]$. With the prime p implicit, let

$$s(e)=\begin{cases}2(p-1)p^{e-1} & \text{if } p \text{ is odd}\\ \max(8,2^{e-1}) & \text{if } p=2.\end{cases}$$

Let $A: M^{n+s(e)}(p^e)\to M^n(p^e)$ denote a map, as introduced by Adams in [1], which induces an isomorphism in K-theory, or, equivalently, in $K(1)_*(-)$, where the latter denotes Morava K-theory. We show in Proposition 2.11 that such a map can be defined provided $n\geq 2e+3$. Then $v_1^{-1}\pi_i(X;\mathbf{Z}/p^e)$ is defined to be $\operatorname{dirlim}_N\,[M^{i+Ns(e)}(p^e),X]$, where the maps A are used to define the direct system. See, for example, [21] for one of the first explicit mentions of a definition of this type. The map A is what Hopkins and Smith would call a v_1-map, and they showed in [17] that any two v_n-maps of a finite complex which admits such a map become homotopic after a finite number of iterations. Note that although $v_1^{-1}\pi_*(X;\mathbf{Z}/p^e)$ is a theory yielding information about the unstable homotopy groups of X, the maps A which define the direct system may be assumed to be stable maps, since the direct limit only cares about large values of $i+Ns(e)$. An alternative way to see that the choice of Adams map does not affect $v_1^{-1}\pi_*(X;\mathbf{Z}/p^e)$ is to use the result of [8] that unique Adams maps can be chosen subject to certain restrictions. Note also that the groups $v_1^{-1}\pi_i(X;\mathbf{Z}/p^e)$ are defined for all spaces X and all integers i and satisfy $v_1^{-1}\pi_i(X;\mathbf{Z}/p^e) \approx v_1^{-1}\pi_{i+s(e)}(X;\mathbf{Z}/p^e)$.

In order to make our new definition of $v_1^{-1}\pi_*(X)$, we use the map $\rho : M^n(p^{e+1}) \to M^n(p^e)$ which has degree p on the top cell, and degree 1 on the bottom cell. This map is entirely canonical; if $f_a : S^{n-1} \to S^{n-1}$ is the canonical map of degree a, satisfying $f_a \circ f_b = f_{ab}$, then

$$\rho : S^{n-1} \cup CS^{n-1}/([0,x] \sim f_{p^{e+1}}(x)) \longrightarrow S^{n-1} \cup CS^{n-1}/([0,x] \sim f_{p^e}(x))$$

is defined by $\rho([t,x]) = [t, f_p(x)]$, $\rho(x) = x$. We need the following lemma, which is a special case of [16, Lemma 4], or is part of [8, 1.1]. It was first proved in [15, p. 633].

Lemma 1.2 *If $A : M^{n+s(e)}(p^e) \to M^n(p^e)$ and $A' : M^{n+s(e+1)}(p^{e+1}) \to M^n(p^{e+1})$ are v_1-maps, then there exists k so that the following diagram commutes.*

$$\begin{array}{ccc} M^{n+ks(e+1)}(p^{e+1}) & \xrightarrow{\rho} & M^{n+ks(e+1)}(p^e) \\ \downarrow A'^k & & \downarrow A^{kp'} \\ M^n(p^{e+1}) & \xrightarrow{\rho} & M^n(p^e) \end{array}$$

Here $p' = p$ unless $p = 2$ and $e < 4$, in which case $p' = 1$.

Here, of course, n is any sufficiently large integer, and all maps A (resp. A') are suspensions of one another.

Thus, after sufficient iteration of the Adams maps, there are morphisms ρ^* between the direct systems used in defining $v_1^{-1}\pi_*(X;\mathbf{Z}/p^e)$ for varying e, and passing to direct limits, we obtain a direct system

$$v_1^{-1}\pi_*(X;\mathbf{Z}/p^e) \xrightarrow{\rho^*} v_1^{-1}\pi_*(X;\mathbf{Z}/p^{e+1}) \xrightarrow{\rho^*} \tag{1.3}$$

defined for any space X. One purpose of this paper is to introduce the following definition and show it equivalent to the definition of [9].

Definition 1.4 *For any space X and any integer i,*

$$v_1^{-1}\pi_i(X) = \operatorname{dirlim}_e v_1^{-1}\pi_{i+1}(X;\mathbf{Z}/p^e),$$

using the direct system in (1.3).

In this paper, we are only concerned with applying this definition to spaces with H-space exponents, but future investigation of its properties for other spaces should be interesting. For example, we propose the following conjecture, the "if" part of which is proved as Corollary 2.10.

Conjecture 1.5 *If X is a finite complex, then $v_1^{-1}\pi_*(X)$ has a finite period if and only if X has an H-space exponent.*

We now state the principal properties of our new definition of $v_1^{-1}\pi_*(-)$, postponing their proofs until the next section. The first two propositions show that our definition satisfies the properties that were used in the calculations of [9], [2], and [3].

Proposition 1.6 *If $F \to E \to B$ is a fibration, then there is a long exact sequence*

$$\to v_1^{-1}\pi_i(F) \to v_1^{-1}\pi_i(E) \to v_1^{-1}\pi_i(B) \to v_1^{-1}\pi_{i-1}(F) \to .$$

Proposition 1.7 *If X has H-space exponent p^e, then there is a split short exact sequence*

$$0 \to v_1^{-1}\pi_i(X) \to v_1^{-1}\pi_i(X;\mathbf{Z}/p^e) \to v_1^{-1}\pi_{i-1}(X) \to 0.$$

Note that p^e in this proposition refers to any sufficiently large p^e. In particular, we obtain a rather circuitous proof of the following result.

Corollary 1.8 *$v_1^{-1}\pi_*(X;\mathbf{Z}/p^e)$ is the same for all e such that p^e is an H-space exponent of X.*

Using the methods developed in the proof of Proposition 1.7, we can then prove the following satisfying result.

Proposition 1.9 *For spaces with H-space exponents, $v_1^{-1}\pi_*(X)$ as defined in this paper agrees with its definition in [9].*

As noted in [9], if X has an H-space exponent, then the space T, defined as the mapping telescope of a certain sequence of maps

$$\Omega^L X \to \Omega^{L+d} X \to \Omega^{L+2d} X \to \cdots,$$

satisfies $v_1^{-1}\pi_i(X) \approx \pi_{i-L}(T)$ for $i > L$. Note that $\Omega^d T \simeq T$. We prefer to think of this infinite loop space T as defining a periodic Ω-spectrum $\mathrm{Tel}(X, v_1)$, whose nth space is T for all $n \equiv -L \bmod d$. Then

$$v_1^{-1}\pi_*(X) \approx \pi_*(\mathrm{Tel}(X, v_1)).$$

There is a map from $\Omega^L X$ into the $-L$th space of the spectrum $\mathrm{Tel}(X, v_1)$ which induces an isomorphism in $v_1^{-1}\pi_*(-)$ in positive gradings. It is natural to ask if $\mathrm{Tel}(X, v_1)$ is some localization in the Bousfield sense ([4]) associated to X. We mention four examples. The proofs of these claims are given in Section 3.

Proposition 1.10 *There is an equivalence of spectra*

$$\mathrm{Tel}(S^{2n+1}, v_1) \approx v_1^{-1}\Sigma^{2n+1}B^{qn}.$$

The spectrum on the right in the above proposition is the telescope of v_1-maps of suspension spectra of the space B^{qn}, which is a skeleton of the p-localization of $B\Sigma_p$. See, e.g., [22] or [12]. As usual, $q = 2(p-1)$. This is the Bousfield localization $\Sigma^{2n+1}L_K(B^{qn})$ in the category of spectra.

One might wonder how $\mathrm{Tel}(S^{2n+1}, v_1)$ is related to $L_K(S^{2n+1})$, the K_*-localization in the category of spaces constructed in [22]. Of course, the former has nonzero homotopy groups in negative dimensions while the latter does not, but this can be circumvented by considering instead the 0th space in the Ω-spectrum, which is $\Omega^{d-L}T$, where T is as above. Another difference is that, whereas the homotopy groups of $v_1^{-1}\Sigma^{2n+1}B^{qn}$ are finite, $\pi_{2n+1}(L_K(S^{2n+1})) \approx \mathbf{Z}$, and $\pi_{2n-1}(L_K(S^{2n+1})) \supset \mathbf{Q}/\mathbf{Z}$, but this difference can be eliminated by applying Ω^{2n+2}. We can relate $\Omega^{2n+2}L_K(S^{2n+1})$ to the 0th space $\Omega^\infty(v_1^{-1}\Sigma^{2n+1}B^{qn})$ in the Ω-spectrum for $v_1^{-1}\Sigma^{2n+1}B^{qn}$, and hence to our $\mathrm{Tel}(S^{2n+1}, v_1)$, as follows.

Proposition 1.11 *There is an equivalence of spaces*

$$\Omega^{2n+2}L_K(S^{2n+1}) \approx \Omega^{2n+2}\Omega^\infty(v_1^{-1}\Sigma^{2n+1}B^{qn}).$$

Note that this implies that $\Omega^{2n+2}L_K(S^{2n+1})$ is an infinite loop space, although $L_K(S^{2n+1})$ is not.

The third example may be the most novel part of the paper. It was the source of much discussion at the conference. Here we localize at $p = 2$.

Proposition 1.12 *There is a finite suspension spectrum X_3 satisfying*

i. There is a cofibration of suspension spectra $\Sigma^3P^2 \to X_3 \to \Sigma^5P^4$;

ii. X_3 admits a v_1-map, and there is an equivalence of spectra

$$v_1^{-1}X_3 \equiv \mathrm{Tel}(SU(3), v_1);$$

iii. $ku_*(X_3)$ contains elements of order 8, but $4 \cdot ko_*(X_3) = 0$;

iv. The identity map of X_3 has order 8, but $4 \cdot v_1^{-1}\pi_*(X_3) = 0$.

Corollary 1.13 *The 2-primary H-space exponent of $SU(3)$ is ≥ 8, but*

$$4 \cdot v_1^{-1}\pi_*(SU(3))_{(2)} = 0.$$

We do not know of any elements in $\pi_*^s(X_3)$ of order 8, but we expect such elements to exist, for if they do not, $4 \cdot 1_{X_3}$ would provide a counterexample to Freyd's Generating Hypothesis ([13]). Similarly, we expect $\pi_*(SU(3))$ to have elements of order 8, but we do not know of any such elements. If such elements exist, they would give a counterexample to the conjecture in [9] that the p-exponent of $SU(n)$ equals the p-exponent of $v_1^{-1}\pi_*(SU(n))$. We still believe in this conjecture when p is odd; the difference in this case is related to the difference between bo and bu when localized at 2, which does not occur at the odd primes. The 2-primary H-space exponent of $SU(3)$ is no greater than the product of that of S^3 and S^5, which is $4 \cdot 8$ ([18], [6]). Using a more careful analysis of $\pi_*(S^5)$, one can show that the 2-primary H-space exponent of $SU(3)$ is ≤ 16.

These examples lead one to wonder for which spaces Y with an H-space exponent do there exist torsion spectra X so that $\mathrm{Tel}(Y, v_1)$ and $v_1^{-1}X$ are equivalent spectra. In an earlier draft of this paper, we conjectured that a finite spectrum X should always exist with this property. The referee pointed out that if Y is a mod p^e Moore space with p odd, then by [27] $\pi_*(\mathrm{Tel}(Y, v_1))$ is infinite dimensional over $\mathbf{Z}/p[v_1, v_1^{-1}]$, and so X satisfying

$$\mathrm{Tel}(Y, v_1) \simeq v_1^{-1}X$$

cannot be finite in this case. We thank the referee for this and other insightful criticism.

We close this section with some comments on stabilization. We shall say that a *generalized homotopy functor* is a covariant functor $\phi_*(-)$ from the homotopy category of spaces to $\mathbf{Z}$-graded abelian groups which is exact on fibrations. Then the functor $\phi_*^s(X) = \operatorname{dirlim}_n \phi_*(\Omega^n\Sigma^n X)$ is a generalized homology theory. For example if $\phi_*(-) = \pi_*(-)$ is ordinary homotopy groups, then of course $\phi_*^s(-) = \pi_*^s(-)$ is the stable homotopy groups. If $\phi_*(Y) = [\Sigma^j X, Y]$ for some finite complex X, then $\phi_*^s(-)$ is the generalized homology theory associated to the Spanier-Whitehead dual DX of X. If this X has a self map $v : \Sigma^d X \to X$, then $v^{-1}\phi_i(Y) = \operatorname{dirlim}_N [\Sigma^{i+Nd}X, Y]$ is a generalized homotopy functor whose associated homology theory is the generalized homology theory associated to the telescope of $DX \to \Sigma^{-d}DX \to \Sigma^{-2d}DX \to \cdots$. An important example of this occurs when v is an Adams map of the mod p^e Moore space, in which case

$$(v_1^{-1}\pi)_*^s(X; \mathbf{Z}/p^e) \approx \pi_*(L_{K/p^e}X).$$

Our final result characterizes the stabilization of the homotopy functor $v_1^{-1}\pi_*(-)$ of Definition 1.4 in terms of the functor M_1 discussed in [24]. Recall that if X is a spectrum, then M_1X is the spectrum $L_K(X_0/X)_{(p)}$, where X_0/X denotes the cofibre of the rationalization of X.

Proposition 1.14 *If $\phi_*(-) = v_1^{-1}\pi_*(-)$, then $\phi_*^s(X) = \pi_{*+1}(M_1 X)$.*

One problem which might be interesting is

Problem 1.15 *Characterize those homology theories which are the stabilization of some generalized homotopy functor.*

2 Proofs related to the definition

In this section, we prove Propositions 1.6, 1.7, 1.9, and a result, Proposition 2.11 regarding desuspensions of Adams maps of mod p^e Moore spaces.

Proof of Proposition 1.6. For each e, there is an exact sequence of $\pi_*(-;\mathbf{Z}/p^e)$. As direct limits preserve exactness, there is a commutative diagram of exact sequences

$$\begin{array}{ccccc} \to v_1^{-1}\pi_i(F;\mathbf{Z}/p^e) \to & & v_1^{-1}\pi_i(E;\mathbf{Z}/p^e) & & \to v_1^{-1}\pi_i(B;\mathbf{Z}/p^e) \to \\ \downarrow \rho^* & & \downarrow \rho^* & & \downarrow \rho^* \\ \to v_1^{-1}\pi_i(F;\mathbf{Z}/p^{e+1}) \to & & v_1^{-1}\pi_i(E;\mathbf{Z}/p^{e+1}) & & \to v_1^{-1}\pi_i(B;\mathbf{Z}/p^{e+1}) \to \\ \downarrow & & \downarrow & & \downarrow \end{array}$$

and again the direct limit of the sequences is exact, implying the result. ■

Proof of Proposition 1.7. Fix n. There is a direct system of direct systems as in the diagram below, in which we have abbreviated $M^n(p^e)$ to M_e^n.

$$\begin{array}{ccccccc} [M_e^{n_{1,1}}, X] \xrightarrow{A^*} \to & [M_e^{n_{1,j}}, X] & \xrightarrow{A^*} \to & [M_e^{n_{1,J}}, X] & \xrightarrow{A^*} \to & v_1^{-1}\pi_n(X;\mathbf{Z}/p^e) \\ & \downarrow \rho^{*e} & & \downarrow \rho^{*e} & & \downarrow \rho^{*e} \\ & [M_{2e}^{n_{2,1}}, X] & \xrightarrow{A^*} \to & [M_{2e}^{n_{2,t}}, X] & \xrightarrow{A^*} \to & v_1^{-1}\pi_n(X;\mathbf{Z}/p^{2e}) \\ & & & \downarrow \rho^{*e} & & \downarrow \rho^{*e} \\ & & & [M_{3e}^{n_{3,1}}, X] & \xrightarrow{A^*} \to & v_1^{-1}\pi_n(X;\mathbf{Z}/p^{3e}) \\ & & & & \downarrow & \downarrow \end{array} \tag{2.1}$$

with each sequence $\langle n_{k,*}\rangle$ a subsequence of $\langle n_{k-1,*}\rangle$. We choose the differences $n_{k,i} - n_{k,i-1}$ to be large enough that Lemma 1.2 applies. We also require that $n_{k,i} - n_{k,i-1}$ is a multiple of $s((k+1)e)$, which is one power of p^e larger than one might naïvely require, and that the maps A^* are in the center of the group of stable self-maps of the appropriate Moore space, as [16] guarantees can be done.

Lemma 2.2 *For every morphism $[M^N(p^{ke}),X]\xrightarrow{\rho^{*e}}[M^N(p^{(k+1)e}),X]$ in diagram (2.1) (with $N=n_{k,\ell}=n_{k+1,s}$ for some ℓ and s), there is a commutative diagram of split short exact sequences*

$$\begin{array}{ccccc} 0\to\pi_N(X) & \xrightarrow{c^*} & [M^N(p^{ke}),X] & \xrightarrow{i^*} & \pi_{N-1}(X)\to 0\\ \downarrow 0 & & \downarrow \rho^{*e} & & \downarrow 1\\ 0\to\pi_N(X) & \xrightarrow{c'^*} & [M^N(p^{(k+1)e}),X] & \xrightarrow{i'^*} & \pi_{N-1}(X)\to 0\end{array}\qquad(2.3)$$

which can be considered to be

$$\begin{array}{ccccc} 0\to\pi_N(X) & \to & \pi_N(X)\oplus\pi_{N-1}(X) & \to & \pi_{N-1}(X)\to 0\\ \downarrow 0 & & 0\ \downarrow\oplus 1 & & \downarrow 1\\ 0\to\pi_N(X) & \to & \pi_N(X)\oplus\pi_{N-1}(X) & \to & \pi_{N-1}(X)\to 0\end{array}\qquad(2.4)$$

Proof. The short exact sequences are just the Barratt-Puppe sequences, together with the observation that, for $*>L$, $p^e=0:\pi_*(X)\to\pi_*(X)$. Since we assume that X has p^e as *H-space* exponent, we obtain sections of i^* and i'^* by choosing null-homotopies of p^{ke} and $p^{(k+1)e}$ on $\Omega^L X$. Moreover, the null-homotopies can be chosen compatibly, under which splittings $\rho^{*e}=0\oplus 1$.

We provide more details on this relationship between null-homotopies and sections. We use n for what was $N-1$ in (2.3), and think of $\Omega^n X$ as $\mathrm{Map}_*(S^n,X)$. We write $\omega\in\Omega^n X$, and $z\in S^n$. If $f:S^n\to S^n$, define $\tilde f:\Omega^n X\to\Omega^n X$ by $\tilde f(\omega)(z)=\omega(f(z))$. If $H:\Omega^n X\times I\to\Omega^n X$ is a null-homotopy of $\tilde f$, i.e., $H(\omega,0)=\tilde f(\omega)$ and $H(\omega,1)=*$, define

$$\widetilde H:\Omega^n X\to\mathrm{Map}_*(S^n\cup CS^n/([0,z]\sim f(z)),X)$$

by $\widetilde H(\omega)(z)=\omega(z)$ and $\widetilde H(\omega)[t,z]=H(\omega,t)(z)$. Let f_d denote the canonical map of degree d, and let H be a null-homotopy of $\tilde f_{p^{ke}}$. Then $\pi_0(\widetilde H)$ yields the splitting of i^* in (2.3). We use $G(\omega,t)(z)=H(\omega,t)(f_{p^e}(z))$ as our null-homotopy of $\tilde f_{p^{(k+1)e}}$. Compatibility of the splittings follows from commutativity of the following diagram, which is an easy consequence of the definitions.

$$\begin{array}{ccc}\Omega^n X & \xrightarrow{\widetilde H} & \mathrm{Map}_*(S^n\cup CS^n/([0,z]\sim f_{p^{ke}}(z)),X)\\ \downarrow 1 & & \downarrow \rho^{*e}\\ \Omega^n X & \xrightarrow{\widetilde G} & \mathrm{Map}_*(S^n\cup CS^n/([0,z]\sim f_{p^{(k+1)e}}(z)),X)\end{array}$$

Lemma 2.5 *The maps A^* in (2.1) respect the splittings of Lemma 2.2.*

Proof. We use that $n_{k,i} - n_{k,i-1}$ is a multiple of $s((k+1)e)$. The advantage of doing this is that the composite

$$\alpha : S^{n_{k,i}-1} \xrightarrow{i} M^{n_{k,i}}(p^{ke}) \xrightarrow{A} M^{n_{k,i-1}}(p^{ke}) \xrightarrow{c} S^{n_{k,i-1}}$$

will be a multiple of p^e. One way to see this most easily is to have chosen our Adams maps *à la* Crabb and Knapp. By their [8, 1.1], there will be a commutative diagram as below.

$$\begin{array}{ccccccc} S^{n_{k,i}-1} & \xrightarrow{i} M^{n_{k,i}}(p^{(k+1)e}) & \xrightarrow{A'} & M^{n_{k,i-1}}(p^{(k+1)e}) & \xrightarrow{c} & S^{n_{k,i-1}} \\ \downarrow 1 & \downarrow p^e & & \downarrow p^e & & \downarrow p^e \\ S^{n_{k,i}-1} & \xrightarrow{i} M^{n_{k,i}}(p^{ke}) & \xrightarrow{A} & M^{n_{k,i-1}}(p^{ke}) & \xrightarrow{c} & S^{n_{k,i-1}} \end{array}$$

Thus the composite

$$\pi_{n_{k,i-1}}(X) \xrightarrow{c^*} [M^{n_{k,i-1}}(p^{ke}), X] \xrightarrow{A^*} [M^{n_{k,i}}(p^{ke}), X] \xrightarrow{i^*} \pi_{n_{k,i}-1}(X)$$

is 0, since it is the morphism induced by $\circ\alpha = \circ p^e c A' i$, and p^e annihilates $\pi_*(X)$ for $* > L$. Therefore, under the same splitting of the groups of (2.3) in which the maps ρ^{*e} are $0 \oplus 1$, the maps A^* yield a commutative diagram

$$\begin{array}{ccccc} 0 \to \pi_{n_{k,i-1}}(X) & \xrightarrow{i_1} & \pi_{n_{k,i-1}}(X) \oplus \pi_{n_{k,i-1}-1}(X) & \xrightarrow{p_2} & \pi_{n_{k,i-1}-1}(X) \to 0 \\ \downarrow A_1 & & \downarrow A^* & & \downarrow A_2 \\ 0 \to \pi_{n_{k,i}}(X) & \xrightarrow{i_1} & \pi_{n_{k,i}}(X) \oplus \pi_{n_{k,i}-1}(X) & \xrightarrow{p_2} & \pi_{n_{k,i}-1}(X) \to 0 \end{array} \quad (2.6)$$

■

Passing to direct limits over A_1, A^*, and A_2, we obtain a direct system of split short exact sequences

$$\begin{array}{ccccc} 0 \to \operatorname{dirlim}_i \pi_{n_{1,i}}(X) & \to & v_1^{-1}\pi_n(X; \mathbf{Z}/p^e) & \to & \operatorname{dirlim}_i \pi_{n_{1,i}-1}(X) \to 0 \\ \downarrow 0 & & \downarrow \rho^{*(k-1)e} & & \downarrow 1 \\ 0 \to \operatorname{dirlim}_j \pi_{n_{k,j}}(X) & \to & v_1^{-1}\pi_n(X; \mathbf{Z}/p^{ke}) & \to & \operatorname{dirlim}_j \pi_{n_{k,j}-1}(X) \to 0 \\ \downarrow 0 & & \downarrow \rho^{*e} & & \downarrow 1 \end{array} \quad (2.7)$$

Thus for each k, the group $\operatorname{dirlim}_i \pi_{n_{k,i}-1}(X)$ on the right side of (2.7) is isomorphic to the direct limit of the morphisms ρ^*, which is our $v_1^{-1}\pi_{n-1}(X)$.

Lemma 2.8 *The groups on the left side of (2.7) are isomorphic to the groups that would appear on the right side of (2.7) done for $n+1$ instead of n.*

Since the above argument (done for $n+1$) shows that each of these groups is $v_1^{-1}\pi_n(X)$, we obtain our desired split short exact sequences, establishing Proposition 1.7. ■

Proof of Lemma 2.8. We use the hypothesis that A is in the center of $[M, M]_*$. In particular, it commutes with the composite

$$M^{n-1} \xrightarrow{c} S^{n-1} \xrightarrow{i} M^n.$$

Thus the following diagram commutes around the outside.

$$\begin{array}{ccccc} [M^{n_{k,i-1}}(p^{ke}), X] & \xrightarrow{i^*} & \pi_{n_{k,i-1}-1}(X) & \xrightarrow{c^*} & [M^{n_{k,i-1}-1}(p^{ke}), X] \\ \downarrow A^* & & \downarrow A_2 \downarrow A_1 & & \downarrow A'^* \\ [M^{n_{k,i}}(p^{ke}), X] & \xrightarrow{i'^*} & \pi_{n_{k,i}-1}(X) & \xrightarrow{c'^*} & [M^{n_{k,i}-1}(p^{ke}), X] \end{array} \qquad (2.9)$$

The left square of (2.9) agrees with the right square of (2:6), which is part of the direct system defining the right side of (2.7), while the right square of (2.9) agrees with the left square of (2.6), with $n-1$ instead of n, which is part of the direct system defining the left side of (2.7) with $n-1$. Thus both squares commute. Therefore,

$$c'^* A_1 i^* = A'^* c^* i^* = c'^* i'^* A^* = c'^* A_2 i^*.$$

Since c'^* is injective and i^* surjective, this implies $A_1 = A_2$. Thus the direct systems on the left and right sides of (2.6) agree under suspension, and passing to direct limits establishes the claim. ■

Proof of Proposition 1.9. The following diagram commutes by (2.9).

$$\begin{array}{ccccccccc} \pi_n X & \xrightarrow{i_2} & \pi_{n+1}X \oplus \pi_n X & \xrightarrow{A^*} & \pi_{n+s+1}X \oplus \pi_{n+s}X & \xrightarrow{p_2} & \pi_{n+s}X & \rightarrow \\ & \searrow 1 & \downarrow p_2 & & \downarrow p_2 & \nearrow 1 & & \\ & & \pi_n X & \xrightarrow{A_2} & \pi_{n+s}X & & & \xrightarrow{A_2} \end{array}$$

The new definition of $v_1^{-1}\pi_n(X)$ is, by the proof of Proposition 1.7, the direct limit of the bottom row, and the old definition is the direct limit of the top row. The direct limits are equal because the diagram shows that the bottom row is cofinal in the top. ■

The following result proves half of Conjecture 1.5.

Corollary 2.10 *If X has an H-space exponent, then $v_1^{-1}\pi_*(X)$ has finite period.*

Proof. $v_1^{-1}\pi_{n-1}(X)$ is any of the groups on the right side of (2.7). The definition of these groups as direct limits over a periodic system imply that they are periodic. ■

We close this section by stating and proving what is easily proved about desuspensions of Adams maps of mod p^e Moore spaces.

Proposition 2.11 *Let p be an odd prime and $e \geq 1$. There is an Adams map of the mod p^e Moore spectrum which desuspends to a map*

$$M^{2e+3+s(e)}(p^e) \to M^{2e+3}(p^e),$$

but none which desuspends to a map

$$M^{2e-1+s(e)}(p^e) \to M^{2e-1}(p^e).$$

Proof. We first prove nonexistence. If there were such a map, then by [1, 12.5] the composite

$$S^{2e-2+s(e)} \to M^{2e-1+s(e)} \to M^{2e-1}(p^e) \to S^{2e-1}$$

would be an element of order p^e in $\pi_*(S^{2e-1})$, but [7] showed that there is no such element.

We now prove existence. It is well-known (e.g., [14]) that there is an element $\alpha \in \pi_{2e+s(e)}(S^{2e+1})$ of order p^e with $e(\alpha) = 1/p^e$. By [29, 3.7], the Toda bracket satisfies

$$0 \in \langle p^e, \Sigma\alpha, p^e \rangle \subset \pi_{2e+s(e)+2}(S^{2e+2}).$$

By [29, 1.8], this implies that a map $S^{2e+s(e)+2} \to M^{2e+3}(p^e)$ which projects to $\Sigma^2\alpha$ extends over $M^{2e+s(e)+3}(p^e)$, as claimed. This is the method employed stably by Adams in [1, 12.4, 12.5].

3 Proofs related to the examples

In this section, we prove 1.10, 1.11, 1.12, 1.13, and 1.14.

Proof of Proposition 1.10. We will show that the Snaith maps Sn ([25]) yield a commutative diagram as below for $k > 0$, and then interpret [21] and [26] to say that the map of telescopes induces an isomorphism of homotopy groups.

$$\begin{array}{ccc}
\Omega^{kd}\Omega^{2n+1}S^{2n+1} & \xrightarrow{\Omega^{kd}\mathrm{Sn}} & \Omega^{kd}QB^{qn} \\
\downarrow{\scriptstyle \Omega^{kd}A'} & & \downarrow{\scriptstyle \Omega^{kd}v} \\
\Omega^{kd}\Omega^{2n+1+d}S^{2n+1} & \xrightarrow{\Omega^{(k+1)d}\mathrm{Sn}} & \Omega^{(k+1)d}QB^{qn} \\
\downarrow & & \downarrow
\end{array}$$

To prove commutativity of the square, we expand it to

$$\begin{array}{ccc}
\Omega^{kd}\Omega^{2n+1}S^{2n+1} & \xrightarrow{\Omega^{kd}\mathrm{Sn}} & \Omega^{kd}QB^{qn} \\
\downarrow c^* & & \downarrow i_* \\
\Omega^{kd}\mathrm{Map}_*(M^{2n+1}, S^{2n+1}) & \xrightarrow{\mathrm{Sn}'} & \Omega^{kd}Q(B^{qn}\wedge M^1) \\
\downarrow \Omega^{kd}A^* & & \downarrow \Omega^{kd}A \\
\Omega^{kd}\mathrm{Map}_*(M^{2n+1+d}, S^{2n+1}) & \xrightarrow{\mathrm{Sn}'} & \Omega^{kd}Q(B^{qn}\wedge M^{1-d}) \\
\downarrow s & & \downarrow s \\
\Omega^{kd}\Omega^{2n+1+d}S^{2n+1} & \xrightarrow{\Omega^{(k+1)d}\mathrm{Sn}} & \Omega^{(k+1)d}QB^{qn}
\end{array}$$

The maps Sn' are defined as the induced map of fibers in the commutative diagram

$$\begin{array}{ccc}
\Omega^{kd}\Omega^{2n+1}S^{2n+1} & \xrightarrow{\Omega^{kd}\mathrm{Sn}} & \Omega^{kd}QB^{qn} \\
\downarrow p^e & & \downarrow p^e \\
\Omega^{kd}\Omega^{2n+1}S^{2n+1} & \xrightarrow{\Omega^{kd}\mathrm{Sn}} & \Omega^{kd}QB^{qn}
\end{array}$$

This diagram commutes because the p^e can be done on loop factors which are not involved in the Snaith maps. Similarly, the middle part of the big diagram (the part involving Sn' and A) commutes because the Adams map can be done on loop factors not involved in either the Snaith map or the p^e. ■

Proof of Proposition 1.11. Recall from [22] that $L_K(S^{2n+1})$ is the universal cover of the fibre of the localized Snaith map

$$s : L_K(QS^{2n+1}) \to L_K(Q\Sigma^{2n+1}B_{(n+1)q-1}),$$

where $QX = \Omega^\infty\Sigma^\infty X$. By [5, 3.1], if X is a suspension spectrum with $\pi_2(X)$ a torsion group, then

$$\Omega^{2n+2}L_K(\Omega^\infty\Sigma^{2n+1}X) = \Omega\Omega^\infty L_K(X) = \Omega L_K\Omega^\infty X.$$

Thus

$$\Omega^{2n+2}L_KS^{2n+1} = \mathrm{fibre}(\Omega L_KQS^0 \xrightarrow{s} \Omega L_KQB_{q(n+1)-1}).$$

By [20], this map s factors as

$$\Omega L_KQS^0 \xrightarrow{s'} \Omega L_KQP_1 \xrightarrow{c} \Omega L_KQB_{q(n+1)-1},$$

with c the pinch map. By [5] and [24], the map s' is an equivalence. Thus

$$\Omega^{2n+2}L_KS^{2n+1} = \mathrm{fibre}(c) = \Omega L_KQB^{qn}.$$

Proof of Proposition 1.12. The existence of a suspension spectrum X_3 as in (i) is suggested by the fibration

$$S^3 \to SU(3) \to S^5 \tag{3.1}$$

and Proposition 1.10. The differentials in the unstable Adams spectral sequence (UASS) for $SU(3)$ suggest that the attaching map in X_3 should be η from the 4-cell of $\Sigma^3 P^2$ to the 6-cell of $\Sigma^5 P^4$, and η^2 from the 5-cell to the 8-cell. We elaborate on this.

The homotopy groups of $SU(3)$ were calculated through a large range in [23] by studying the exact homotopy sequence of (3.1). We like to visualize this in the UASS by combining the charts for S^3 and S^5, similarly to the method of [11]. In the first chart below, the groups for S^3 are indicated by •'s and those of S^5 by o's. Below it, we present the resulting E_∞-term.

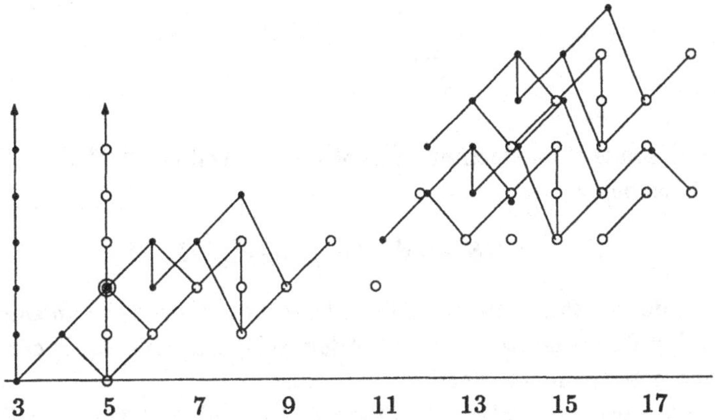

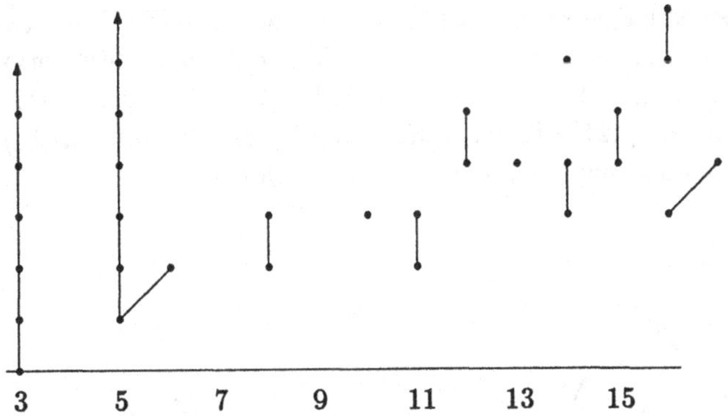

The pattern of differentials near the upper edge of the top chart will be repeated with period $(t-s, s) = (8, 4)$. In order that the chart for $\pi_*^s(X_3)$ have this pattern of differentials near its upper edge, the attaching maps must be as stated above, with the d_1's implying η, and the d_2's implying η^2. We now show that such a stable complex X_3 can be formed. The ASS for $\pi_*^s(\Sigma^3 P^2)$ begins as indicated below. Using this, we see that there are two maps from $S^5 \cup_2 e^6 \to \Sigma^3 P^2$ which send the bottom cell nontrivially. We choose the one which sends $\pi_7(-)$ trivially, and map

$$(S^5 \cup_2 e^6) \vee S^7 \xrightarrow{f} \Sigma^3 P^2$$

by sending S^7 by the circled class in the chart below.

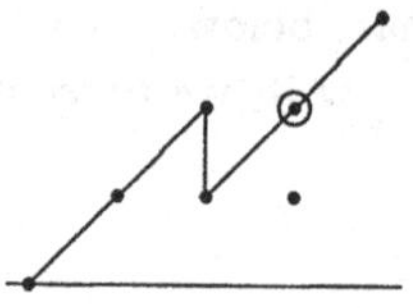

Then f_* sends the attaching map of the top cell of $\Sigma^4 P^4$ trivially, and so extends as below.

$$\Sigma^4 P^4 = ((S^5 \cup_2 e^6) \vee S^7) \cup_{\eta,2} e^8 \xrightarrow{\tilde{f}} \Sigma^3 P^2$$

Letting X_3 denote the cofibre of $\tilde{f}$ establishes (i). We note as an aside that $\mathrm{Sq}^4 H^5(X_3) \neq 0$ in order to satisfy an Adem relation. This Sq^4 is carried by the null-homotopy on the 9-cell.

Next we prove (iii). The ASS's for $ko_*(X_3)$ and $ku_*(X_3)$ have E_2-term $\mathrm{Ext}_{A_1}(H^* X_3, \mathbf{Z}_2)$ and $\mathrm{Ext}_{E_1}(H^* X_3, \mathbf{Z}_2)$, respectively. Here A_1 is the subalgebra of the mod 2 Steenrod algebra A generated by Sq^1 and Sq^2, and E_1 is the exterior subalgebra generated by Sq^1 and $\mathrm{Sq}^3 + \mathrm{Sq}^2\mathrm{Sq}^1$. These are easily calculated to be as in the charts below, extended with period (8,4) in the first starting in $t - s = 6$, and period (2,1) in the second starting in $t - s = 8$. See, e.g., [21]. The d_2-differential in the ASS for $ko_*(X_3)$ is due to the η^2-attaching map. These charts establish (iii).

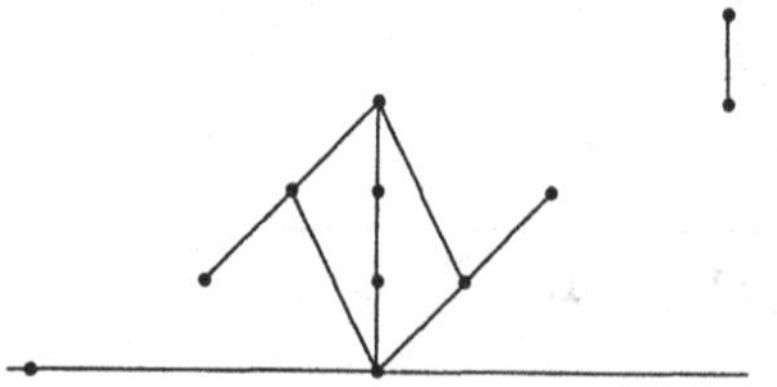

The first part of (iii) shows that the order of 1_{X_3} must be at least 8. To see that $8 \cdot 1_{X_3} \simeq *$, we note that the ASS converging to $\pi_*^s(X_3)$ has $E_\infty^{s,t}(X_3) = 0$ if $s > 2$ and $t - s \leq 9$. Since $8 \cdot 1_{X_3}$ has filtration 3, and, if nontrivial, must be detected by an element in $\pi_j^s(X_3)$ with $j \le 9$, it must be trivial, establishing the first part of (iv).

Since X_3 is a torsion complex with $ku_*(X_3) \neq 0$, it admits a v_1-map by [16]. To prove the second part of (iv), we note that by [21, 6.2], $v_1^{-1}\pi_*(X_3) \approx v_1^{-1}J_*(X_3)$, which in this case is just $v_1^{-1}ko_*(X_3) \oplus v_1^{-1}ko_{*+1}(X_3)$, with the filtration of the second summand 2 less than that of the first, and hence is given by the following chart, extended with period (8,4), which establishes the result.

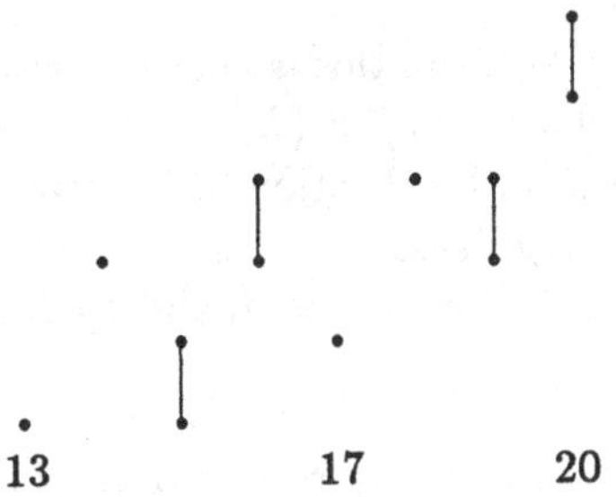

Finally, we must prove part (ii). First we note that the above chart shows that $\pi_*(v_1^{-1}X_3)$ is isomorphic to the groups $\pi_*(\mathrm{Tel}(SU(3), v_1)) = v_1^{-1}\pi_*(SU(3))$ computed in [2]. Moreover, the chart of UASS($SU(3)$) given earlier shows that the filtrations of the homotopy classes agree. But we still must construct a map between the two spectra. Using the UASS chart for $SU(3)$ and the attaching maps in X_3, we will construct a commutative diagram of maps of filtration 5 as below.

$$\begin{array}{ccccccc}
\Sigma^{12}P^4 & \xrightarrow{\tilde f} & \Sigma^{11}P^2 & \longrightarrow & \Sigma^8 X_3 & \longrightarrow & \Sigma^{13}P^4 \\
\big\downarrow{\scriptstyle \Omega\beta} & & \big\downarrow{\scriptstyle \alpha} & & \big\downarrow{\scriptstyle \gamma} & & \big\downarrow{\scriptstyle \beta} \\
\Omega S^5 & \xrightarrow{h} & S^3 & \longrightarrow & SU(3) & \longrightarrow & S^5
\end{array}$$

Since the top row is a cofibre sequence and the bottom row a fibre sequence, it is enough to show that α and β can be chosen to make the left square commute. The map α can be seen in the chart of UASS($SU(3)$) given earlier as the map which sends the 12-cell to the • in position $(t-s, s) = (12, 5)$. Similarly, β sends the 14- and 16-cells to ∘'s in filtration 5. Both $\alpha \circ \tilde f$ and $h \circ \Omega\beta$ have filtration ≥ 6. An ASS calculation shows that the group of maps $\Sigma^{12}P^4 \to S^3$ of filtration ≥ 6 is $\mathbf{Z}_2 \oplus \mathbf{Z}_2 \oplus \mathbf{Z}_2$, corresponding to the •'s, x_1, x_2, and x_3, in positions (13,6), (15,6), and (15,7), respectively. We claim that both of our composites correspond to $x_1 + x_3$. The construction of the maps $\tilde f$ and α showed this is true for their composite. Since differentials in the chart of UASS($SU(3)$) correspond to h_*, the d_1- and d_2-differentials hitting x_1 and x_3, respectively, show that $[h \circ \Omega\beta] = x_1 + x_3$.

The maps α and β induce isomorphisms in $v_1^{-1}\pi_*(-)$ by [10], and hence so does γ, proving (ii). ■

Proof of Corollary 1.13. The second part is immediate from [2]. To prove the first part, we note first that if γ is the map constructed in the proof of Proposition 1.12(ii), and q the inverse of the equivalence constructed there, then the composite

$$\Sigma^{8-L}X_3 \xrightarrow{\gamma} \Omega^L SU(3) \xrightarrow{4} \Omega^L SU(3) \to \Omega^L \mathrm{Tel}(SU(3), v_1) \xrightarrow{q} \Sigma^{-L} v_1^{-1} X_3$$

is the v_1-localization of $4\cdot 1_{X_3}$, and this is nonzero, essentially by Proposition 1.12. Indeed, using the chart for $v_1^{-1}\pi_*(X_3)$ given in the proof of Proposition 1.12, the multiples of the v_1-localization of 1_{X_3} can be viewed as forming a $\mathbf{Z}/8$ with the generator g detected by the top class in dimension 12, $2g$ detected by the top class in dimension 14, and $4g$ detected by the top class in dimension 16. ■

Proof of Proposition 1.14. Using results from [24, §8], we have

$$\begin{aligned} \pi_i(M_1 X) &\approx \operatorname*{dirlim}_{e} \pi_i(L_K(X/p^e)) \approx \operatorname*{dirlim}_{e,L} \pi_i(X \wedge M^{1-Ls(e)}(p^e)) \\ &\approx \operatorname*{dirlim}_{e,L} \pi_{i+Ls(e)}(X; \mathbf{Z}/p^e) \approx (v_1^{-1}\pi)^s_{i-1}(X) \end{aligned}$$

References

[1] J. F. Adams, *On the groups J(X), IV*, Topology **5** (1966) 21-71.

[2] M. Bendersky and D. M. Davis, *The 2-primary v_1-periodic homotopy groups of $SU(n)$*, to appear in Amer Jour Math.

[3] M. Bendersky, D. M. Davis, and M. Mimura, *v_1-periodic homotopy groups of exceptional Lie groups—torsion-free cases*, to appear.

[4] A. K. Bousfield, *Localization of spectra with respect to homology*, Topology **18** (1979) 257-281.

[5] ———, *K-localizations and K-equivalences of infinite loop spaces*, Proc London Math Soc **44** (1982) 291-311.

[6] F. R. Cohen, *Unstable decomposition of $\Omega^2\Sigma^2X$ and its applications*, Math Zeit **182** (1983) 553-568.

[7] F. R. Cohen, J. C. Moore, and J. Neisendorfer, *The double suspension and exponents of the homotopy groups of spheres*, Annals of Math **110** (1979) 549-565.

[8] M. Crabb and K. Knapp, *Adams periodicity in stable homotopy*, Topology **24** (1985) 475-486.

[9] D. M. Davis, *v_1-periodic homotopy groups of $SU(n)$ at odd primes*, to appear in Proc London Math Soc.

[10] D. M. Davis and M.Mahowald, *v_1-periodicity in the unstable Adams spectral sequence*, Math Zeit **204** (1990) 319-339.

[11] ———, *The $SO(n)$-of origin*, Forum Math **1** (1989) 239-250.

[12] D. M. Davis, M. Mahowald, and H. R. Miller, *Mapping telescopes and K_*-localization*, Annals of Math Studies **113** (1987) Princeton Univ Press, 152-167.

[13] P. Freyd, *Stable homotopy*, Proc Conf on Categorical Algebra, La Jolla, Springer-Verlag 1966.

[14] B. Gray, *On the sphere of origin of infinite families in the homotopy groups of spheres*, Topology **8** (1969) 219-232.

[15] P. Hoffman, *Relations in the stable homotopy rings of Moore spaces*, Proc. London Math Society **18** (1968) 621-634.

[16] M. J. Hopkins, *Global methods in homotopy theory*, London Math Soc Lecture Notes **117** (1987) 73-96.

[17] M. Hopkins and J. Smith, to appear.

[18] I. M. James, *On the suspension sequence*, Annals of Math **65** (1957) 74-107.

[19] ———, *On Lie groups and their homotopy groups*, Proc Camb Phil Soc **55** (1959) 244-247.

[20] N. J. Kuhn, *The geometry of James-Hopf maps*, Pac Jour Math **102** (1982) 397-412.

[21] M. Mahowald, *The image of J in the EHP sequence*, Annals of Math **116** (1982) 65-112.

[22] M. Mahowald and R. D. Thompson, *The K-theory localization of an unstable sphere*, to appear.

[23] M. Mimura and H. Toda, *Homotopy groups of $SU(3)$, $SU(4)$, and $Sp(2)$*, J Math Kyoto Univ **3** (1964) 217-250.

[24] D. C. Ravenel, *Localization with respect to certain periodic homology theories*, Amer Jour Math **106** (1984) 351-414.

[25] V. P. Snaith, *A stable decomposition of* $\Omega^n\Sigma^n X$, Jour London Math Soc **7** (1974) 577-583.

[26] R. D. Thompson, *The* v_1*-periodic homotopy groups of an unstable sphere at an odd prime*, Trans Amer Math Soc **319** (1990) 535-559.

[27] ———, v_1-periodic homotopy groups of a Moore space, Proc Amer Math Soc **107** (1989) 833-845.

[28] H. Toda, *On the double suspension* E^2, Jour Inst Polytech Osaka City Univ **7** (1956) 103-145.

[29] ———, *Composition methods in the homotopy groups of spheres*, Annals of Math Studies, **49** (1962) Princeton Univ Press.

The unstable Novikov spectral sequence for $Sp(n)$, and the power series $\sinh^{-1}(x)$ *

Martin Bendersky
CUNY, Hunter College
New York 10021

Donald M. Davis
Lehigh University
Bethlehem, PA 18015 †

This paper is dedicated to the memory of Frank Adams.

1 Statement of results

The unstable Novikov spectral sequence (UNSS) converging to the homotopy groups of a space X was introduced in [4], and has recently been shown to be very useful in computing the v_1-periodic homotopy groups of certain spaces. ([11], [6],[3]). In this paper, we calculate the 1-line of the UNSS, $E_2^{1,*}(Sp(n))$, for the symplectic groups $Sp(n)$, and the entire UNSS for their union, Sp. We hope to apply these results in a future paper to the determination of $v_1^{-1}\pi_*(Sp(n))$. We also discuss several results about the power series $\sinh^{-1}(x)$. The most interesting, Theorem 1.5, was suggested by a comparison of $E_2^{1,*}(Sp(n))$ with $E_2^{1,*}(SU(2n))$, and was then proved by Michael Crabb.

We use the UNSS based upon the complex cobordism spectrum MU. Our first main result, which will be proved in Section 2, gives the entire spectral sequence for Sp.

*AMS Subject Classification 55T15. Key words: unstable Novikov spectral sequence, symplectic groups, power series

†Both authors were partially supported by the National Science Foundation. The second author was also supported by the S.E.R.C.

Theorem 1.1 *The E_2-term of the UNSS for Sp is given by*

$$E_2^{s,t}(Sp) = \begin{cases} \mathbf{Z} & \text{if } s=0,\ t\equiv 3 \bmod 4, \text{ and } t>0 \\ \mathbf{Z}/2 & \text{if } s>0,\ t-2s\equiv 3 \bmod 4, \text{ and } t-2s>0 \\ 0 & \text{otherwise.} \end{cases}$$

The only nonzero differentials are $d_3 : E_3^{s,t} \to E_3^{s+3,t+2}$ with $t-2s \equiv 7 \bmod 8$, and these are always nonzero when $t-2s>0$.

The picture for this UNSS, with coordinates $(t-s, s)$, begins as below, yielding the homotopy groups

$$\pi_i(Sp) \approx \begin{cases} \mathbf{Z} & i\equiv 3 \bmod 4,\ i>0 \\ \mathbf{Z}/2 & i\equiv 4,5 \bmod 8,\ i>0 \\ 0 & \text{otherwise,} \end{cases}$$

as is known from Bott periodicity.

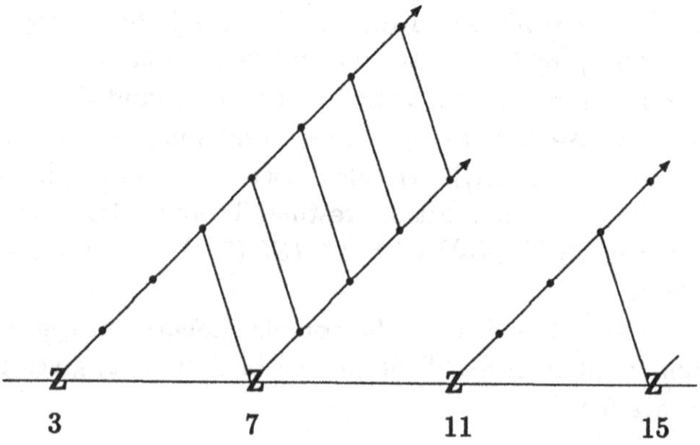

Our second main result is a general result comparing the 1-lines for $Sp(n)$ and $SU(2n)$. The proof, which will be given in Section 2, is easy once an appropriate exact sequence is derived.

Theorem 1.2 $E_2^{1,4k-1}(SU(2n)) \approx E_2^{1,4k-1}(Sp(n)) \approx E_2^{1,4k-1}(SU(2n+1))$.

The groups $E_2^{1,*}(SU(m))$ were determined in [2], and will be reviewed later in this section. Our result appears not to have been noticed before, even the part involving just the groups for $SU(2n)$ and $SU(2n+1)$. Groups very closely related to these were studied in [6], [11], [9], and [16]. Indeed, [9, 3.8] comes very close to our Theorem 1.2, but the James numbers with which it deals sometimes differ by one power of 2 from the 1-line numbers.

Our third main result, which will be proved in Section 3, is a calculation of the 1-line for $Sp(n)$, independent of Theorem 1.2. It involves numbers $e_{Sp}(k,n)$, which we now define along with their complex analogues.

$$e_U(k,n) = \gcd\{\mathrm{coef}(\frac{x^k}{k!}, (e^x - 1)^j) : j \geq n\}$$

$$e_{Sp}(k,n) = \gcd\{\mathrm{coef}(\frac{x^{2k}}{(2k)!}, \frac{(2\sinh(\frac{x}{2}))^{2j}}{2j}) : j > n\}$$

Another form for these numbers, namely $\gcd\{\sum_{i=0}^{j-1}(-1)^i\binom{j}{i}(j-i)^k : j \geq n\}$ and $\gcd\{\frac{1}{j}\sum_{i=0}^{j-1}(-1)^i\binom{2j}{i}(j-i)^{2k} : j > n\}$, respectively, can be obtained by expanding powers of $e^x - 1$ or $e^{x/2} - e^{-x/2}$ in terms of powers of e^x or $e^{x/2}$.

Theorem 1.3 $E_2^{1,j}(Sp(n)) \approx \begin{cases} \mathbf{Z}/e_{Sp}(k,n) & j = 4k-1,\ k > n \\ \mathbf{Z}/2 & j = 4k+1,\ k > 0 \\ 0 & \textit{otherwise.} \end{cases}$

The following corollary is immediate from Theorems 1.2 and 1.3 together with the analogous result for $SU(n)$,

$$E_2^{1,2k+1}(SU(n)) \approx \mathbf{Z}/e_U(k,n),$$

which was proved in [2].

Corollary 1.4 $e_U(2k-1, 2n) = e_{Sp}(k,n) = e_U(2k-1, 2n+1)$.

This corollary is itself a purely number-theoretic result which appears to be new, but reinterpreting it with a bit of combinatorics suggested a more striking statement about power series. Define rational power series $f(x)$ and $g(x)$ to be equivalent ($\sim$) if for all integers n and positive integers k

$$\text{l.c.m.}\{\mathrm{den}(\mathrm{coef}(x^j, f(x)^n)) : j < k\} = \text{l.c.m.}\{\mathrm{den}(\mathrm{coef}(x^j, g(x)^n)) : j < k\}.$$

Thus $f(x)$ and $g(x)$ are equivalent if, for any power of the series, through any range of powers of x, the least common multiples of denominators of the coefficients are equal for f and g. Equivalently, $f(x) \sim g(x)$ if and only if, for all integers $k > 0$, n, and N, $N \cdot f(x)^n \bmod x^k$ is integral if and only if $N \cdot g(x)^n \bmod x^k$ is integral. The following statement was suggested by Corollary 1.4, and was subsequently proved by Michael Crabb.

Theorem 1.5 *(*Crabb*)* $(\log(1+x)/x)^2 \sim (2\sinh^{-1}(\frac{x}{2})/x)^2$.

In Section 4, we discuss the relationship between Corollary 1.4 and Theorem 1.5, and give Crabb's proof of Theorem 1.5. Further investigation of this equivalence relation between power series should be interesting.

2 Proofs of Theorems 1.1 and 1.2

A key role in this section is played by the following well-known result. See e.g., [16]. From now on, HP^n denotes quaternionic projective space, and Q^{n+1} denotes the quaternionic quasiprojective space. Q^{n+1} is naturally embedded in $Sp(n+1)$. (See e.g. [12].)

Proposition 2.1 *Q^{n+1} is the Thom space of a 3-plane bundle ς over HP^n with sphere bundle CP^{2n+1}.*

The cofibre sequence of $(D(\varsigma), S(\varsigma))$ extends to $Q^{n+1} \to \Sigma CP^{2n+1} \to \Sigma HP^n$, which induces short exact sequences in $MU_*(-)$ and $MU^*(-)$. There is a commutative diagram as below, where the first two vertical maps are well-known, and the third one is induced by them.

$$\begin{array}{ccccc} Q^{n+1} & \to & \Sigma CP^{2n+1} & \to & \Sigma HP^n \\ \downarrow & & \downarrow & & \downarrow \\ Sp(n+1) & \to & SU(2n+2) & \to & SU(2n+2)/Sp(n+1) \end{array}$$

For each of these vertical maps $X \to Y$, the primitives $P(MU_*(Y))$ are isomorphic to $MU_*(X)$, and $MU_*(Y)$ is a cofree coalgebra, so that by [2, 2.2] $E_2^{s,t}(Y) \approx \text{Ext}^s_{\mathcal{U}}(A[t], MU_*(X))$. Here $\mathcal{U}$ is the category of unstable $MU_*(MU)$-comodules, and $A[t]$ is a free MU_*-module on a generator of degree t. The short exact sequence in $MU_*(-)$ of the top row induces a long exact sequence in $\text{Ext}_{\mathcal{U}}(A[t], MU_*(-))$.

We can now prove Theorem 1.2. For $\epsilon = 0$ or 1, we obtain a commutative diagram of exact sequences as below. The case $\epsilon = 0$ is obtained by restriction from the case $\epsilon = 1$. All groups have second superscript $4k-1$. $SU(n)$ and $Sp(n)$ are written as SU_n and Sp_n in order to conserve space. HP_{n-1} is a stunted projective space, and $\text{Ext}^s_{\mathcal{U}}(A[t], MU_*(-))$ has been abbreviated to $\text{Ext}^{s,t}(-)$.

$$\begin{array}{ccccccc} & & & & & & \text{Ext}^0(HP_{n-1}) \\ & & & & & & \downarrow \\ \text{Ext}^0(\Sigma HP^n) & \to & E_2^1(Sp_{n+\epsilon}) & \xrightarrow{h} & E_2^1(SU_{2n+1+\epsilon}) & \to & \text{Ext}^1(\Sigma HP^n) \to \\ & & \downarrow & & \downarrow & & \downarrow \\ & & E_2^1(Sp) & \to & E_2^1(SU) & \to & \text{Ext}^1(\Sigma HP^\infty) \to \end{array}$$

The relevant groups $\text{Ext}^{0,4k-1}(HP_{n-1})$, $\text{Ext}^{0,4k-1}(\Sigma HP^n)$, and $E_2^{1,4k-1}(SU)$ are all 0, and an easy diagram chase then implies that h is bijective. ■

A similar argument suggests, but does not prove, the following conjecture, which would be useful in our study of $v_1^{-1}\pi_*(Sp(n))$.

Conjecture 2.2 $E_2^{1,4k+1}(SU(2m+1)) \approx \mathrm{Ext}^{1,4k+1}(\Sigma HP^m)$.

This conjecture has been supported by a large amount of computer calculation. It is related to a conjecture about power series similar to, but more complicated than, Theorem 1.5.

The following result is central to our proof of Theorem 1.1.

Proposition 2.3 $E_2^{s,t}(SU/Sp) \approx E_2^{s,t-2}(Sp)$.

Proof. We prove that each is isomorphic to $E_2^{s,t-1}(BSp)$. For $X = BSp$ or Sp, $H^*(X)$ is a free algebra, and so $MU_*(X)$ is a cofree coalgebra, and hence we may apply [2, 2.2] to say that $E_2^{s,t}(X) \approx \mathrm{Ext}_{\mathcal{U}}^s(A[t], P(MU_*(X)))$. The primitives of each agree after a dimension shift, and so we obtain $E_2^{s,t-1}(BSp) \approx E_2^{s,t-2}(Sp)$. It is important here that the primitives for BSp are even-dimensional, so that the unstable condition of [4, 6.6(1)], $2\ell(J) < |x|$, applies to a primitive x of $MU_*(BSp)$ if and only if it applies to its desuspension.

For the second part, we use Bott periodicity $BSp \simeq \Omega(SU/Sp)$, and the result of [7] that $H^*(SU/Sp)$ is a free algebra, so that again we can apply [2, 2.2]. Thus we must show that applying Ω to SU/Sp just shifts $\mathrm{Ext}_{\mathcal{U}}^s(A[t], P(-))$. Similarly to the first case, we have that

$$P(MU_*(SU/Sp)) \approx P(MU_{*-1}(\Omega SU/Sp))$$

as MU_*-comodules, but in this case the shift alters the unstable condition, and so a more elaborate argument is required. The situation here is analogous to that of [5, 6.1], which proved that $E_2^{s,t-1}(\Omega S^{2n+1}) \approx E_2^{s,t}(S^{2n+1})$. The key point there was that if N is a free MU_*-module on odd dimensional generators, then

$$\sigma^{-1}(U(N)) \approx Q(G(\sigma^{-1}(N))),$$

where the functors U, Q, and G are defined in [5]. This applies to our case as well, and the rest of the argument on [5, p. 387] proceeds almost verbatim. The fact that $MU_*(BSp)$ is a free algebra, as was $MU_*(\Omega S^{2n+1})$, is another important ingredient in this argument. ■

Proof of Theorem 1.1. Letting n go to ∞ in the exact sequences derived earlier yields an exact sequence

$$\to E_2^{s,t}(SU) \to E_2^{s,t}(SU/Sp) \to E_2^{s+1,t}(Sp) \to E_2^{s+1,t}(SU) \to .$$

By [2, 3.1], $E_2^{s,t}(SU)$ for $s > 0$. Thus the exact sequence, together with 2.3, implies

$$E_2^{s,t-2}(Sp) \approx E_2^{s+1,t}(Sp)$$

for $s > 0$, and $E_2^{0,t-2}(Sp) \to E_2^{1,t}(Sp)$ is surjective. It follows easily from [2, 2.2] that $E_2^{0,t}(Sp)$ is $\mathbf{Z}$ when $t \equiv 3 \bmod 4$ and $t > 0$, and is 0 otherwise. These remarks imply that the only nonzero groups $E_2^{s,t}(Sp)$ occur when $t - 2s \equiv 3 \bmod 4$ and $t - 2s > 0$, and that for fixed $t - 2s$ (corresponding to a diagonal of slope 1 in the usual chart) all groups for $s \geq 1$ are isomorphic cyclic groups. That the groups and differentials are as claimed in the theorem is now a consequence of Bott's result for $\pi_*(Sp)$. For example, $E_2^{1,5}$ must be $\mathbf{Z}/2$ to give the correct $\pi_4(Sp)$. Thus $E_2^{s,2s+3}(Sp) \approx \mathbf{Z}/2$ for all $s > 0$. Next, we must have $E_2^{1,9} \approx \mathbf{Z}/2$ to kill $E_2^{3,9}$. It cannot be larger than that since $\pi_8(Sp) = 0$, and there is nothing else for it to kill.

3 The 1-line for $Sp(n)$

In this section, we prove Theorem 1.3. If t is even, then $E_2^{s,t}(Sp(n)) = 0$ since the analogue is true for odd-dimensional spheres by [4, p. 245]. The exact sequence

$$E_2^{0,t}(Sp/Sp(n)) \to E_2^{1,t}(Sp(n)) \to E_2^{1,t}(Sp) \to E_2^{1,t}(Sp/Sp(n))$$

together with Theorem 1.1 and the following facts imply the remaining 0 and $\mathbf{Z}/2$-groups in Theorem 1.3.

i. $E_2^{0,t}(Sp/Sp(n))$ is zero unless $t \equiv 3 \bmod 4$ and $t \geq 4n + 3$,

ii. $E_2^{1,t}(Sp/Sp(n))$ is zero if $t < 4n + 5$,

iii. $\pi_i(Sp(1)) \to \pi_i(Sp)$ is surjective if $i \equiv 4, 5 \bmod 8$, and

iv. The element in $E_2^{4,8i+11}(Sp(1))$ which is divisible by $[h_1]^3$ is killed by a d_3-differential

Now we work toward the groups of order $e_{Sp}(k, n)$ in Theorem 1.3. The first step in the proof is the following lemma. It involves the morphism $\bar{e} : MU_*(MU) \to Q$ defined in [2]. The properties that we need are

i. $\bar{e}$ is a ring homomorphism,

ii. $\bar{e}(\eta_R(x)) = 0$ if $|x| > 0$,

iii. $\bar{e}(m_i) = 1/(i + 1)$, and

iv. $\bar{e}$ passes to an injection $E_1^{1,2k+1}(S^{2n+1}) \to Q/\mathbf{Z}$.

Lemma 3.1 *Let* $\Psi : MU_*(Sp(n)) \to MU_*(MU) \otimes_{MU_*} MU_*(Sp(n))$ *denote the coaction. Let* $\{z_i : i \geq 1\}$ *denote the natural basis for the primitives in* $MU_*(Sp(n))$, *and* B *the matrix whose* i*th row is the coefficients of* $\bar{e}(\Psi(z_i))$, *expressed in terms of* $\{z_1, z_2, \ldots\}$. *Let* $P(x) = (2\sinh^{-1}(\sqrt{x}/2))^2 = x - \frac{1}{12}x^2 + \frac{1}{90}x^3 - \cdots$. *Then the entries* $b_{i,j}$ *of* B *are related to* P *and its derivative* P' *by*

$$P(x)^{j-1}P'(x) = \sum_{i\geq 1} b_{i,j}x^{i-1}.$$

Moreover, if $C = B^{-1}$, *then the entries* $c_{i,j}$ *of* C *satisfy*

$$\frac{(2\sinh(x/2))^{2j}}{2j} = \sum_{i\geq 1} \frac{1}{2i} c_{i,j} x^{2i}.$$

We now begin the lengthy proof of this lemma. There is some overlap of our method with [14] and [15]. We thank Andy Baker for helpful comments on this proof.

Let $\Phi : MU^*(X) \to MU_*(MU) \otimes_{MU_*} MU^*(X)$ be dual to Ψ. It is a ring homomorphism. For $X = CP^\infty$ with generator $x \in MU^2(CP^\infty)$, Φ is determined by $\Phi(x) = \sum B_i x^{i+1}$, where $B_i \in MU_{2i}(MU)$ is as in [14]. This formula is dual to [17, 1.48e].

In this proof, we use both the log series for MU and the ordinary rational log series. We shall denote the former by $\log^{MU} x$. Note that

$$\bar{e}(\log^{MU} x) = \sum \bar{e}(m_i)x^{i+1} = \sum \tfrac{1}{i+1}x^{i+1} = -\log(1-x).$$

Let $\bar{x} \in MU^2(CP^\infty)$ be induced from the conjugate of the canonical line bundle. Then $\bar{x} = \exp^{MU}(-\log^{MU} x)$, and $\Phi(\bar{x}) = \sum B_i \bar{x}^{i+1}$.

Proposition 3.2 $\bar{e}(\Phi(x)) = -\log(1-x)$.

Proof. The formula

$$m_n = \sum \eta_R(m_i)(B)^{i+1}_{n-i}$$

is obtained by conjugating ([17, 1.48b]), which says

$$\eta_R(m_n) = \sum m_i (m^{MU})^{i+1}_{n-i}.$$

Here the unsubscripted m and B refer to the sum of all m_i (resp. B_i), the $i+1$ superscript refers to a power of this sum, and the $n-i$ subscript refers to the component in grading $2(n-i)$. Applying $\bar{e}$ to this yields $\bar{e}(m_n) = (\bar{e}(B))_n$, since $\bar{e}(\eta_R(-)) = 0$. Thus

$$\bar{e}(\Phi(x)) = \sum \bar{e}(B_i)x^{i+1} = \sum \frac{1}{i+1}x^{i+1}.$$

Proposition 3.3 $\bar{e}(\bar{x}) = -x/(1-x)$.

Proof. Let $\ell(x) = -\log(1-x)$. Note that $\ell^{-1}(x) = 1 - e^{-x}$. Thus

$$\bar{e}(\bar{x}) = \bar{e}(\exp^{MU}(-\log^{MU} x)) = \ell^{-1}(-\ell(x)) = 1 - e^{-\log(1-x)} = -x/(1-x).$$

As noted earlier, there is a short exact sequence

$$0 \to MU^*(\Sigma HP^\infty) \to MU^*(\Sigma CP^\infty) \to MU^*(Q^\infty) \to 0.$$

As in [15], $MU^*(HP^\infty)$ is generated over MU_* by a 4-dimensional class y that maps to $-x\bar{x} \in MU^*(CP^\infty)$. We may choose the suspensions of $\{x(-x\bar{x})^i : i \geq 0\}$ as an MU_*-basis of $MU^*(Q^\infty)$. We do not write the suspension.

Proposition 3.4 *In* $MU^*(HP^\infty)$, $\bar{e}\Phi(y) = (2\sinh^{-1}(\frac{\sqrt{y}}{2}))^2$.

Proof. $\bar{e}\Phi(-x\bar{x}) = -\ell(x)\cdot\ell(\ell^{-1}(-\ell(x))) = \ell(x)^2 = (\log(1-x))^2$. Using the formula $\sinh^{-1}(x) = \log(x + \sqrt{1+x^2})$, one readily verifies that

$$2\sinh^{-1}\Big(\frac{x}{2\sqrt{1-x}}\Big) = -\log(1-x). \tag{3.5}$$

In writing $\bar{e}\Phi(y)$, we must also apply $\bar{e}$ to y, which yields $-x\cdot\frac{-x}{1-x} = \frac{x^2}{1-x}$. Thus $\bar{e}\Phi(y) = P(y)$, where

$$P(\frac{x^2}{1-x}) = (\log(1-x))^2. \tag{3.6}$$

By (3.5), $P(y) = (2\sinh^{-1}(\frac{\sqrt{y}}{2}))^2$ is the desired series. ∎

To obtain $\bar{e}\Phi$ in $MU^*(Q^\infty)$, we must write $\bar{e}\Phi(x)$ in terms of $x(\frac{x^2}{1-x})^i$ and $(\frac{x^2}{1-x})^i$, and then ignore the latter terms. Differentiating (3.6) yields

$$-2\log(1-x) = P'(\frac{x^2}{1-x})\frac{2x-x^2}{1-x} = P'(\frac{x^2}{1-x})(2x + \frac{x^2}{1-x}),$$

and so, using Propositions 3.2 and 3.4, we have in $MU^*(Q^\infty)$

$$\bar{e}(\Phi(x)) = xP'(y) \quad\text{and}\quad \bar{e}(\Phi(xy^i)) = xP'(y)P(y)^i. \tag{3.7}$$

Since Ψ is dual to Φ, we can now write the matrix B of Lemma 3.1. Similarly to [2, 3.4], $P(MU_*(Sp)) \approx MU_*(Q^\infty)$, with the same coaction. Our basis $\{z_1, z_2, \ldots\}$ for $MU_*(Q^\infty)$ has z_i dual to xy^{i-1}. Then

$$B = (P'(y),\ P'(y)P(y),\ P'(y)P(y)^2,\ P'(y)P(y)^3,\ \cdots),$$

where each power series yields a column of the matrix, with the numbers in that column being the coefficients of the power series, beginning with the coefficient of y^0. We have omitted here the writing of the x-factor which appears on every term in $MU^*(Q^\infty)$. This yields the first part of Lemma 3.1.

Continuing to denote matrices by a list of power series which could be thought of as generating functions for the columns of the matrix, we let the inverse matrix $C = (f_1(y),\ f_2(y),\ \cdots)$. Then

$$(P'(y),\ P'(y)P(y),\ P'(y)P(y)^2,\ \cdots)\cdot(f_1(y),\ f_2(y),\ \cdots) = (1, y, y^2, \cdots).$$

The jth column of this matrix equation says that if, as in the last part of Lemma 3.1, $f_j(y) = c_{1,j} + c_{2,j}y + c_{3,j}y^2 + \cdots$, then

$$c_{1,j}P'(y) + c_{2,j}P'(y)P(y) + c_{3,j}P'(y)P(y)^2 + \cdots = y^{j-1}.$$

This says that $P'(y)f_j(P(y)) = y^{j-1}$. Letting $x = P(y)$, this becomes

$$f_j(x) = \frac{P^{-1}(x)^{j-1}}{P'(P^{-1}(x))} = \frac{d}{dx}\left(\frac{P^{-1}(x)^j}{j}\right).$$

Integrating yields

$$\sum c_{i,j}\frac{x^i}{i} = \frac{(2\sinh^{-1}(\sqrt{x}/2))^{2j}}{j},$$

which is the last part of Lemma 3.1. ∎

In order to deduce the main part of Theorem 1.3 from Lemma 3.1, we shall discuss a general result which contains that needed here and the main result of [2]. For $L \le \infty$, let $\{X_i : 0 \le i \le L\}$ be a sequence of spaces satisfying

i. $X_0 = *$.

ii. There are fibrations $X_{i-1} \to X_i \to S^{n_i}$ with n_i odd and $n_1 < n_2 < \cdots$.

iii. $H^*(X_i;\mathbf{Z})$ is an exterior algebra $\Lambda(x_1, \ldots, x_i)$ with $|x_i| = n_i$.

Then $MU_*(X_k) = \Lambda_{MU_*}(y_1, \ldots, y_k)$ with $|y_i| = n_i$, and

$$E_2(X_k) = \mathrm{Ext}_{\mathcal{U}}(MU_*\{y_1, \ldots, y_k\}),$$

where $MU_*\{y_1, \ldots, y_k\}$ denotes a free MU_*-module on the y_i.

Definition 3.8 $\overline{E}^{1,t}(X_k) = \ker(E_2^{1,t}(X_k) \to E_2^{1,t}(X_L))$.

In the case of interest to us, $X_k = Sp(k)$, $L = \infty$, and $X_L = Sp$.

The exact sequence

$$\to E_2^0(S^{n_i}) \to E_2^1(X_{i-1}) \to E_2^1(X_i) \to$$

together with the fact that $E_2^{0,t}(S^n) = \mathbf{Z}$ if $t = n$, and is 0 otherwise, implies

Proposition 3.9 *$\overline{E}^{1,t}(X_k) = 0$ unless $t = n_i$ for some $i > k$, in which case $\overline{E}^{1,t}(X_k)$ is cyclic.*

By mimicking the proof in [2] with X_i corresponding to $SU(i+1)$, we can prove the following general result.

Theorem 3.10 *For $n \le k$, the cokernel of*

$$\overline{E}^{1,n_k}(X_{n-1}) \to \overline{E}^{1,n_k}(X_{k-1})$$

is cyclic of order $\omega_k(n)$, where $\omega_k(n)$ is defined as follows. Let $\gamma_{k,j} \in MU_*(MU)$ be defined in terms of the coaction in $MU_*(X_i)$ by

$$\Psi(y_k) = \sum_{j=1}^{k} \gamma_{k,j} \otimes y_j.$$

Note that this is independent of i. Define rational numbers $b_{k,j}$ by $b_{k,j} = \bar{e}(\gamma_{k,j})$. Then the matrix $B = (b_{k,j})$ is lower triangular with 1's on the diagonal. Let $C = (c_{k,j})$ be the inverse of B. Then

$$\omega_k(n) = \text{l.c.m.}\{\text{den}(c_{k,j}) : n \le j \le k\}.$$

The proof, although by no means easy, is such a direct adaptation of the proof of [2, 4.7] that we shall omit it.

In our case, $X_n = Sp(n)$, and the matrix B described in Theorem 3.10 is the same as the matrix B described in Lemma 3.1. We deduce from these two results that

$$\begin{aligned} \omega_k(n) &= \text{l.c.m.}\{\text{den}(\text{coef}(\frac{x^{2k}}{2k}, \frac{(2\sinh(\frac{x}{2}))^{2j}}{2j})) : j \ge n\} \\ &= (2k-1)!/\gcd\{\text{coef}(\frac{x^{2k}}{(2k)!}, \frac{(2\sinh(\frac{x}{2}))^{2j}}{2j}) : j > n-1\} \\ &= (2k-1)!/e_{Sp}(k, n-1) \end{aligned}$$

On the other hand, $\omega_k(n)$ is the order of the cyclic cokernel of

$$\overline{E}^{1,4k-1}(Sp(n-1)) \to \overline{E}^{1,4k-1}(Sp(k-1)).$$

Note that, using Theorem 1.2 and [2, 4.2iii],

$$\overline{E}^{1,4k-1}(Sp(k-1)) = E_2^{1,4k-1}(Sp(k-1)) \approx E_2^{1,4k-1}(SU(2k-1)) \approx \mathbf{Z}/(2k-1)!.$$

Thus $\overline{E}^{1,4k-1}(Sp(n-1))$ is cyclic of order $(2k-1)!/\omega_k(n) = e_{Sp}(k, n-1)$, as claimed.

4 Power series

In this section, we discuss Theorem 1.5 and some related results about the power series $\sinh^{-1}(x)$.

The motivation for Theorem 1.5 came from applying the method of [10] to Corollary 1.4, yielding the following result, which says that the theorem is true for negative exponents in a limited range of coefficients. Of course, this proposition is rendered inconsequential by Theorem 1.5, whose proof we give after discussing this weaker result.

Proposition 4.1 *If $m \le k-2$, then*

$$\begin{aligned} & \text{l.c.m.}\{\text{den}(\text{coef}(x^j, (\log(1+x)/x)^{-2k})) : j \le m\} \\ = \; & \text{l.c.m.}\{\text{den}(\text{coef}(x^j, (2\sinh^{-1}(\tfrac{x}{2})/x)^{-2k})) : j \le m\}. \end{aligned}$$

Proof. For $\epsilon = 0$ or 1, we have the following, where the last equality is a result of [1], which was also used in [10].

$$\begin{aligned} & e_U(2k-1, 2n+\epsilon) \\ = \; & \gcd\{\text{coef}(\frac{x^{2k-1}}{(2k-1)!}, (e^x-1)^j) : j \ge 2n+\epsilon\} \\ = \; & (2k-1)!/\text{l.c.m.}\{\text{den}(\text{coef}(x^{2k-1}, (e^x-1)^j)) : j \ge 2n+\epsilon\} \\ = \; & (2k-1)!/\text{l.c.m.}\{\text{den}(\text{coef}(x^j, (\log(1+x)/x)^{-2k})) : \\ & \qquad j \le 2k-2n-1-\epsilon\} \end{aligned}$$

On the other hand, using the Lagrange inversion formula ([8, p. 148]) at the middle equality, we have

$$\begin{aligned} & e_{Sp}(k,n) \\ = \; & (2k)!/\text{l.c.m.}\{\text{den}(\tfrac{1}{2i}\text{coef}(x^{2k}, (2\sinh(\tfrac{x}{2}))^{2i})) : i \ge n+1\} \\ = \; & (2k)!/\text{l.c.m.}\{\text{den}(\tfrac{1}{2i}\tfrac{2i}{2k}\text{coef}(x^{2k-2i}, (2\sinh^{-1}(\tfrac{x}{2})/x)^{-2k})) : i \ge n+1\} \\ = \; & (2k-1)!/\text{l.c.m.}\{\text{den}(\text{coef}(x^{2j}, (2\sinh^{-1}(\tfrac{x}{2})/x)^{-2k})) : j \le k-n-1\} \end{aligned}$$

By Corollary 1.4, both of these strings of equalities are equal, and the equality of the l.c.m.'s in the denominators at the end of the two strings yields the proposition. ■

Crabb's proof of Theorem 1.5 was motivated by his considerations of K-theory of Thom spaces. The subspace of $K^0((CP^{k-1})^{2n(H-1)})_{(2)}$ invariant under ψ^3 leads to information about integrality of $(\log(1+x)/x)^{2n}$, while similar analysis for the bundle $n(H \oplus \overline{H} - 2)$ leads to information about integrality of $(2\sinh^{-1}(x/2)/x)^{2n}$. Since these bundles are equivalent as real bundles, the information must be the same in the two cases. See [9] for related arguments.

This analysis can be reformulated in purely number-theoretic terms as follows.

Proof of Theorem 1.5. We begin by noting that for any rational power series $f(-)$ and positive integers N and k

$$N\cdot f(y)\in \mathbf{Z}[y]/y^{[(k+1)/2]} \quad \text{iff} \quad N\cdot f(\tfrac{x^2}{1+x})\in \mathbf{Z}[x]/x^k.$$

We apply this to $f(y) = (2\sinh^{-1}(\frac{1}{2}\sqrt{y})/\sqrt{y})^{2n}$, where n is any integer. Since (3.5) implies

$$2\sinh^{-1}(\tfrac{1}{2}\sqrt{x^2/(1+x)}) = \log(1+x),$$

we obtain the middle "iff" below.

$$\begin{aligned} & N(\log(1+x)/x)^{2n}\in \mathbf{Z}[\![x]\!]/x^k \\ \text{iff}\quad & N(1+x)^n(\log(1+x)/x)^{2n}\in \mathbf{Z}[\![x]\!]/x^k \\ \text{iff}\quad & N(2\sinh^{-1}(\tfrac{1}{2}\sqrt{y})/\sqrt{y})^{2n}\in \mathbf{Z}[\![y]\!]/y^{[(k+1)/2]} \\ \text{iff}\quad & N(2\sinh^{-1}(\tfrac{z}{2})/z)^{2n}\in \mathbf{Z}[\![z]\!]/z^k, \end{aligned}$$

where we have let $y = z^2$ in the last line. ∎

While performing some computer verifications of 1.5, we noted that the exponent of 2 in the denominator of the coefficient of x^{2k} in $(2\sinh^{-1}(\frac{x}{2})/x)^{-2}$ seemed to be always $2k$. Replacing $x/2$ by x led us to conjecture the following result, which was then proved by Nigel Byott.

Proposition 4.2 *The coefficients in* $(\sinh^{-1}(x)/x)^{-2}$ *are all 2-adic units.*

Byott's proof involves using calculus to show

$$(\sinh^{-1}(x)/x)^2 = \sum(-4)^j\frac{1}{(2j+1)(j+1)}\binom{2j}{j}^{-1}x^{2j},$$

and deducing from this the result about its reciprocal.

We list the first few terms of some of these series, to make the point that the denominators of coefficients are not always equal. It is only their least common multiple through a range that are equal.

$$\begin{aligned}(\log(1+x)/x)^2 \;=\;& 1-x+\tfrac{11}{12}x^2-\tfrac{5}{6}x^3+\tfrac{137}{180}x^4-\tfrac{7}{10}x^5+\tfrac{363}{560}x^6\\ &-\tfrac{761}{1260}x^7+\tfrac{7129}{12600}x^8-\tfrac{671}{1260}x^9+\tfrac{83711}{166320}x^{10}+\cdots\end{aligned}$$

$$(2\sinh^{-1}(\tfrac{x}{2})/x)^2 = 1-\tfrac{1}{12}x^2+\tfrac{1}{90}x^4-\tfrac{1}{560}x^6+\tfrac{1}{3150}x^8-\tfrac{1}{16632}x^{10}+\cdots$$

$$(\log(1+x)/x)^{-2} = 1+x+\tfrac{1}{12}x^2-\tfrac{1}{240}x^4+\tfrac{1}{240}x^5-\tfrac{221}{60480}x^6+\tfrac{19}{6048}x^7 -\tfrac{9829}{3628800}x^8+\tfrac{407}{172800}x^9-\tfrac{330157}{159667200}x^{10}+\cdots$$

$$(2\sinh^{-1}(\tfrac{x}{2})/x)^{-2} = 1+\tfrac{1}{12}x^2-\tfrac{1}{240}x^4+\tfrac{31}{60480}x^6-\tfrac{289}{3628800}x^8+\tfrac{317}{22809600}x^{10}+\cdots$$

We close with another amusing result about the $\sinh^{-1}$ power series which fell out from our work. It can be proved by comparing our computation of the coaction in $MU_*(Q^\infty)$ with that obtained using the method of [15]. See (3.6).

Proposition 4.3 *For $k \geq 0$, let $s_k(y)$ be the polynomial defined by*

$$s_k(\tfrac{x^2}{1-x}) = \tfrac{x^2}{1-x}(x^k + (\tfrac{-x}{1-x})^k),$$

and let

$$f_k(y) = \sum_{i=0}^{\infty}(-1)^i \frac{1}{(i+1)(i+1+k)} y^i.$$

Then

$$(2\sinh^{-1}(\frac{\sqrt{y}}{2}))^2 = \tfrac{1}{2}s_0(y)f_0(y) + \textstyle\sum_{k=1}^{\infty} s_k(y)f_k(y).$$

We omit the proof, but exemplify slightly as follows. The polynomials $s_k(y)$ for $0 \leq k \leq 4$ are $2y$, $-y^2$, $2y^2+y^3$, $-3y^3-y^4$, and $2y^3+4y^4+y^5$. The series $f_k(y)$ can be expressed in terms of the log series as follows. Let

$$\ell_j(y) = (-1)^j(\log(1+y) - \sum_{i=1}^{j}(-1)^{i+1}\frac{y^i}{i})/y^{j+1}.$$

Then $f_k(y) = \frac{1}{k}(\ell_0(y) - \ell_k(y))$.

References

[1] M. F. Atiyah and J. A. Todd, *On complex Stiefel manifolds*, Proc Camb Phil Soc **56** (1960) 342-353.

[2] M. Bendersky, *Some calculations in the unstable Adams-Novikov spectral sequence*, Publ RIMS **16** (1980) 529-542.

[3] ———, *The v_1-periodic unstable Novikov spectral sequence*, to appear.

[4] M. Bendersky, E. B. Curtis, and H. R. Miller, *The unstable Adams spectral sequence for generalized homology*, Topology **17** (1978) 229-248.

[5] M. Bendersky, E. B. Curtis, and D. Ravenel, *EHP sequences in BP theory*, Topology **21** (1982) 373-391.

[6] M. Bendersky and D. M. Davis, *2-primary v_1-periodic homotopy groups of $SU(n)$*, to appear in Amer Jour Math.

[7] A. Borel, *Cohomologie des espaces fibres principaux*, Annals of Math **57** (1953) 115-207.

[8] L. Comtet, *Advanced combinatorics*, Reidel (1974).

[9] M. C. Crabb and K. Knapp, *James numbers*, Math Ann **282** (1988) 395-422.

[10] D. M. Davis, *Divisibility of generalized exponential and logarithmic coefficients*, Proc Amer Math Soc **109** (1990) 553-558.

[11] ———, *v_1-periodic homotopy groups of $SU(n)$ at an odd prime*, to appear in Proc London Math Soc.

[12] I. M. James, *The topology of Stiefel manifolds*, London Math Society Lecture Notes Series **24** (1976).

[13] K. Morisugi, *Homotopy groups of symplectic groups and the quaternionic James numbers*, Osaka Jour Math **23** (1986) 867-880.

[14] D. M. Segal, *The cooperation on $MU_*(CP^\infty)$ and $MU_*(HP^\infty)$ and the primitive generators*, Jour Pure Appl Alg **14** (1979) 315-322.

[15] T. Sugawara, *Landweber-Novikov operations on the complex cobordism ring of HP^n*, Mem Fac Sci Kyushu Univ **29** (1975) 193-202.

[16] G. Walker, *Estimates for the complex and quaternionic James numbers*, Quar Jour Math Oxford **32** (1981) 467-489.

[17] W. S. Wilson, *Brown-Peterson homology, an introduction and sampler*, Regional Conference Series in Math, Amer Math Soc **48** (1980).

UNSTABLE ADAMS SPECTRAL SEQUENCE CHARTS

DIANNE BARNES, DAVID PODUSKA AND PAUL SHICK

Dedicated to the memory of J. Frank Adams

ABSTRACT. We present charts for the E_2 term of the unstable Adams spectral sequence for $\pi_* S^n$, complete out to the 60-stem, for $3 \leq n \leq 25$ at the prime 2, and complete out to the 81-stem for $3 \leq n \leq 15$ at the prime 3. The charts are produced from the computer calculations of Curtis, et al [2], extended by Mahowald [4] (p=2) and Tangora [6] (p=3), using a *C* language program to convert the Lambda algebra files into LaTEX typesetting code.

The purpose of this paper is to present the computer calculations of the unstable Adams spectral sequences for spheres of Curtis, et al [2] (extended by Mahowald [4]) and Tangora [6] in the conventional $(t-s, s)$ chart presentation, thereby making these results more accessible to homotopy theorists. The main tool is a *C* language program which inputs the ASCII files of Lambda algebra calculations and has as its output ASCII files written in the LaTEX typesetting code ([3]).

The Unstable Adams spectral sequence (UASS) converges to the p-component of the homotopy groups of the n-sphere, S^n. The E_2 term of the UASS can be calculated by homological algebra techniques, using [1], where it was shown that the E_2 term of the classical Adams spectral sequence (clASS) converging to the p-component of the stable homotopy groups of spheres, $\pi_* S^0$, is isomorphic to the homology of the differential graded algebra Λ. A companion paper ([5]) showed that the E_2 term for the UASS for S^n is isomorphic to the homology of a submodule of Λ, denoted by $\Lambda(n)$. For each prime p, the algebra Λ is defined by generators and relations, although for p odd the description is somewhat more complicated that at the prime 2. For this reason, we will describe the Lambda algebra only for $p = 2$, referring to [6] for the odd primary case.

1980 *Mathematics Subject Classification* (1985 *Revision*). Primary 55Q45; Secondary 55P42, 55N22.

Project supported by NSF grant DMS-8714043.

Let Λ be the $\mathbb{F}_2$-algebra with generators λ_i, for all $i \geq 0$, with the following relations:

$$\lambda_i \lambda_j = \sum_{k \geq 0} \binom{j-2i-2-k}{k} \lambda_{j-i-k-1} \lambda_{2i+k+1},$$

where the binomial coefficient is, of course, reduced mod 2. We make Λ a differential algebra by defining

$$d(\lambda_i) = \sum_{k \geq 1} \binom{i-k}{k} \lambda_{i-k} \lambda_{k-1}.$$

The bigrading is given by assigning λ_i bidegree $(1, i+1)$, where the first coordinate is the homological degree and the second is the internal degree.

For a sequence of nonnegative integers $I = (i_1, i_2, \ldots, i_r)$, λ_I denotes the monomial $\lambda_{i_1} \lambda_{i_2} \ldots \lambda_{i_r}$. A sequence I is admissible if $2i_j \geq i_{j+1}$ for all j. The Lambda algebra Λ has as a basis over $\mathbb{F}_2$ given by $\{\lambda_I : I \text{ is admissible }\}$, as one can see from the relations above. For each $n > 0$, we define $\Lambda(n)$ to be the differential subalgebra of Λ spanned by the set of λ_I with I admissible and $i_1 < n$. The main theorem of [5] is that $\Lambda(n)$ with its differential is the E_1 term for the UASS for $\pi_* S^n$, so that

$$E_2^{*,*}(S^n) = H_*(\Lambda(n), d).$$

The work of [2] and [6] computes the homology of Λ, displaying the results in Curtis tables. The basic idea is to lexicographically order the admissible monomials in each bidegree, and order all polynomials in decreasing order of terms. The Curtis tables list the leading term of the minimal representative of each homology class of polynomials. The tables are further simplified by considering only terms λ_I which end with an odd index, eliminating the towers from the resulting homology. The notation used by [2] and (at the prime 2 only) by [6] is basically this:

$$(i_1, i_2, \ldots, i_s) \longleftarrow (j_1, j_2, \ldots, j_{s-1})$$

means that d of any polynomial homologous to the minimal representative whose leading term is $\lambda_{j_1} \lambda_{j_2} \ldots \lambda_{j_{s-1}}$ is a polynomial homologous to a minimal representative having leading term $\lambda_{i_1} \lambda_{i_2} \ldots \lambda_{i_s}$.

The main program used in this project was written in C to convert the ASCII files of Lambda algebra calculations into LaTEX typesetting code [3]. Both C and LaTEX were chosen because of their great portability. The program has three main sections. The first reads the Curtis table and sets up a series of lists, one in each bigrading. The second section checks for h_0 and h_1 multiplications (a_0 and h_0 at the prime 3). The third section writes the ASCII files in LaTEX which are then used to print the chart.

To print a chart of $E_2^{*,*}(S^n)$, the program looks for all classes where the leading term is less than or equal to $n-1$ and where, if the term is the target of a differential, the source of the differential has leading term greater than or equal to n. Once the table is set up from such classes, the h_0 and h_1 multiplications are determined. The basic algorithm is quite simple: if $\lambda_k a$ is hit by $\lambda_{k+1} b$ for all k (or for all even k), then $h_0 b = a$. If $\lambda_k a$ is hit by $\lambda_{k+2} b$ for all k (or for all even k), then $h_1 b = a$. This algorithm detects most of the known h_0 and h_1 extensions, but fails in some cases, particularly cases involving classes which live only on S^n and "die" on S^{n+1}. Any extensions which were known to exist but failed to show up using this method were added to the charts using auxiliary files. Because of this, there may be some missing h_1 extensions in the unstable charts. Because of the complexity of the charts to be printed, the LaTEX files were printed (usually) on several 8.5 by 11 inch pages, which were manually taped together and then reduced to form the charts found below.

The charts are displayed in the usual $(t-s, s)$ presentation, so that a given column $t-s=k$ provides the E_2 term approximation for $\pi_k S^n$. For the prime 2, we provide charts for S^n for n odd, for $3 \leq n \leq 25$, complete out to the 60-stem. We also include a chart for the stable Adams spectral sequence E_2 term, $\mathrm{Ext}_A(\mathbb{Z}/2, \mathbb{Z}/2)$ which includes labels of classes, and a second chart including all known differentials. For the prime 2 charts, vertical lines indicate multiplication by h_0, diagonal lines multiplication by h_1. Similar charts are provided at the prime 3, complete out to the 81 stem, for S^n, $3 \leq n \leq 15$, and for $\mathrm{Ext}_A(\mathbb{Z}/3, \mathbb{Z}/3)$. At the prime 3, vertical lines indicate multiplication by a_0, diagonal lines multiplication by h_0. An "open dot" in a given bigrading indicates that there are at least 4 classes there.

Acknowledgements: We wish to thank Mark Mahowald and Martin Tangora for graciously providing us with the ASCII files of their calculations, and to thank Mahowald also for his input throughout the project. We also wish to acknowledge the National Science Foundation for their support during this project.

References.

1. A. Bousfield, E. Curtis, D. Kan, D. Quillen, D. Rector and J. Schlesinger, *The mod p lower central series and the Adams spectral sequence,*, Topology **5** (1966), 331-342.

2. E. Curtis, P. Goerss, M. Mahowald and J. Milgram, *Calculations of the unstable Adams E_2 terms for spheres*, Algebraic Topology, Proc. Seattle Emphasis Year, Springer Lec. Notes Math. 1286, 1987, pp. 208-266.
3. L. Lamport, *LaTEX, A Document Preparation System*, Addison-Wesley, 1986.
4. M. Mahowald, *Lambda algebra calculations*, unpublished.
5. D. Rector, *An unstable Adams spectral sequence*, Topology **5** (1966), 342-346.
6. M. Tangora, *Computing the Homology of the Lambda Algebra*, A.M.S. Memoir **337** (1985).

NORTHWESTERN UNIVERSITY, EVANSTON IL 60208

CASE WESTERN RESERVE UNIVERSITY, CLEVELAND OH 44106

JOHN CARROLL UNIVERSITY, UNIVERSITY HTS OH 44118

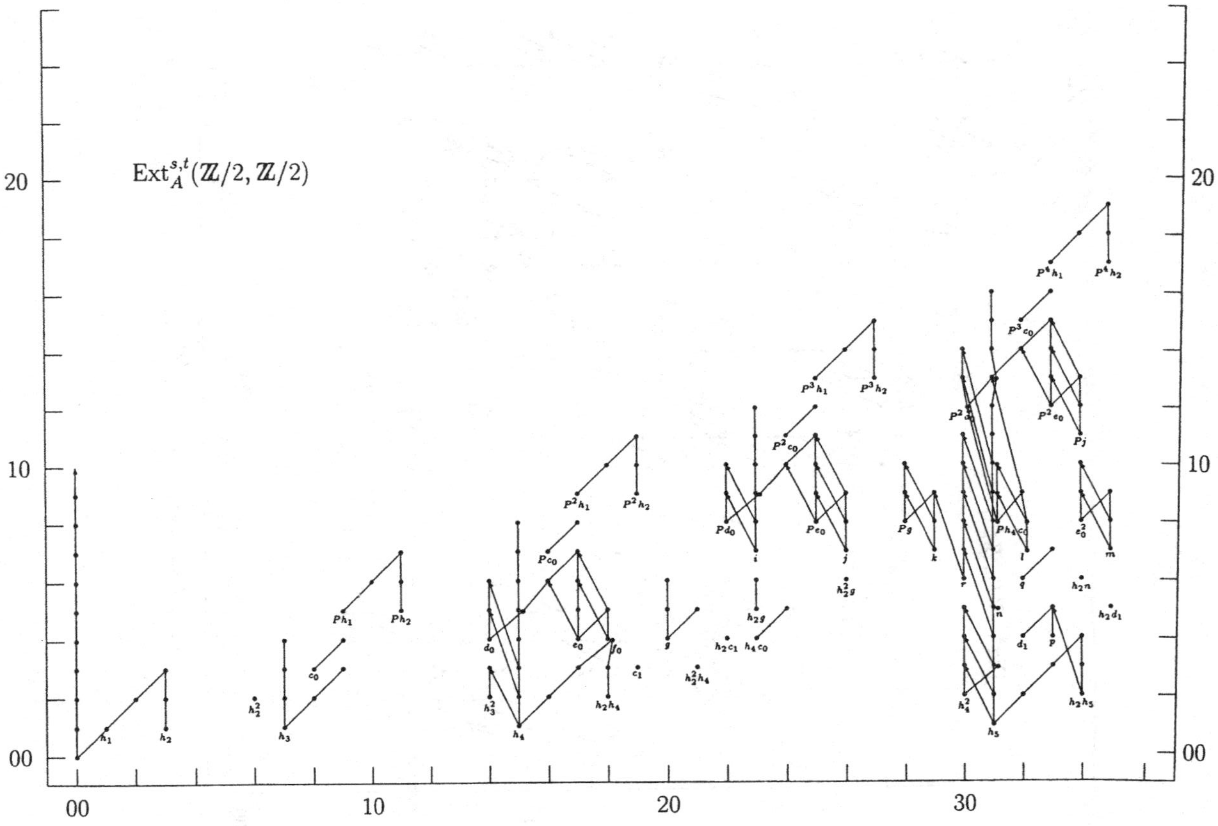
$\mathrm{Ext}_A^{s,t}(\mathbb{Z}/2, \mathbb{Z}/2)$
00
10
20
30
h_1
h_2
h_2^2
h_3
c_0
Ph_1
Ph_2
h_3^2
h_4
d_0
Pc_0
e_0
f_0
h_2h_4
P^2h_1
P^2h_2
c_1
g
$h_2^2h_4$
h_2c_1
h_4c_0
Pd_0
i
h_2g
P^2c_0
Pe_0
j
h_2^2g
P^3h_1
P^3h_2
Pg
k
r
h_4^2
h_5
n
P^2d_0
Ph_4c_0
l
q
d_1
p
P^3c_0
P^2e_0
Pj
e_0^2
h_2n
h_2h_5
m
h_2d_1
P^4h_1
P^4h_2

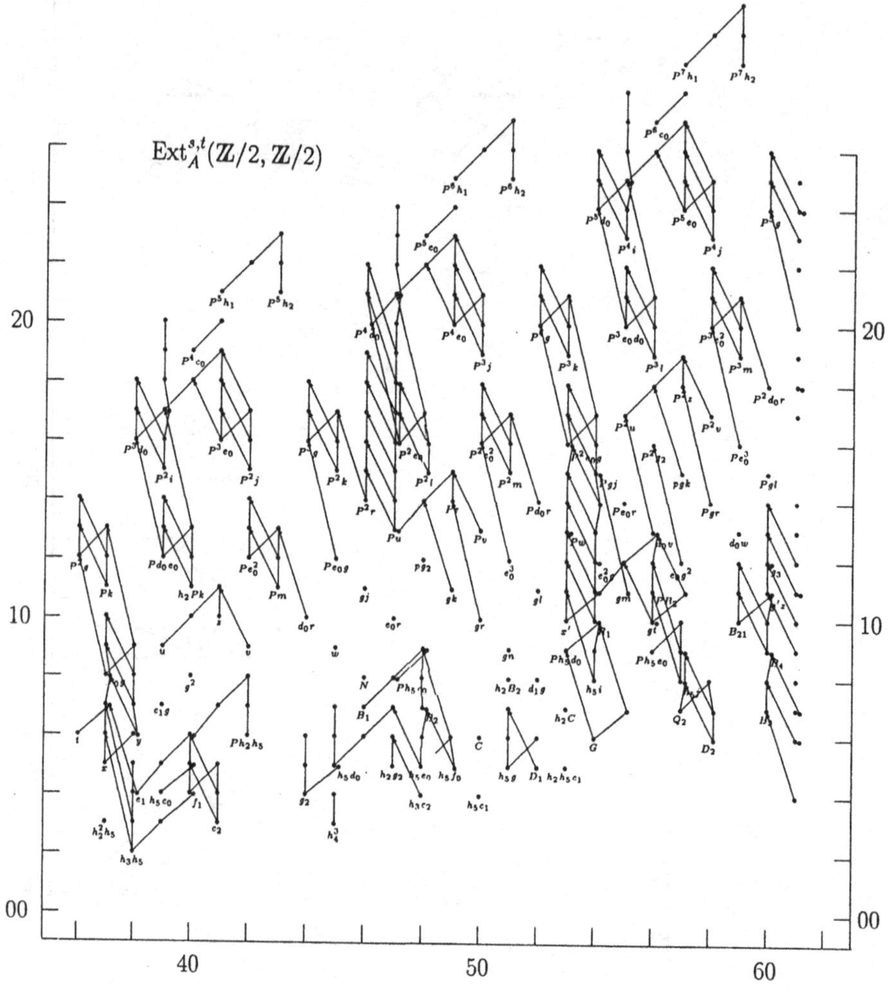

$\mathrm{Ext}_A^{s,t}(\mathbb{Z}/2, \mathbb{Z}/2)$
20
10
00
40
50
60
P^7h_1
P^7h_2
P^6c_0
P^6h_1
P^6h_2
P^5d_0
P^4i
P^5e_0
P^4j
P^5c_0
P^5h_1
P^5h_2
P^4c_0
P^4d_0
P^4e_0
P^3j
P^3k
$P^3e_0d_0$
P^3l
P^3m
P^3d_0
P^2i
P^3e_0
P^2j
P^2k
P^2e_0
P^2l
P^2m
P^2r
P^2u
P^2v
P^2d_0r
Pe_0^3
P^2g_2
pgk
Pgl
Pgr
Pd_0r
Pe_0r
Pgj
Pu
Pv
Pk
Pd_0e_0
h_2Pk
Pe_0^2
Pm
Pe_0g
pg_2
d_0r
e_0r
gj
gk
gr
gl
e_0^3
Pw
gm
gt
d_0w
B_{21}
u
z
v
w
g^2
c_1g
N
B_1
Ph_5n
B_2
gn
h_2B_2
d_1g
Ph_5d_0
h_5i
Ph_5e_0
Q_2
D_2
h_2C
C
G
t
x
y
Ph_2h_5
e_1
h_5c_0
f_1
c_2
g_2
h_5d_0
h_2g_2
h_5e_0
h_5f_0
h_3c_2
h_5g
D_1
$h_2h_5c_1$
h_5c_1
h_4^3
$h_2^2h_5$
h_3h_5

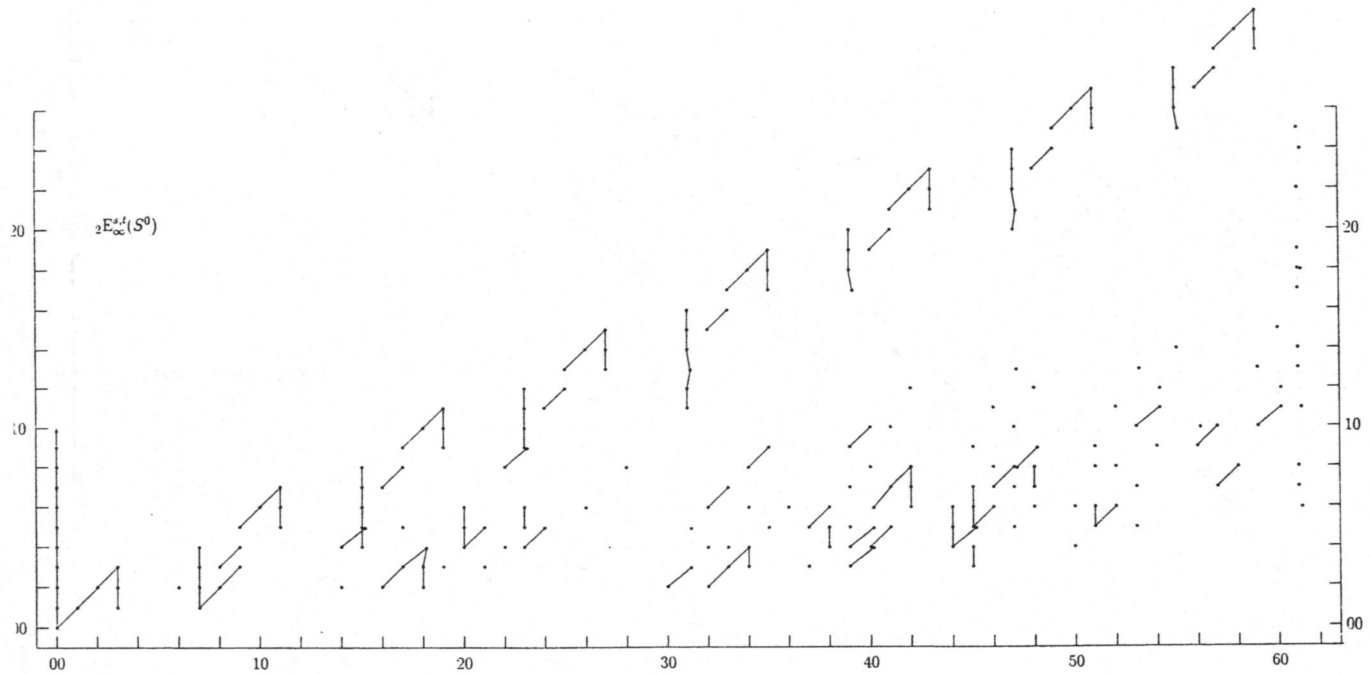

$_2E_\infty^{s,t}(S^0)$
00
10
20
30
40
50
60

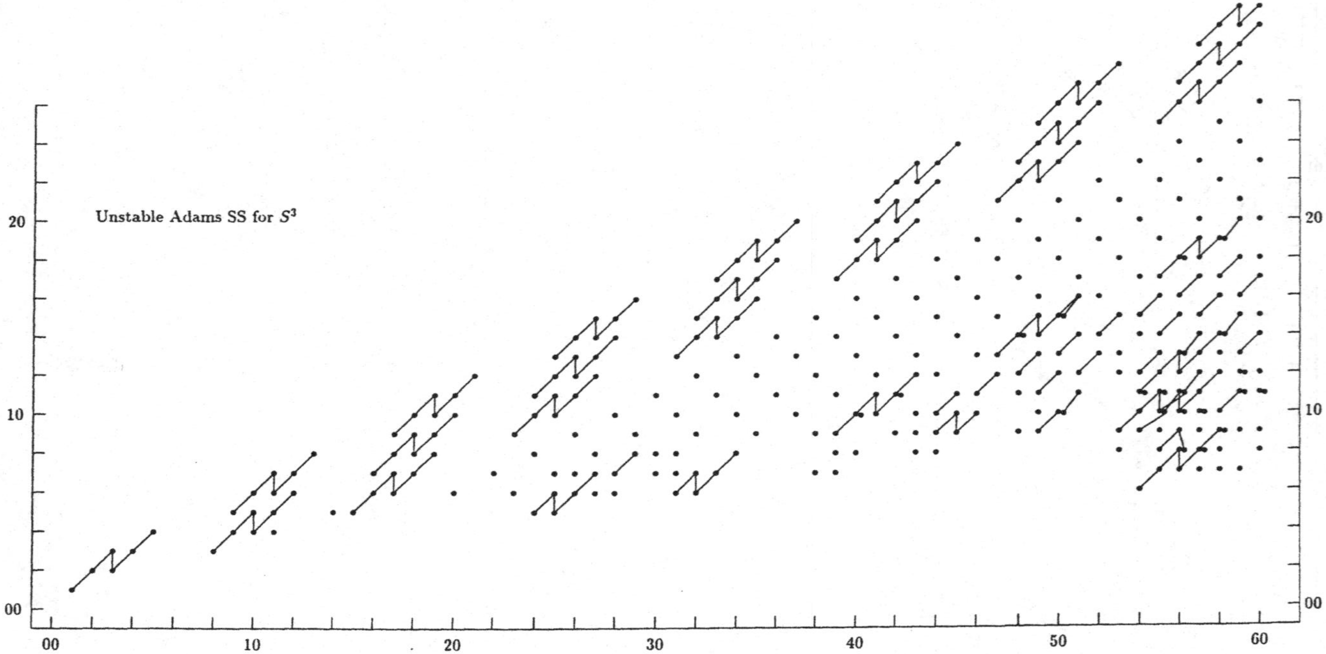
Unstable Adams SS for S^3
00
10
20
30
40
50
60
00
10
20

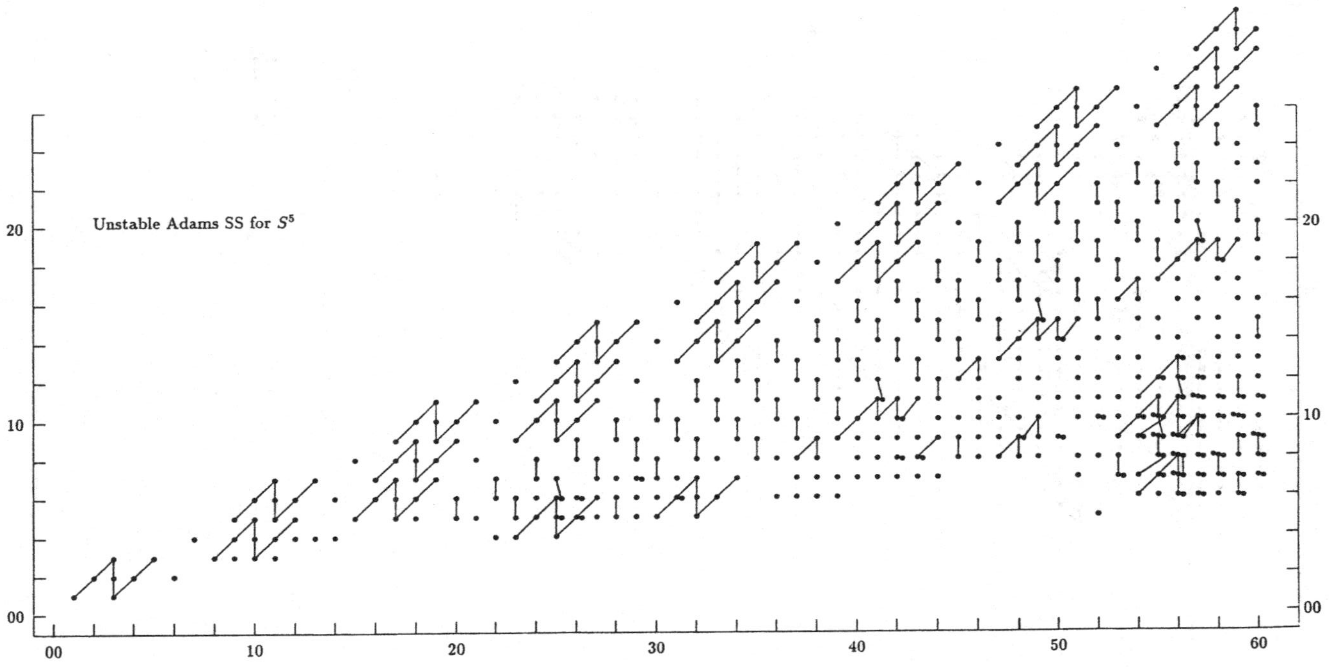
Unstable Adams SS for S^5
00
10
20
30
40
50
60

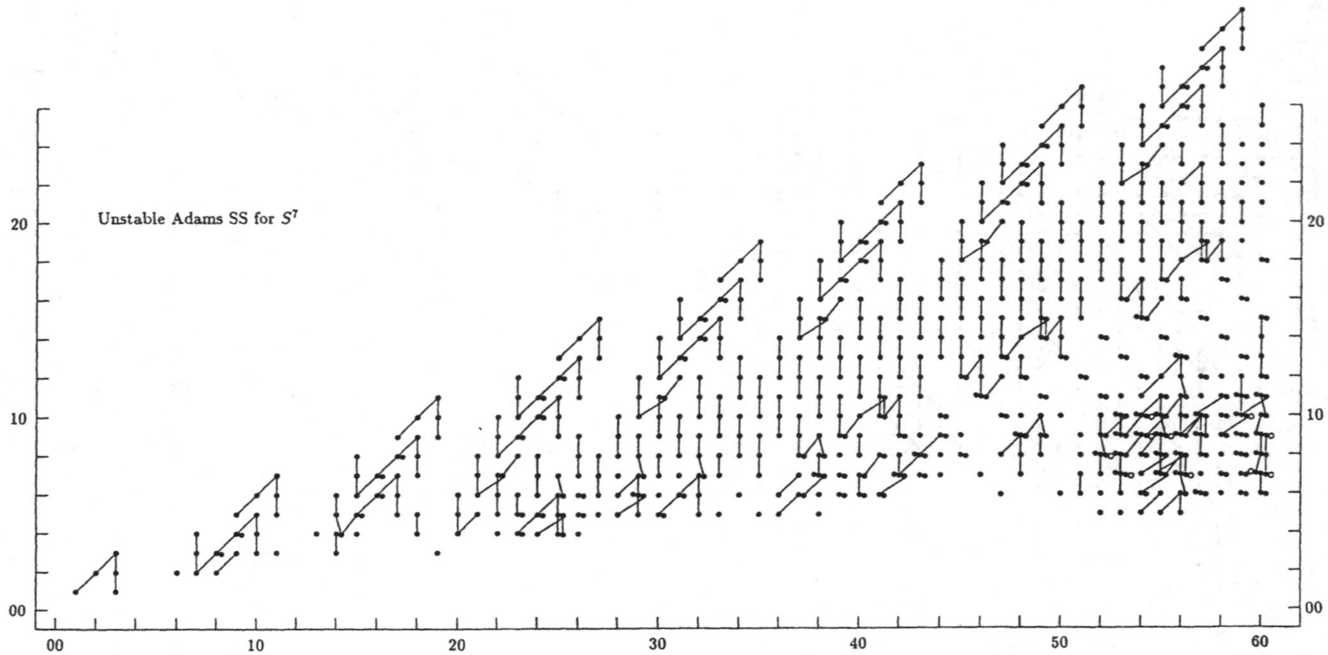
Unstable Adams SS for S^7
00
10
20
30
40
50
60

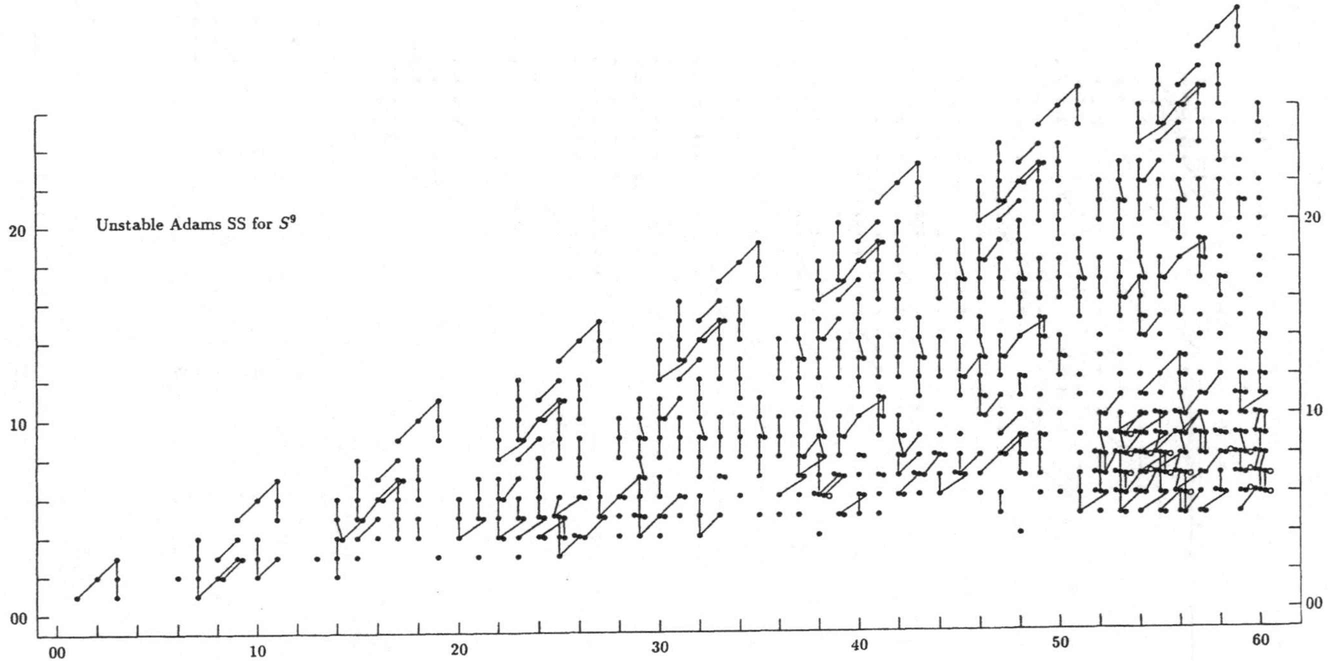
Unstable Adams SS for S^9
00
10
20
30
40
50
60

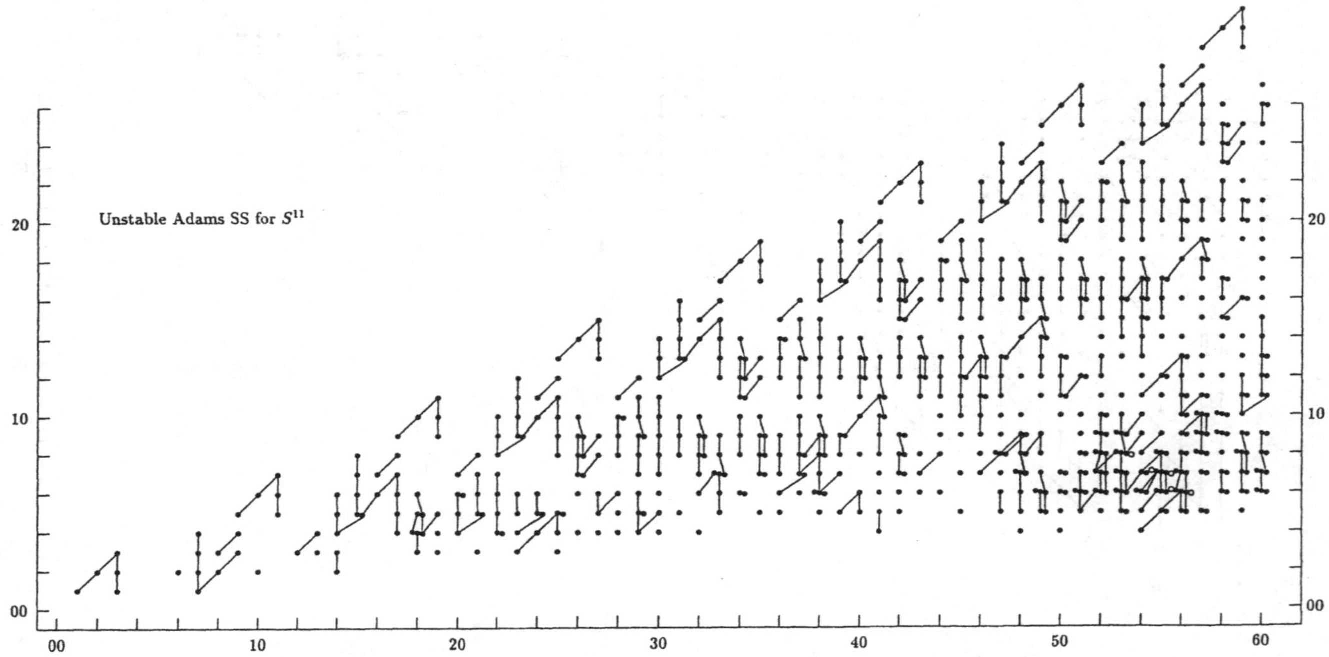

Unstable Adams SS for S^{11}
00
10
20
30
40
50
60
00
10
20

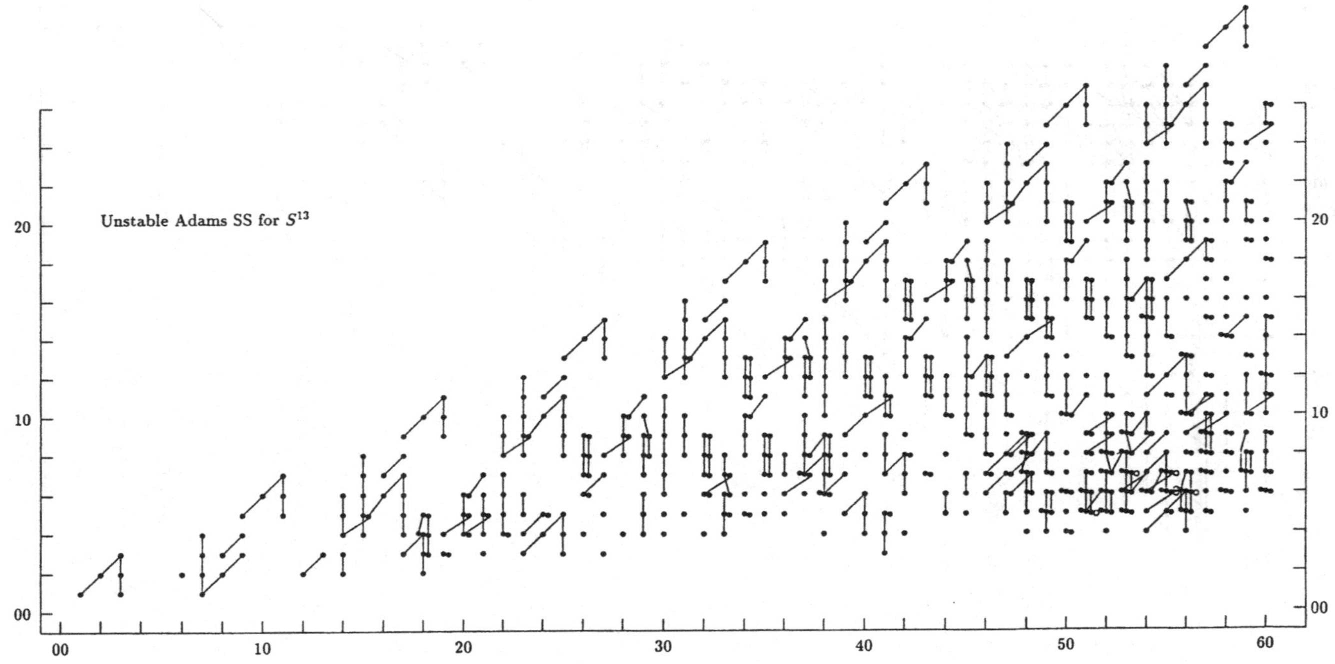
Unstable Adams SS for S^{13}
00
10
20
30
40
50
60
00
10
20

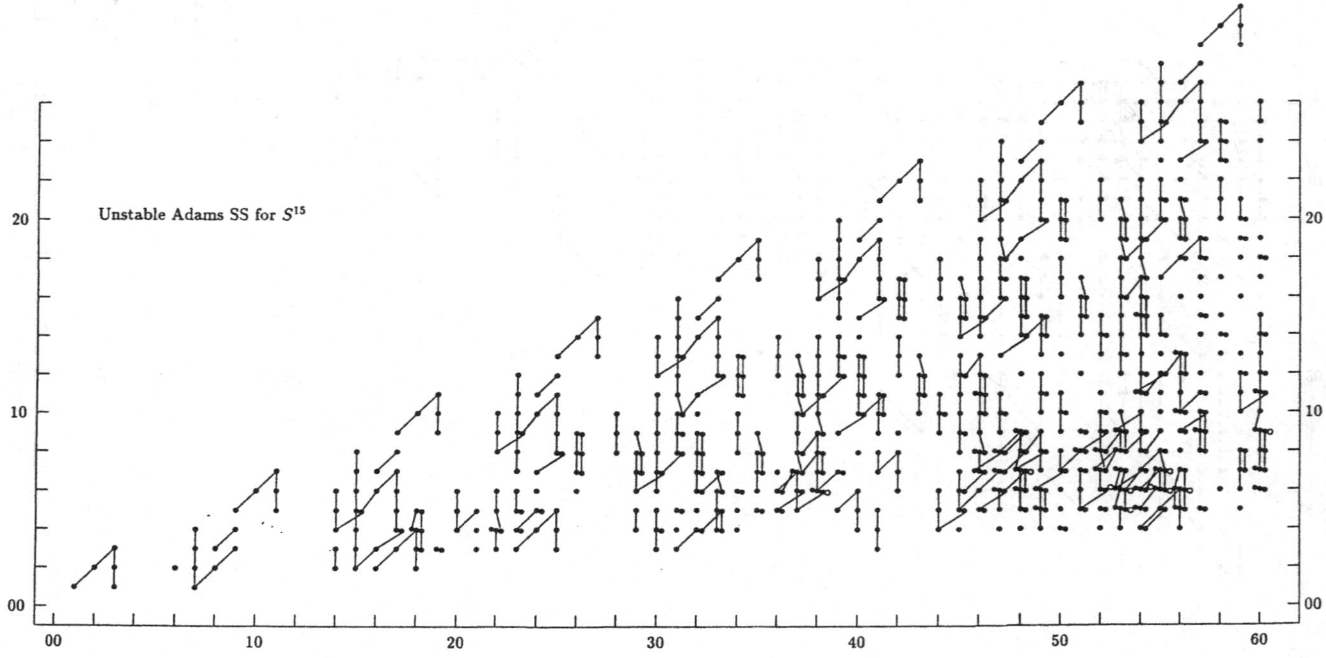

Unstable Adams SS for S^{15}
00
10
20
30
40
50
60

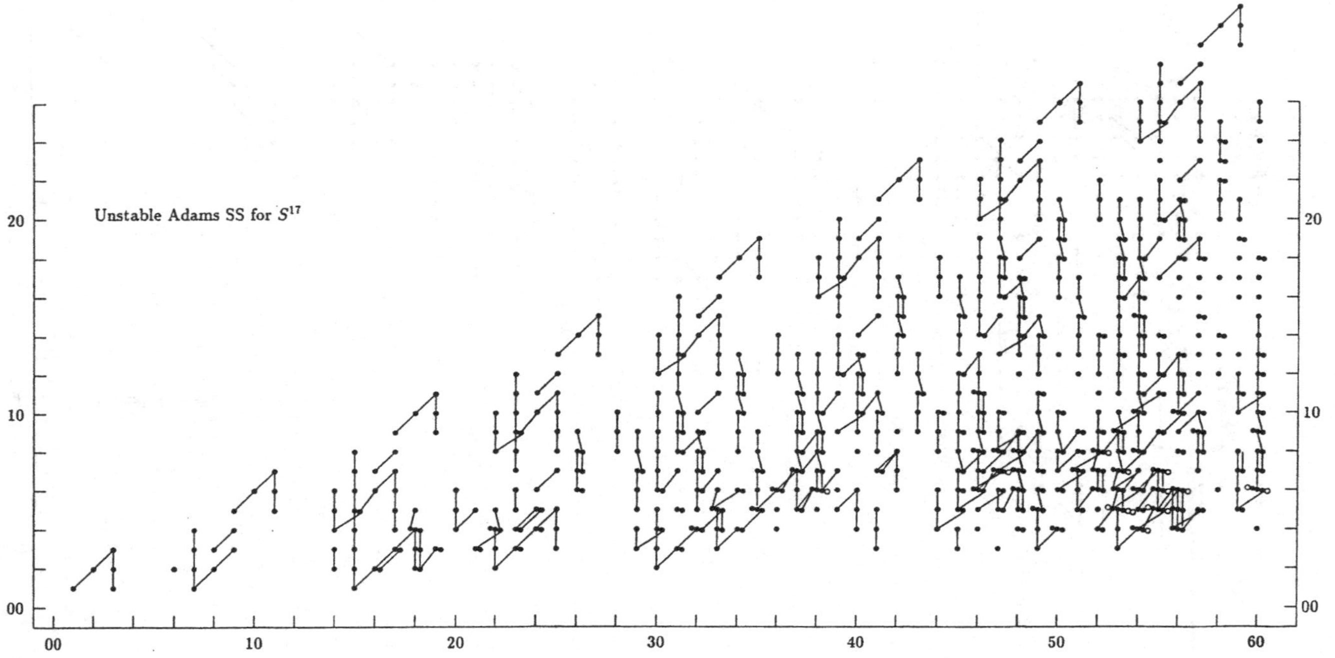
Unstable Adams SS for S^{17}
20
10
00
00
10
20
30
40
50
60
20
10
00

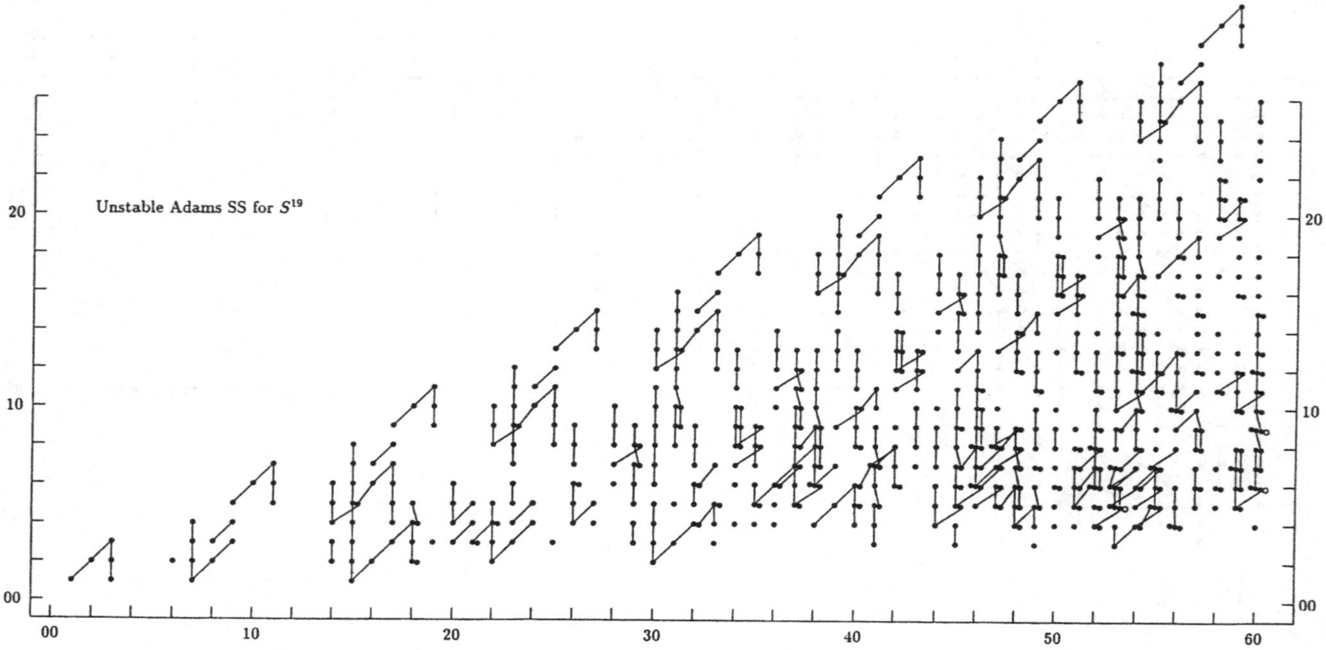

Unstable Adams SS for S^{19}
00
10
20
30
40
50
60
00
10
20

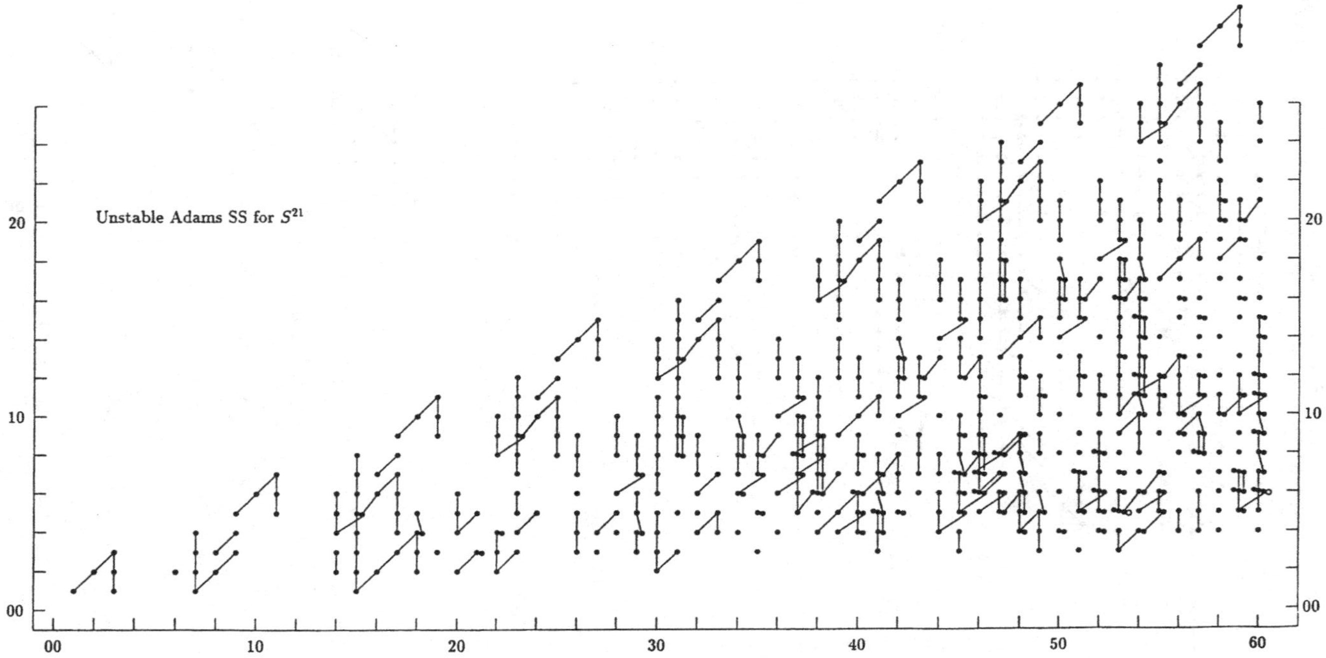
Unstable Adams SS for S^{21}
00
10
20
00
10
20
30
40
50
60
00
10
20

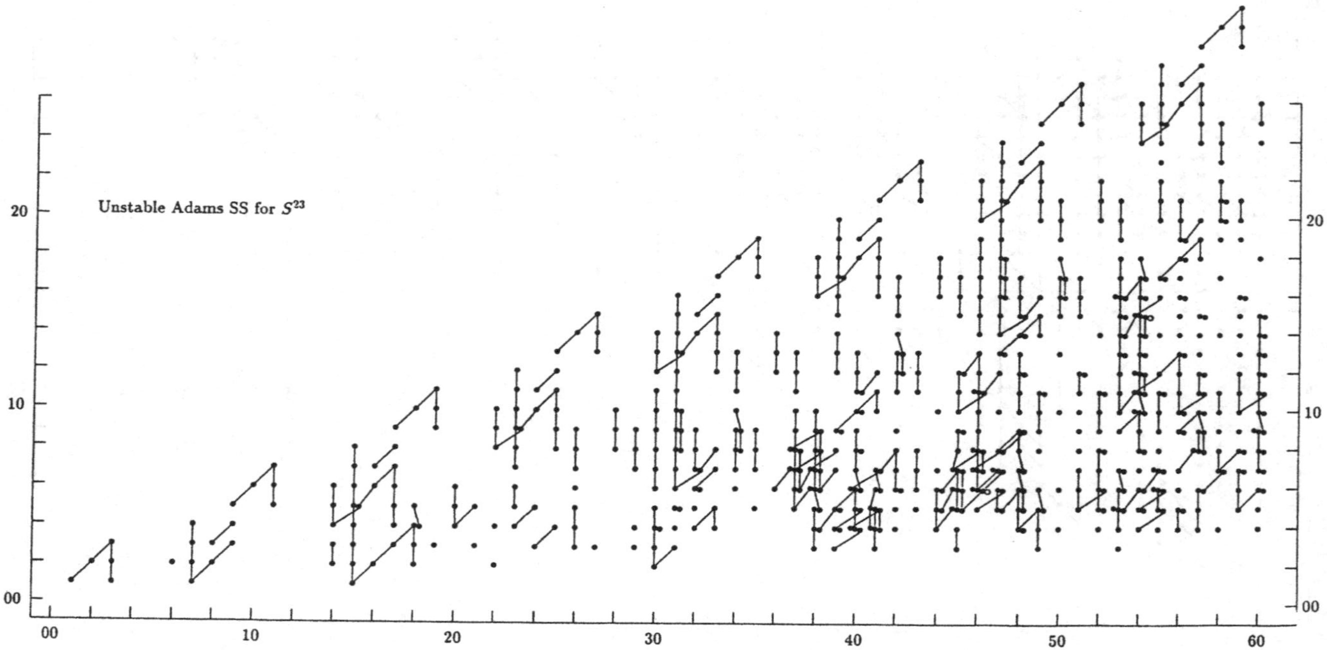
Unstable Adams SS for S^{23}
00
10
20
30
40
50
60

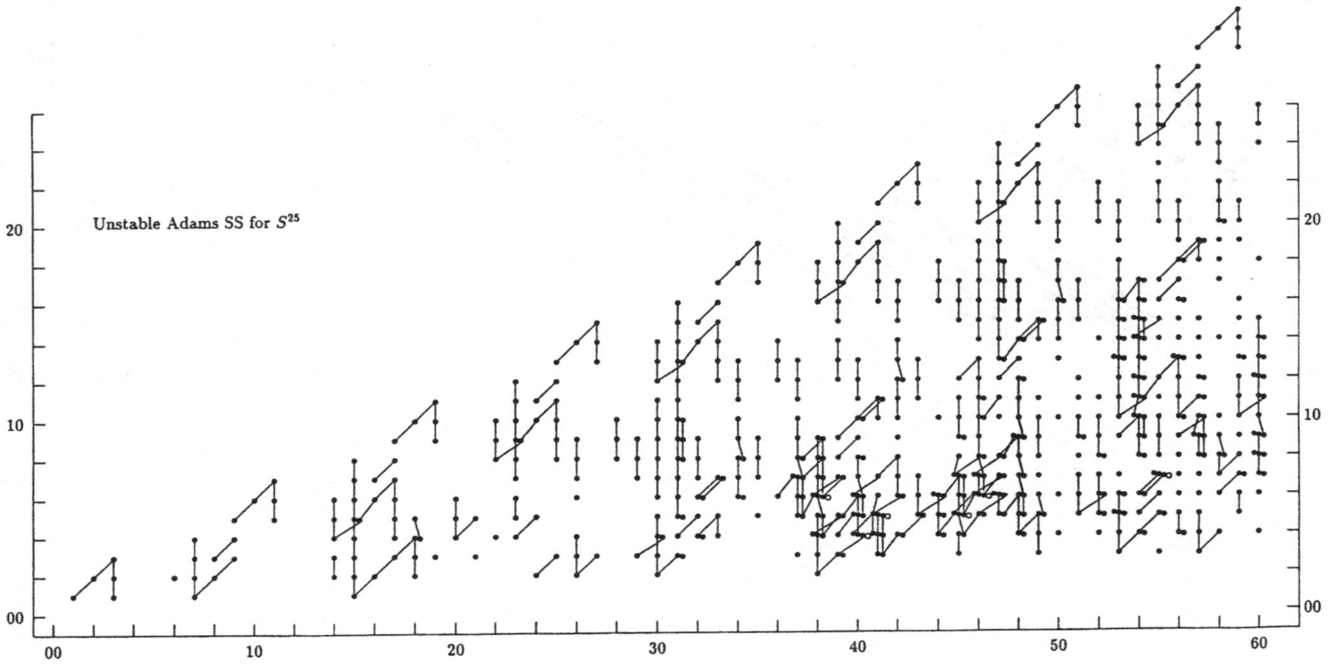
Unstable Adams SS for S^{25}
20
10
00
00
10
20
30
40
50
60
20
10
00

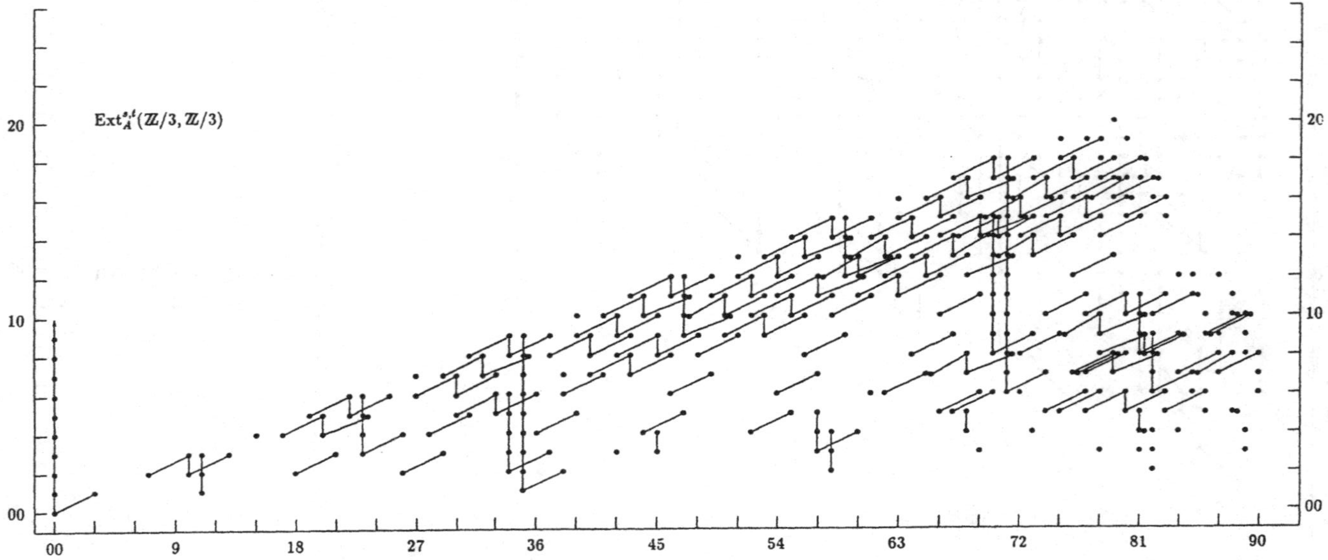
$Ext_A^{s,t}(\mathbb{Z}/3, \mathbb{Z}/3)$
00
10
20
9
18
27
36
45
54
63
72
81
90

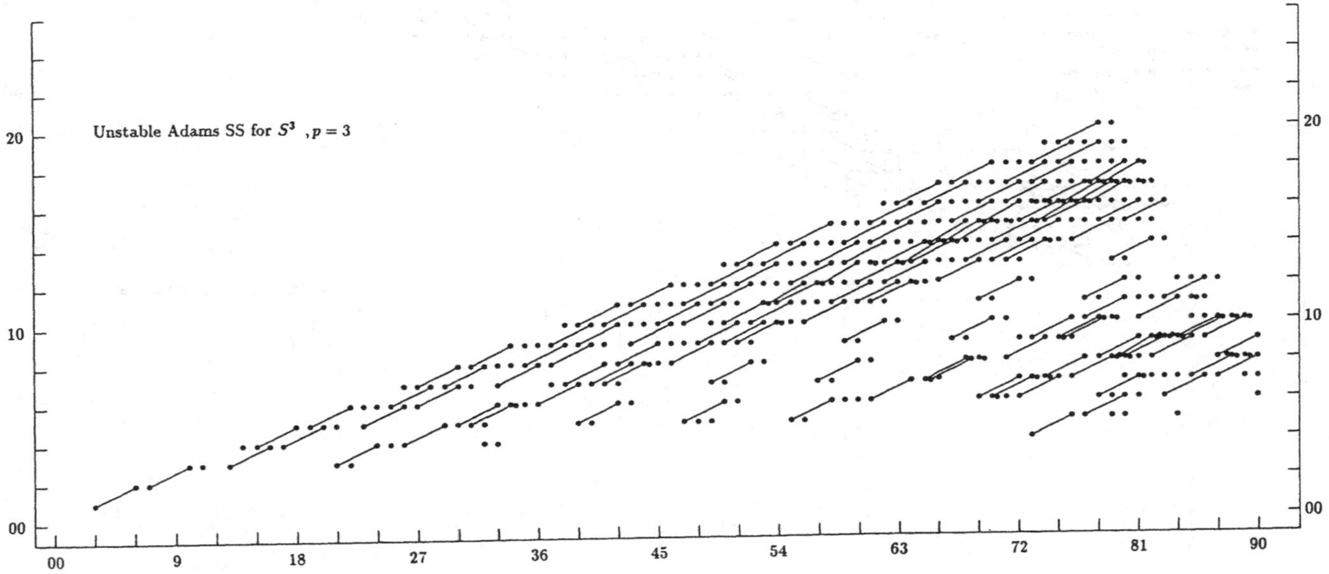
Unstable Adams SS for S^3 , $p = 3$
00
9
18
27
36
45
54
63
72
81
90
00
10
20

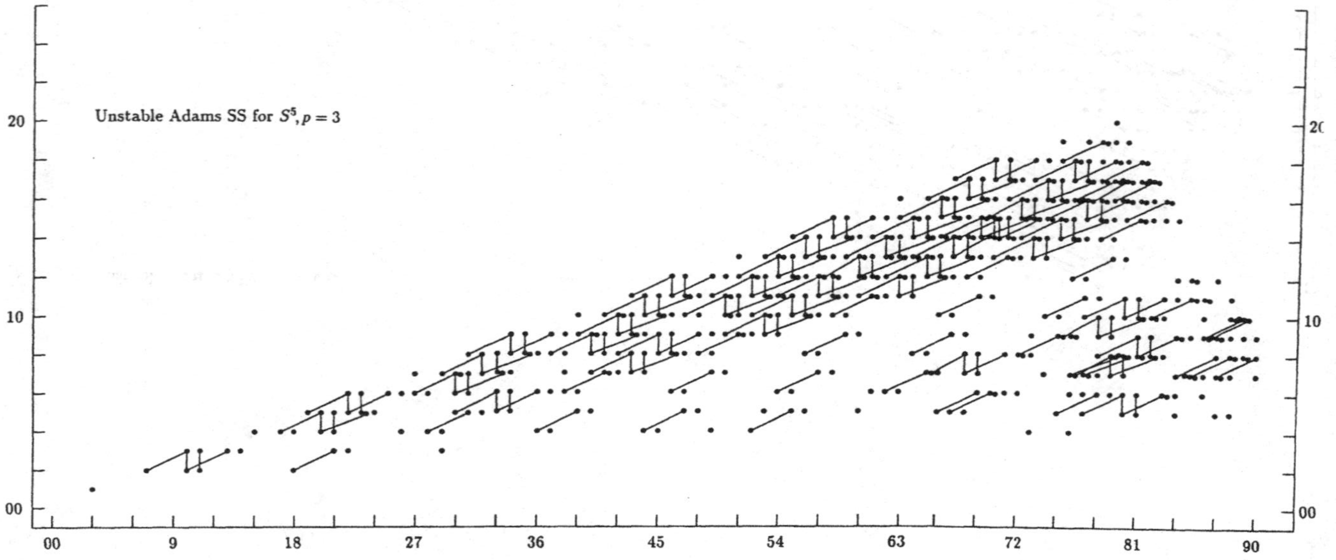
Unstable Adams SS for $S^5, p = 3$
20
10
00
00
9
18
27
36
45
54
63
72
81
90
20
10
00

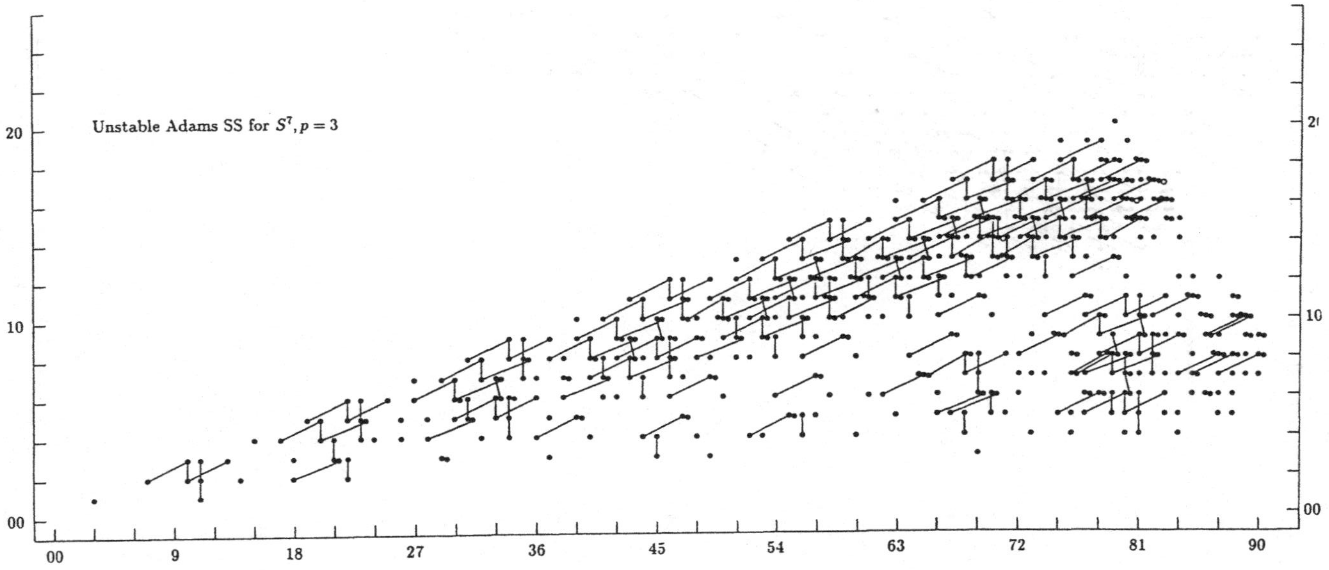

Unstable Adams SS for $S^7, p = 3$
00
9
18
27
36
45
54
63
72
81
90
00
10
20

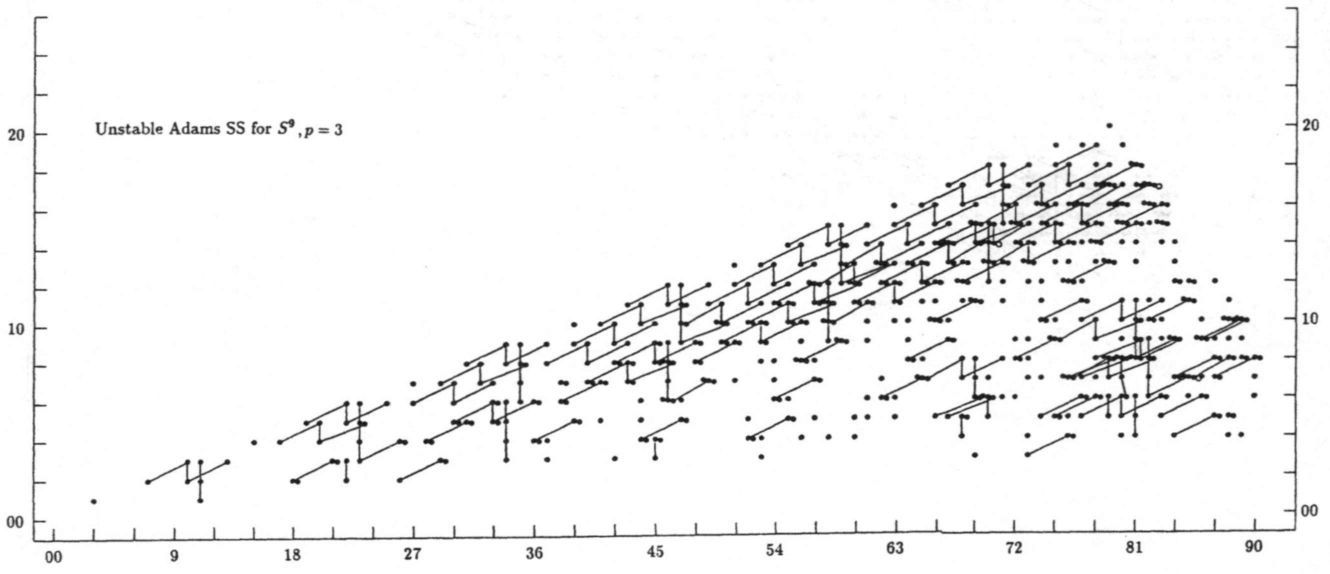

Unstable Adams SS for $S^9, p = 3$
20
10
00
00
9
18
27
36
45
54
63
72
81
90
20
10
00

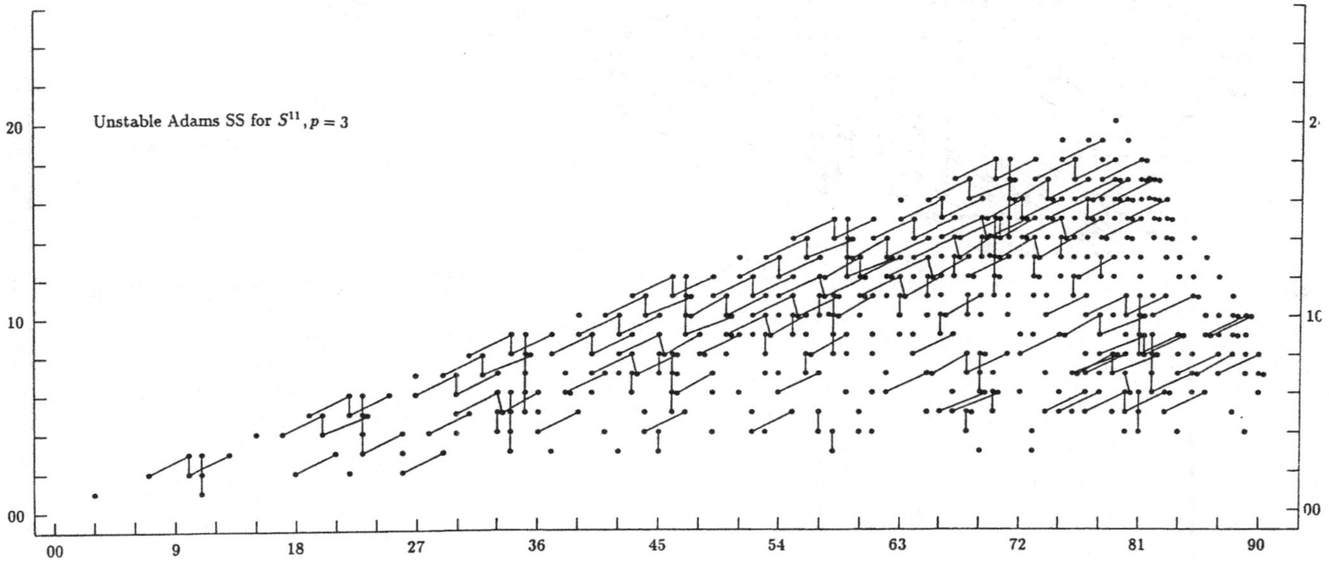

Unstable Adams SS for $S^{11}, p = 3$

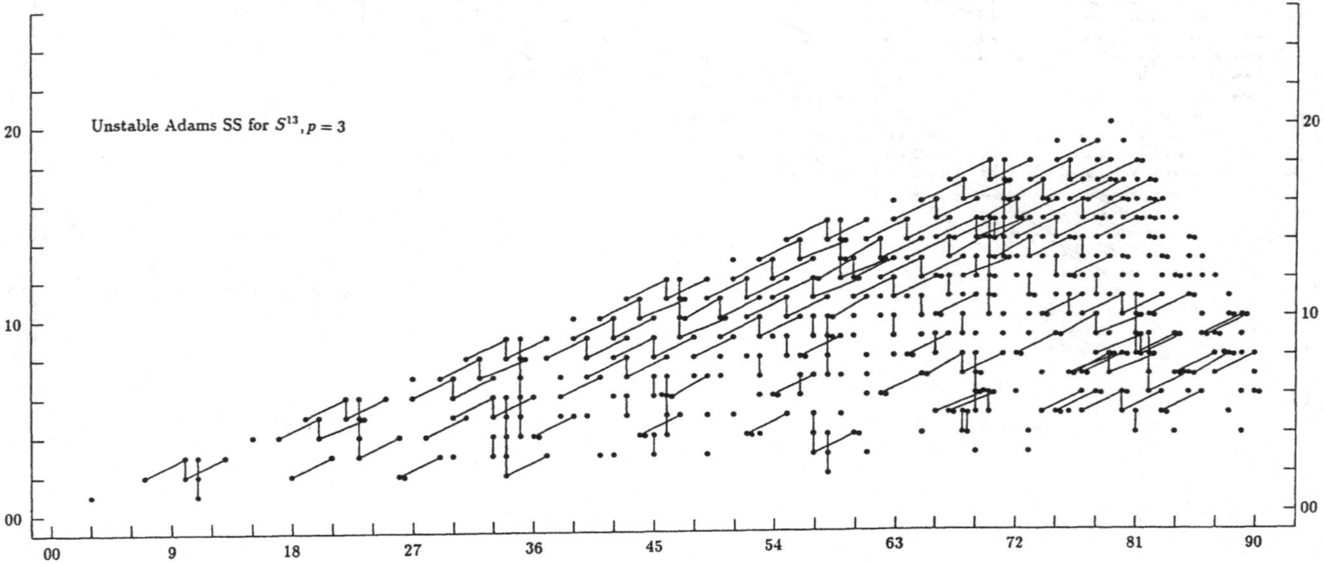

Unstable Adams SS for $S^{13}, p = 3$

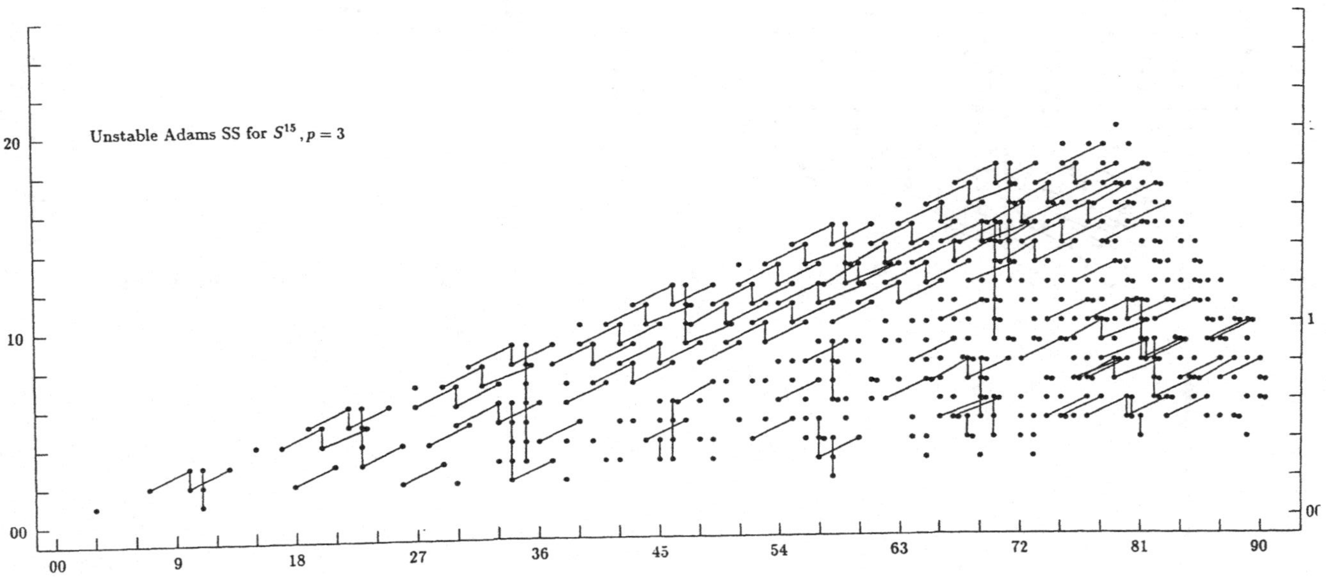

Unstable Adams SS for $S^{15}, p = 3$
20
10
00
00
9
18
27
36
45
54
63
72
81
90
1

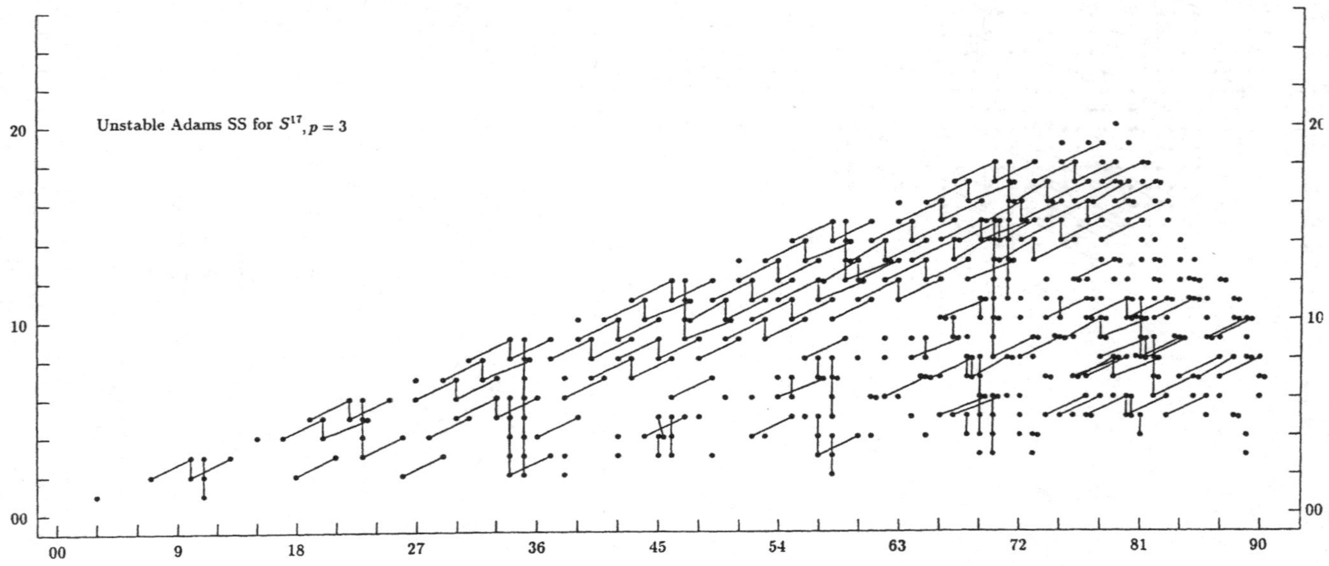
Unstable Adams SS for $S^{17}, p = 3$
20
10
00
00
9
18
27
36
45
54
63
72
81
90
20
10
00

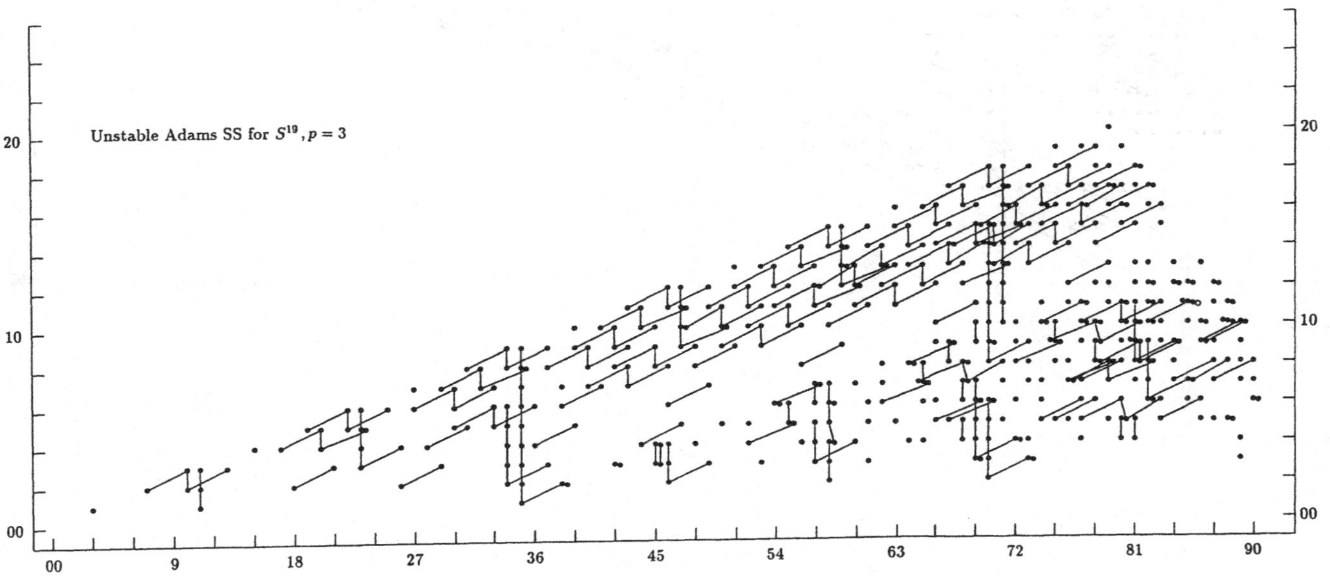

Unstable Adams SS for $S^{19}, p = 3$
00
9
18
27
36
45
54
63
72
81
90
00
10
20

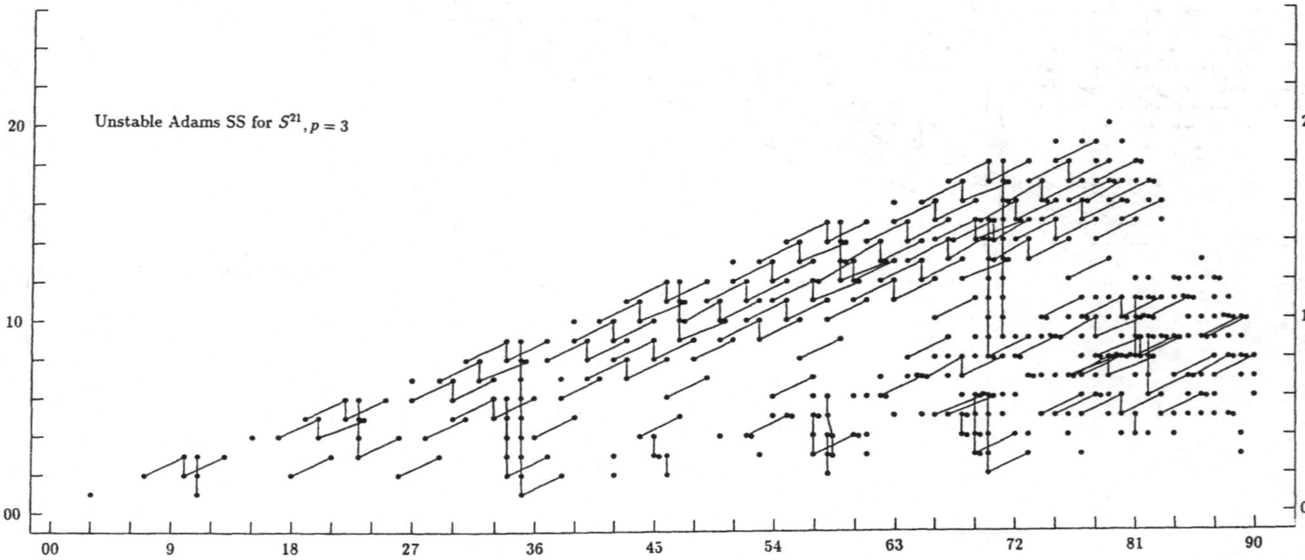
Unstable Adams SS for $S^{21}, p = 3$
20
10
00
00
9
18
27
36
45
54
63
72
81
90
20
10
00

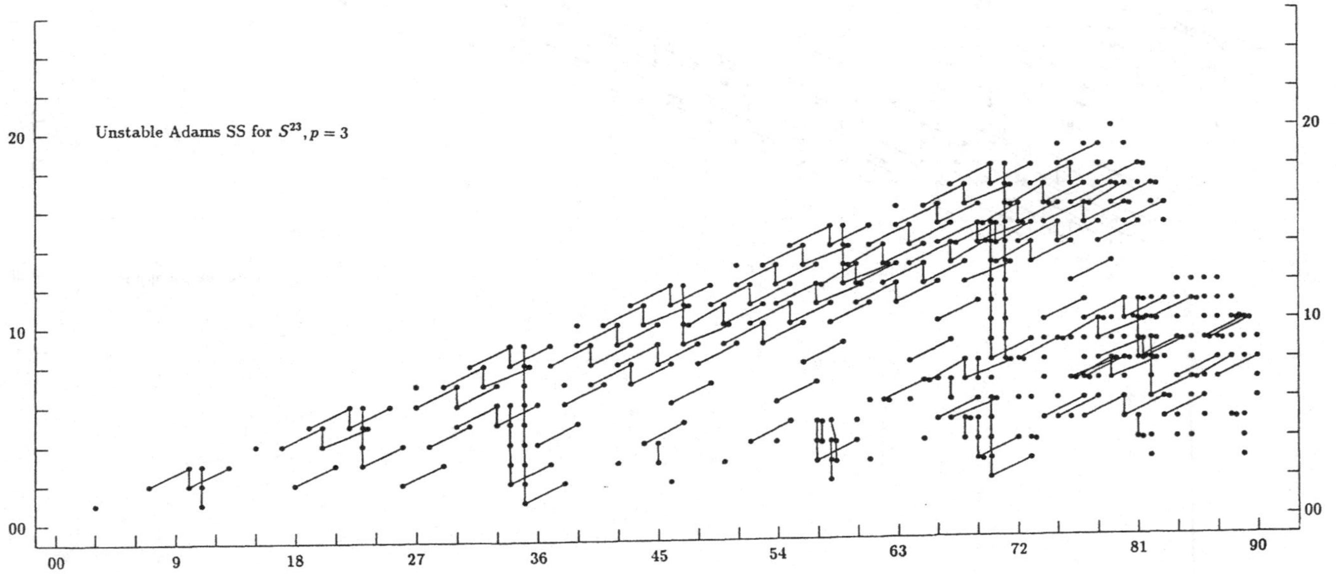
Unstable Adams SS for $S^{23}, p = 3$
20
10
00
00
9
18
27
36
45
54
63
72
81
90
20
10
00

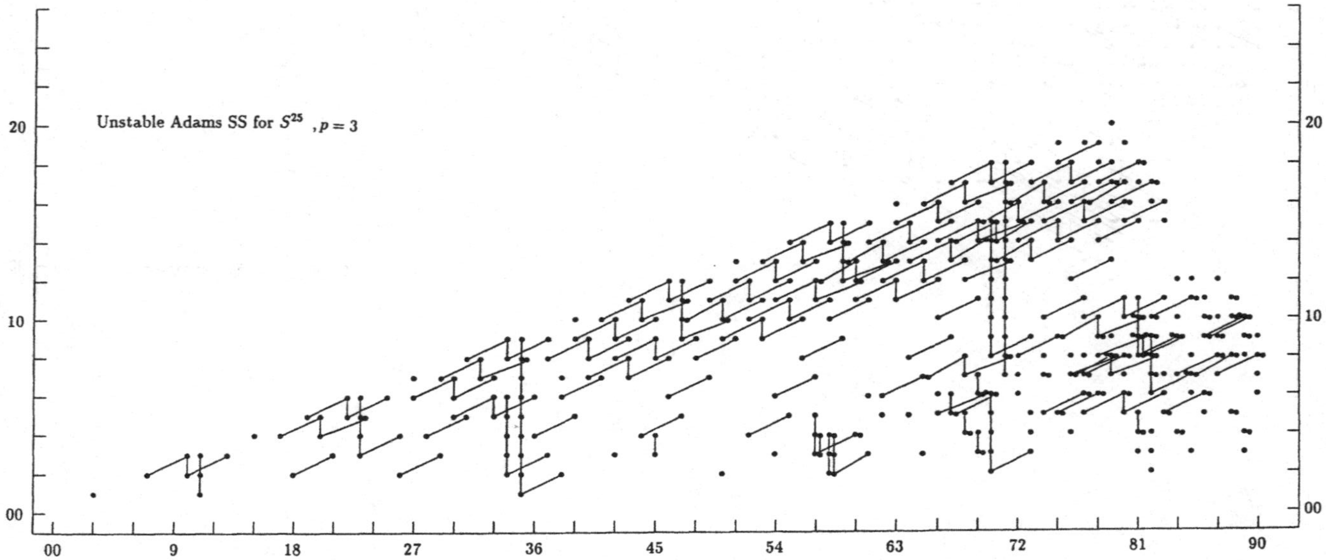

Unstable Adams SS for S^{25} , $p = 3$
20
10
00
00
9
18
27
36
45
54
63
72
81
90

On a certain localization of the stable homotopy of the space X_Γ

Goro Nishida

Kyoto University

§1. Introduction.

In [2], we have defined a space $X_\Gamma = B(\Gamma \cdot T^2)$ for a modular group $\Gamma \subset SL_2\mathbf{Z}$. It was shown there that the stable real homotopy group $\pi_*^s(X_\Gamma) \otimes \mathbf{R}$ is identified with a certain subspace of the vector space of modular forms for Γ *via* the Eichler-Shimura isomorphism. In this note we shall study what we can say about the mod p^r homotopy group

$$\pi_*^s(X_\Gamma; \mathbf{Z}/p^r\mathbf{Z}) = \pi_*^s(X_\Gamma \wedge M_{p^r})$$

for a prime $p \geq 5$, where M_{p^r} is the Moore spectrum. The computation of $\pi_*^s(X_\Gamma \wedge M_{p^r})$ itself is quite difficult, but we can define a stable self map α_r of $X_\Gamma \wedge M_{p^r}$ of degree $4p^{r-1}(p-1)$ and we shall show that the stable homotopy group of the telescope $(X_\Gamma \wedge M_{p^r})[\alpha_r^{-1}]$ can be described in a satisfactory way. In the final section we mention about an analogy of this construction and the definition of p-adic modular forms in the sense of Serre [4].

§2. A localization of $\pi_*^s(BT^2)$.

Let p be a prime. We recall the notion of a (p-local) numerical polynomial [1]. Let $f \in \mathbf{Q}[x, y]$ be a *homogeneous* polynomial over $\mathbf{Q}$. We call f p-local numerical if $f(n, m)$ is p-integral for any integers n and m. We regard $\mathbf{Q}[x, y]$ as a graded ring with $\deg x = \deg y = 2$. We denote by N_* the graded subring of $\mathbf{Q}[x, y]$ consisting of all p-local numerical polynomials. Let $\alpha = x^{2p-2} - x^{p-1}y^{p-1} + y^{2p-2}$, then we have the following

LEMMA 2.1. *For any $f \in N_{4d}$,*

$$\frac{1}{p}(f^p - f\alpha^d) \in N_{4pd}.$$

PROOF: Let $\alpha = \alpha(x, y)$, and let

$$g(n, m) = \frac{1}{p}(f(n, m)^p - f(n, m)\alpha(n, m)^d)$$

for integers n and m. If $n = n'p$ and $m = m'p$, then $g(n,m) = p^{2pd}g(n',m')$ and is p-integral. If $n \not\equiv 0 \bmod p$ or $m \not\equiv 0 \bmod p$, then $\alpha(n,m) \equiv 1 \bmod p$ and it is clear that $g(n,m)$ is p-integral. This completes the proof.

Now let $h_0 = \frac{1}{p}(x^p y - xy^p)$, then clearly $h_0 \in N_{2(p+1)}$. We define h_n inductively by

$$h_n = \frac{1}{p}(h_{n-1}^p - h_{n-1}\alpha^d)$$

where $d = \frac{1}{2}p^{n-1}(p+1)$. Then by the lemma, $h_n \in N_*$ and we have

THEOREM 2.2 ([1]). *A $\mathbf{Z}_{(p)}$-basis of N_* is given by the elements of the form $x^i y^j h_0^{k_0} \cdots h_r^{k_r}$, where $j = 0$, or $j > 0, 0 \le i \le p-1$, and $0 \le k_i \le p-1$.*

Let BT^2 be the classifying space of the 2-torus T^2. We consider the K-theory Hurewicz homomorphism

$$H\colon \pi_*^s(BT^2_{(p)}) \to K_*(BT^2_{(p)})$$

where $BT^2_{(p)}$ denotes the localization at p. We denote the standard basis of $\pi_2^s(BT^2)$ and their image under H by x and y. Let PK_* denote the submodule of primitive elements under the coaction of K_*K.

PROPOSITION 2.3 ([1]). *There is an isomorphism*

$$PK_*(BT^2_{(p)}) \cong N_*.$$

By this proposition we may regard H as a ring homomorphism

$$H\colon \pi_*^s(BT^2_{(p)}) \to N_*.$$

Let R_* be a $\mathbf{Z}$ graded ring and a a homogeneous element of degree d. The localization of R_* with respect to the multiplicative set $\{a^r\}$ is denoted by $R_*[a^{-1}]$. It is naturally $\mathbf{Z}$ graded and $R_n[a^{-1}]$ denotes the submodule of homogeneous elements of degree $n \in \mathbf{Z}$.

THEOREM 2.4. *The Hurewicz homomorphism induces an isomorphism*

$$H\colon \pi_*^s(BT^2_{(p)})[\alpha^{-1}] \to N_*[\alpha^{-1}].$$

PROOF: For any $u \in \pi_n^s(BT^2)$, the composite $\Sigma^n BT^2 \xrightarrow{u\wedge 1} BT^2 \wedge BT^2 \xrightarrow{m} BT^2$ is also denoted by u, where m is the stable map induced from the product. We denote the telescope of the sequence

$$BT^2 \xrightarrow{u} \Sigma^{-n} BT^2 \xrightarrow{u} \cdots$$

by $BT^2[u^{-1}]$. Note that $BT^2[x^{-1}] \simeq BT[x^{-1}] \wedge (BT_+)$ and $BT[x^{-1}] \simeq \mathbf{K}$ by the theorem of Snaith [6], where $\mathbf{K}$ denotes the periodic K spectrum,and similarly for $BT^2[y^{-1}]$. Hence we have

$$\pi_*^s(BT^2)[x^{-1}] \cong K_*(BT)$$

and similarly

$$\pi_*^s(BT^2)[y^{-1}] \cong K_*(BT).$$

Since $K_*(BT)$ is torsion free, we see that for a torsion element $u \in \pi_*^s(BT^2)$, $x^k u = y^l u = 0$ for some k and l, and hence $\alpha^r u = 0$ for some r. This shows that

$$\pi_*^s(BT^2)[\alpha^{-1}] \cong (\pi_*^s(BT^2)/\mathrm{Tor})[\alpha^{-1}]$$

and we see that $\pi_*^s(BT^2)[\alpha^{-1}]$ is torsion free. Since

$$\pi_*^s(BT^2) \otimes \mathbf{Q} \to N_* \otimes \mathbf{Q}$$

is an isomorphism, we see that

$$\pi_*^s(BT^2_{(p)})[\alpha^{-1}] \to N_*[\alpha^{-1}]$$

is a monomorphism. Now a polynomial $g(w) \in \mathbf{Q}[w]$ is called p-local numerical polynomial if $g(n)$ is p-integral for any $n \in \mathbf{Z}$, and the ring of those polynomials is denoted by A. It is shown [1] that $K_0(BT_{(p)}) \cong A$ and hence

$$K_*(BT_{(p)}) \cong K_* \otimes A.$$

Let $f(x,y) \in N_{2k}[x^{-1}]$, then there exists a unique polynomial $g(w)$ such that $f(x,y) = x^k g(y/x)$. Since $x^N f(x,y)$ is numerical for some positive integer N, we see that $f(1,m) = g(m)$ is p-integral for any $m \in \mathbf{Z}$ and hence $g(w) \in A$. Then identifying K_* with $\mathbf{Z}[x, x^{-1}]$ we have a homomorphism $N_*[x^{-1}] \to K_*(BT_{(p)})$. A $\mathbf{Z}_{(p)}$-basis of A is given by

$$\binom{w}{n} = \frac{1}{n!} w(w-1)\cdots(w-n+1), \quad n = 0, 1, \cdots.$$

It is easy to see that

$$\frac{1}{n!} y(y-x)\cdots(y-(n-1)x)x^N$$

is p-local numerical for large N and we see that $N_*[x^{-1}] \cong K_*(BT_{(p)})$. Hence we see that

$$H\colon \pi_*^s(BT^2_{(p)})[x^{-1}] \to N_*[x^{-1}]$$

is an isomorphism and similarly for y. Then for any $f \in N_*$ we can choose $r \geq 0$ such that $x^r f \in \mathrm{Im}[H\colon \pi_*^s(BT^2_{(p)}) \to N_*]$ and similarly for $y^r f$. Hence $\alpha^s f \in \mathrm{Im} H$ for some s and

$$H\colon \pi_*^s(BT^2_{(p)})[\alpha^{-1}] \to N_*[\alpha^{-1}]$$

is an epimorphism. This completes the proof.

§3. Reduction mod p^r.

Let $\gamma = \begin{pmatrix} a & b \\ c & d \end{pmatrix}$ be a matrix over $\mathbf{Z}$. We put $\gamma x = ax + cy$ and $\gamma y = bx + dy$. This defines an action of the semigroup $M_2\mathbf{Z}$ on N_* as ring endomorphisms. Now in the following let p be a prime ≥ 5 and let $\alpha = x^{2p-2} - x^{p-1}y^{p-1} + y^{2p-2}$ as before.

PROPOSITION 3.1. *$\alpha^{p^{r-1}}$ is an $SL_2\mathbf{Z}$-invariant element in $N_* \otimes \mathbf{Z}/p^r\mathbf{Z}$.*

PROOF: Let $S = \begin{pmatrix} 0 & -1 \\ 1 & 0 \end{pmatrix}, T = \begin{pmatrix} 1 & -1 \\ 1 & 0 \end{pmatrix}$ be the standard generators of $SL_2\mathbf{Z}$. Then $S\alpha = \alpha$ and $T\alpha = (x+y)^{2p-2} - (x+y)^{p-1}x^{p-1} + x^{2p-2}$. Note that $x^p y \equiv xy^p \bmod p$ in N_*. Now

$$\begin{aligned}(x+y)^{2p-2} =& \sum \binom{2p-2}{i} x^i y^{2p-2-i} \\ \equiv & x^{2p-2} + \left(\binom{2p-2}{1} + \binom{2p-2}{p}\right) x^{2p-3}y + \cdots \\ & + \left(\binom{2p-2}{p-2} + \binom{2p-2}{2p-3}\right) x^p y^{p-2} \\ & + \binom{2p-2}{p-1} x^{p-1}y^{p-1} + y^{2p-2}.\end{aligned}$$

Note that $\binom{2p-2}{p-1} \equiv 0$ and

$$\binom{2p-2}{i} + \binom{2p-2}{p+i-1} \equiv \binom{p-2}{i} + \binom{p-2}{i-1} = \binom{p-1}{i} \quad \text{mod } p$$

for $1 \leq i \leq p-2$. Therefore we see that $T\alpha \equiv \alpha \bmod p$, and this shows the proposition for $r = 1$. The proof for $r > 1$ is obvious.

By this proposition we can define an $SL_2\mathbf{Z}$-action on the localisation $(N_* \otimes \mathbf{Z}/p^r\mathbf{Z})[(\alpha^{p^{r-1}})^{-1}] \cong (N_* \otimes \mathbf{Z}/p^r\mathbf{Z})[\alpha^{-1}]$. Note that as a graded $SL_2\mathbf{Z}$-module, this is periodic with period $\deg \alpha^{p^{r-1}} = 2p^{r-1}(p-1)$. Now our key result is the following

PROPOSITION 3.2. *$H_0(SL_2\mathbf{Z}; (N_* \otimes \mathbf{Z}/p\mathbf{Z})[\alpha^{-1}]) = 0$.*

In order to prove the proposition we consider the module $N_*[x^{-1}]$. Let $\gamma = \begin{pmatrix} 1 & 1 \\ 0 & 1 \end{pmatrix}$ and Γ_0 the subgroup of $SL_2\mathbf{Z}$ generated by γ.

LEMMA 3.3. *$H_0(\Gamma_0; N_*[x^{-1}]) = 0$*

PROOF: As in the proof of Theorem 2.4, we have an isomorphism

$$\varphi \colon N_*[x^{-1}] \to \mathbf{Z}[x, x^{-1}] \otimes A$$

where φ is given by $\varphi(f) = x^k \otimes g(w)$ for $f \in N_{2k}[x^{-1}]$ and the polynomial $g(w)$ is determined by $f(x,y) = x^k g(y/x)$. We can define a Γ_0 -action on A by $(\gamma g)(w) = g(w+1)$. Then we have

$$\gamma f(x,y) = f(x, x+y) = x^k g((y/x)+1) = x^k(\gamma g)(y/x)$$

and hence φ is a Γ_0-isomorphism. For the basis of A we have

$$\gamma\binom{w}{n} - \binom{w}{n} = \binom{w+1}{n} - \binom{w}{n} = \binom{w}{n-1}.$$

This implies $H_0(\Gamma_0; A) = 0$ and the lemma follows.

PROOF OF PROPOSITION 3.2: By the lemma we see that for any $f \in N_*$, $x^k f$ is a γ-boundary, i.e., there exists a $u \in N_*$ such that $x^k f = \gamma u - u$. Let $\gamma' = \begin{pmatrix} 1 & 0 \\ 1 & 1 \end{pmatrix}$, then similarly there exist k' and u' such that $y^{k'} f = \gamma' u' - u'$. Note that

$$\alpha^m \equiv x^{2m(p-1)} - x^{(2m-1)(p-1)} y^{p-1} + y^{2m(p-1)} \mod p$$

for $\alpha x^2 \equiv x^{2p}$ and $\alpha y^2 \equiv y^{2p} \bmod p$. Then for any $f \in N_*$ we can choose m so that $x^{2m(p-1)} f$ and $x^{(2m-1)(p-1)} y^{p-1} f$ are γ-boundaries and $y^{2m(p-1)} f$ is γ'-boundary. Hence $\alpha^m f = 0$ in $H_0(SL_2\mathbf{Z}; N_* \otimes \mathbf{Z}/p\mathbf{Z})$ and this completes the proof.

COROLLARY 3.4. *$H_0(SL_2\mathbf{Z}; (N_* \otimes \mathbf{Z}/p^r\mathbf{Z})[\alpha^{-1}]) = 0$ for any $r \geq 1$.*

COROLLARY 3.5. *Let $r \geq s$, then the homomorphism*

$$H_1(SL_2\mathbf{Z}; (N_* \otimes \mathbf{Z}/p^r\mathbf{Z})[\alpha^{-1}]) \to H_1(SL_2\mathbf{Z}; (N_* \otimes \mathbf{Z}/p\mathbf{Z})[\alpha^{-1}])$$

induced from the reduction is an epimorphism.

PROOF: Consider the short exact sequence of $SL_2\mathbf{Z}$-modules

$$0 \to (N_* \otimes \mathbf{Z}/p\mathbf{Z})[\alpha^{-1}] \to (N_* \otimes \mathbf{Z}/p^r\mathbf{Z})[\alpha^{-1}] \to (N_* \otimes \mathbf{Z}/p^{r-1}\mathbf{Z})[\alpha^{-1}] \to 0$$

induced from the extension $0 \to \mathbf{Z}/p\mathbf{Z} \to \mathbf{Z}/p^r\mathbf{Z} \to \mathbf{Z}/p^{r-1}\mathbf{Z} \to 0$. Note [2] that $H_i(SL_2\mathbf{Z}; M) = 0$ for $i \geq 2$ and for a $\mathbf{Z}_{(p)}[SL_2\mathbf{Z}]$- module M for $p \geq 5$. Then by Proposition 3.2, the exact sequence for group homology turns out to be

$$0 \to H_1(SL_2\mathbf{Z}; (N_* \otimes \mathbf{Z}/p\mathbf{Z})[\alpha^{-1}]) \to H_1(SL_2\mathbf{Z}; (N_* \otimes \mathbf{Z}/p^r\mathbf{Z})[\alpha^{-1}]) \to H_1(SL_2\mathbf{Z}; (N_* \otimes \mathbf{Z}/p^{r-1}\mathbf{Z})[\alpha^{-1}]) \to 0$$

and

$$H_0(SL_2\mathbf{Z}; (N_* \otimes \mathbf{Z}/p^r\mathbf{Z})[\alpha^{-1}]) \cong H_0(SL_2\mathbf{Z}; (N_* \otimes \mathbf{Z}/p^{r-1}\mathbf{Z})[\alpha^{-1}].$$

This shows Corollary 3.4 and 3.5.

§4. A localization of the space X_Γ.

Let us recall [2] the definition of the space X_Γ. The group $SL_2\mathbf{Z}$ acts on the 2-torus T^2 as group automorphisms and we can define the semi-direct product group $SL_2\mathbf{Z}\cdot T^2$. Then X_Γ is defined to be the classifying space $B(SL_2\mathbf{Z}\cdot T^2)$. The center of $SL_2\mathbf{Z}$ is the cyclic group of order 2 generated by $\begin{pmatrix} -1 & 0 \\ 0 & -1 \end{pmatrix}$, which we denote by C_0.

Let C_1 and C_2 be the subgroups of $SL_2\mathbf{Z}$ generated by $S = \begin{pmatrix} 0 & -1 \\ 1 & 0 \end{pmatrix}$ and $T = \begin{pmatrix} 1 & -1 \\ 1 & 0 \end{pmatrix}$, respectively. Then $C_1 \cong \mathbf{Z}/4\mathbf{Z}$ and $C_2 \cong \mathbf{Z}/6\mathbf{Z}$, and $SL_2\mathbf{Z} \cong C_1 *_{C_0} C_2$ the amalgamed product of C_1 and C_2. We write the semi-direct products $C_i\cdot T^2$ as $\widetilde{T}_i^2$, $i = 0, 1$ and 2, respectively. Let $l_i\colon \widetilde{T}_0^2 \to \widetilde{T}_i^2, i = 1, 2,$ denote the canonical inclusions. Then it is easy to see the following

LEMMA 4.1. *X_Γ is homotopy equivalent to the double mapping cylinder of the maps*

$$B\widetilde{T}_1^2 \overset{Bl_1}{\leftarrow} B\widetilde{T}_0^2 \overset{Bl_2}{\rightarrow} B\widetilde{T}_2^2.$$

By this lemma we see that there exists a cofiber sequence

$$B\widetilde{T}_0^2 \to B\widetilde{T}_1^2 \vee B\widetilde{T}_2^2 \to X_\Gamma$$

in the stable category. Let $M_{p^r} = S^0 \cup_{p^r} e^1$ be the Moore spectrum. The graded module $\{BT^2 \wedge M_{p^r}, BT^2 \wedge M_{p^r}\}_*$ of stable maps is a left and right $SL_2\mathbf{Z}$-module in the usual way. For any subgroup $G \subset SL_2\mathbf{Z}$ we call a stable map f a G-map if $\gamma f = f\gamma$ as stable homotopy classes for any $\gamma \in G$. We now suppose that $p \geq 5$. Note that the natural maps $BT^2 \to B\widetilde{T}_i^2, i = 0, 1$ and 2 are all finite coverings of degree prime to p. Then it is easy to see the following

LEMMA 4.2. *Let $f\colon \Sigma^k BT^2 \wedge M_{p^r} \to BT^2 \wedge M_{p^r}$ be an $SL_2\mathbf{Z}$-map. Then for any i, there is a stable map $\tilde{f}\colon \Sigma^k B\widetilde{T}_i^2 \wedge M_{p^r} \to B\widetilde{T}_i^2 \wedge M_{p^r}$, unique up to homotopy, such that the diagram*

$$\begin{array}{ccc} \Sigma^k BT^2 \wedge M_{p^r} & \longrightarrow & \Sigma^k B\widetilde{T}_i^2 \wedge M_{p^r} \\ {\scriptstyle f}\downarrow & & \downarrow{\scriptstyle \tilde{f}} \\ BT^2 \wedge M_{p^r} & \longrightarrow & B\widetilde{T}_i^2 \wedge M_{p^r} \end{array}$$

is homotopy commutative.

Now for $u \in \pi_k^s(BT^2)$ we have a stable map $u\colon \Sigma^k BT^2 \to BT^2$ as in §2. We denote $u \wedge 1\colon \Sigma^k BT^2 \wedge M_{p^r} \to BT^2 \wedge M_{p^r}$ also by u. Consider

$\alpha = x^{2p-2} - x^{p-1}y^{p-1} + y^{2p-2} \in \pi^s_{4p-4}(BT^2)$. By Proposition 3.1, we easily see that

$$\alpha^{p^{r-1}} : \Sigma^k BT^2 \wedge M_{p^r} \to BT^2 \wedge M_{p^r}$$

is an $SL_2\mathbf{Z}$-map for any r, where $k = 4p^{r-1}(p-1)$. Then by Lemma 4.2, we have a stable map

$$\alpha_r = \tilde{\alpha}^{p^{r-1}} : \Sigma^k B\widetilde{T}^2_i \wedge M_{p^r} \to B\widetilde{T}^2_i \wedge M_{p^r},$$

and it is easy to see that the diagram

$$\begin{array}{ccc} \Sigma^k B\widetilde{T}^2_0 \wedge M_{p^r} & \xrightarrow{Bl_i} & \Sigma^k B\widetilde{T}^2_i \wedge M_{p^r} \\ {\scriptstyle \alpha_r}\downarrow & & \downarrow{\scriptstyle \alpha_r} \\ B\widetilde{T}^2_0 \wedge M_{p^r} & \xrightarrow{Bl_i} & B\widetilde{T}^2_i \wedge M_{p^r} \end{array}$$

is homotopy commutative. Therefore we have obtained

PROPOSITION 4.3. *There is a stable map $\alpha_r \colon \Sigma^k X_\Gamma \wedge M_{p^r} \to X_\Gamma \wedge M_{p^r}, k = 4p^{r-1}(p-1)$ such that the diagram*

$$\begin{array}{ccccc} \Sigma^k(B\widetilde{T}^2_1 \vee B\widetilde{T}^2_2) \wedge M_{p^r} & \longrightarrow & \Sigma^k X_\Gamma \wedge M_{p^r} & \longrightarrow & \Sigma^{k+1} B\widetilde{T}^2_0 \wedge M_{p^r} \\ {\scriptstyle \alpha_r \vee \alpha_r}\downarrow & & {\scriptstyle \alpha_r}\downarrow & & \downarrow{\scriptstyle \alpha_r} \\ (B\widetilde{T}^2_1 \vee B\widetilde{T}^2_2) \wedge M_{p^r} & \longrightarrow & X_\Gamma \wedge M_{p^r} & \longrightarrow & \Sigma B\widetilde{T}^2_0 \wedge M_{p^r} \end{array}$$

is homotopy commutative, where the horizontal sequences are the cofiber sequences.

REMARK: The map $\alpha_r \colon \Sigma^k X_\Gamma \wedge M_{p^r} \to X_\Gamma \wedge M_{p^r}$ is not uniquely determined. It depends on a choice of a homotopy between the maps

$$\alpha_r Bl_i, Bl_i \alpha_r \colon \Sigma^k B\widetilde{T}^2_0 \wedge M_{p^r} \to B\widetilde{T}^2_i \wedge M_{p^r}.$$

Now we can state our main theorem. We fix a map α_r, and denote the telescope of the sequence

$$X_\Gamma \wedge M_{p^r} \xrightarrow{\alpha_r} \Sigma^{-k} X_\Gamma \wedge M_{p^r} \xrightarrow{\alpha_r} \cdots$$

by $(X_\Gamma \wedge M_{p^r})[\alpha_r^{-1}]$.

THEOREM 4.4. *For any n there are isomorphisms*

$$\pi^s_{2n}((X_\Gamma \wedge M_{p^r})[\alpha_r^{-1}]) \cong 0$$

and

$$\pi^s_{2n+1}((X_\Gamma \wedge M_{p^r})[\alpha_r^{-1}]) \cong H_1(SL_2\mathbf{Z}; (N_{2n} \otimes \mathbf{Z}/p^r\mathbf{Z})[\alpha^{-1}]).$$

PROOF: Let $C_i, i = 0, 1$ and 2 be as before. Then $BSL_2\mathbf{Z}$ is the double mapping cylinder of the maps $Bl_{i*}: BC_0 \to BC_i, i = 1, 2$. Let M be a $\mathbf{Z}_{(p)}[SL_2\mathbf{Z}]$-module, and consider the homomorphism

$$l_{1*} \oplus l_{2*}: H_0(C_0; M) \to H_0(C_1; M) \oplus H_0(C_2; M).$$

Since $p \geq 5$, we easily see that

$$H_0(SL_2\mathbf{Z}; M) \cong \mathrm{Cok}(l_{1*} \oplus l_{2*})$$

and

$$H_1(SL_2\mathbf{Z}; M) \cong \mathrm{Ker}(l_{1*} \oplus l_{2*}).$$

Now the natural map $BT^2 \to B\widetilde{T}^2_i$ is a finite covering of degree prime to p, we see that

$$\pi^s_*(B\widetilde{T}^2_i \wedge M_{p^r}) \cong H_0(C_i; \pi^s_*(BT^2 \wedge M_{p^r})$$

for $i = 0, 1$ and 2. Then the homotopy exact sequence of the cofiber sequence in Proposition 4.3 gives a short exact sequence

$$0 \to H_0(SL_2\mathbf{Z}; \pi^s_*(BT^2 \wedge M_{p^r})) \to \pi^s_*(X_\Gamma \wedge M_{p^r}) \\ \to H_1(SL_2\mathbf{Z}; \pi^s_{*-1}(BT^2 \wedge M_{p^r})) \to 0.$$

By Theorem 2.4, we have the isomorphism $H: \pi^s_*(BT^2)[\alpha^{-1}] \cong N_*[\alpha^{-1}]$, and hence we see that $\pi^s_*(BT^2)[\alpha^{-1}]$ is torsion free. Then we have an isomorphism

$$\pi^s_*(BT^2 \wedge M_{p^r})[\alpha^{-1}] \cong (N_* \otimes \mathbf{Z}/p^r\mathbf{Z})[\alpha^{-1}]$$

as $SL_2\mathbf{Z}$-modules. Then localizing the above exact sequence with respect to $\alpha^{p^{r-1}}$, the theorem follows from Corollary 3.4.

Next we define a reduction map

$$\rho: (X_\Gamma \wedge M_{p^{r+1}})[\alpha_{r+1}^{-1}] \to (X_\Gamma \wedge M_{p^r})[\alpha_r^{-1}]$$

for each r. For each r, we choose and fix a homotopy to define

$$\alpha_r: \Sigma^k X_\Gamma \wedge M_{p^r} \to X_\Gamma \wedge M_{p^r}$$

as in the remark. Let $X_{i,r}$ denote $B\widetilde{T}_i^2 \wedge M_{p^r}$ for $i = 0, 1$ and 2. Then the chosen homotopy defines a map

$$\lambda_i\colon X_{0,r}[\alpha_r^{-1}] \to X_{i,r}[\alpha_r^{-1}]$$

of the telescopes and we easily see that $(X_\Gamma \wedge M_{p^r})[\alpha_r^{-1}]$ is homotopy equivalent to the cofiber of

$$\lambda_1 \vee \lambda_2\colon X_{0,r}[\alpha_r^{-1}] \to X_{1,r}[\alpha_r^{-1}] \vee X_{2,r}[\alpha_r^{-1}].$$

The reduction map $M_{p^{r+1}} \to M_{p^r}$ induces a map $X_{i,r+1} \to X_{i,r}$ and it is clear that the diagram

$$\begin{array}{ccc} \Sigma^k X_{i,r+1} & \xrightarrow{\alpha_{r+1}} & X_{i,r+1} \\ \downarrow & & \downarrow \\ \Sigma^k X_{i,r} & \xrightarrow{\alpha_r^p} & X_{i,r} \end{array}$$

is homotopy commutative, where α_r^p is the p-times composition of α_r. Therefore we have a reduction map (not unique)

$$\rho\colon X_{i,r+1}[\alpha_{r+1}^{-1}] \to X_{i,r}[\alpha_r^{-1}]$$

for each r and $i = 0, 1$ and 2.

LEMMA 4.5. *For each r and $i = 1, 2$ we can choose ρ such that the diagram*

$$\begin{array}{ccc} X_{0,r+1}[\alpha_{r+1}^{-1}] & \xrightarrow{\lambda_i} & X_{i,r+1}[\alpha_{r+1}^{-1}] \\ \rho\downarrow & & \downarrow\rho \\ X_{0,r}[\alpha_r^{-1}] & \xrightarrow{\lambda_i} & X_{i,r}[\alpha_r^{-1}] \end{array}$$

is homotopy commutative.

PROOF: Let $Y = X_{0,r+1}[\alpha_{r+1}^{-1}]$. Then Y is a homotopy colimit of a countable direct system of finite complexes Y_β, and we have a cofiber sequence $\bigvee Y_\beta \xrightarrow{f} Y \xrightarrow{g} \bigvee \Sigma Y_\beta$. Now we first choose arbitrary ρ, then it is clear that the diagram in the proposition is homotopy commutative when restricted to Y_β for any β. Hence by the Milnor exact sequence there is a map $\gamma\colon \bigvee \Sigma Y_\beta \to X_{i,r}[\alpha_r^{-1}]$ such that

$$\lambda_i\rho - \rho\lambda_i = g^*(\gamma) \in \{Y, X_{i,r}[\alpha_r^{-1}]\}.$$

It is easy to see that

$$\lambda_{i*}\colon \{Y_\beta, X_{0,r}[\alpha_r^{-1}]\} \to \{Y_\beta, X_{i,r}[\alpha_r^{-1}]\}$$

is an epimorphism for a finite complex Y_β, and hence so is for $\bigvee Y_\beta$. Then there is a map $\gamma'\colon \bigvee \Sigma Y_\beta \to X_{0,r}[\alpha_r^{-1}]$ such that $\lambda_i\gamma' = \gamma$, and we can replace the left hand ρ in the diagram with $\rho - \gamma' g$ so that the diagram is homotopy commutative. This completes the proof.

By the lemma and by the above argument we can define a map (though not unique)

$$\rho\colon (X_\Gamma \wedge M_{p^{r+1}})[\alpha_{r+1}^{-1}] \to (X_\Gamma \wedge M_{p^r})[\alpha_r^{-1}].$$

We denote the homotopy inverse limit $\mathrm{holim}(X_\Gamma \wedge M_{p^r})[\alpha_r^{-1}]$ of the sequence of reduction maps by $(X_\Gamma)^\wedge_\alpha$. Then by Corollary 3.5 and Theorem 4.4 we easily obtain

THEOREM 4.6. *For any $n \in \mathbf{Z}$ we have isomorphisms*

$$\pi^s_{2n}((X_\Gamma)^\wedge_\alpha) \cong 0$$

and

$$\pi^s_{2n+1}((X_\Gamma)^\wedge_\alpha) \cong \varprojlim H_1(SL_2\mathbf{Z}; (N_{2n} \otimes \mathbf{Z}/p^r\mathbf{Z})[\alpha^{-1}]).$$

REMARK: By the Theorem of Eichler and Shimura [2],[5], we see that

$$\begin{aligned} H_1(SL_2\mathbf{Z}; N_*) \otimes \mathbf{R} &\cong H_1(SL_2\mathbf{Z}; \mathbf{R}[x,y]) \\ &\cong \bigoplus \mathbf{R}\{E_k\} \oplus S_k \end{aligned}$$

where E_k is the Eisenstein function of weight k and S_k is the complex vector space of cusp forms for $SL_2\mathbf{Z}$ of weight k. Hence $H_1(SL_2\mathbf{Z}; N_*)$ can be identified with a certain $\mathbf{Z}_{(p)}$-submodule of the space of modular forms. Therefore

$$\pi^s_*((X_\Gamma)^\wedge_\alpha) = \varprojlim H_1(SL_2\mathbf{Z}; (N_* \otimes \mathbf{Z}/p^r\mathbf{Z})[(\alpha^{p^{r-1}})^{-1}])$$

could be described in terms of modular forms. It may be worth while to compare this with the p-adic modular forms in the sense of Serre [4]. Let M_* be the ring of modular forms $f = \sum a_n q^n$ such that the Fourier coefficients $a_n \in \mathbf{Z}_{(p)}$ for all n. Then we can show that the ring $M_*^\wedge$ of p-adic modular forms is given by

$$M_*^\wedge \cong \varprojlim (M_* \otimes \mathbf{Z}/p^r\mathbf{Z})[(E_{p-1}^{p^{r-1}})^{-1}].$$

References

1 A. Baker, F.Clarke, N.Ray and L. Schwartz, *On the Kummer congruence and the stable homotopy of BU*, Trans. Amer. Math. Soc. **316** (1989), 385–432.
2 G.Nishida, *Modular forms and the double transfer for BT^2*, to appear in Japan J.Math..
3 G.Nishida, *On the mod p cohomology of the space X_Γ and the mod p trace formula for Hecke operators*, to appear in J.Math.Kyoto Univ..
4 J.-P. Serre, *Formes modulaires et fonction zeta p-adiques*, Springer Lecture Notes in Mathematics **350** (1973), 191-268.
5 G.Shimura, *Introduction to the arithmetic theory of automorphic functions*, Iwanami Shoten Tokyo, University Press;Princeton,1971.
6 V.P.Snaith, *Algebraic cobordism and K-theory*, Memoirs of the AMS **221** (1979).
7 A.Baker, *Elliptic cohomology, p-adic modular forms and Atkin's operator U_p*, Contemporary Math. **96** (1989), 33–38.

Department of Mathematics, Kyoto University, Kyoto 606, Japan.

COOPERATIONS IN ELLIPTIC HOMOLOGY

FRANCIS CLARKE AND KEITH JOHNSON

University College of Swansea, Dalhousie University

INTRODUCTION

In this paper we study the ring $\mathrm{Ell}_*(\mathrm{Ell})$ of cooperations in elliptic homology. In the first section we discuss the ring $E_*(E)$ for a general ring spectrum which has a Conner-Floyd type isomorphism. Sections 2 and 3 are devoted to K-theory and elliptic homology respectively. In an appendix we discuss the exponential series for the formal group law of elliptic cohomology.

1. GENERALITIES ON HOPF ALGEBROIDS OF COOPERATIONS

Let E be a commutative ring spectrum. Adams [2] considered the ring of cooperations $E_*(E) = \pi_*(E \wedge E)$. If $E_*(E)$ is flat over the coefficient ring $E_* = \pi_*(E)$ then one can define a number of algebraic structure maps which make $E_*(E)$ into what is today called a Hopf algebroid [21] or bilateral Hopf algebra [20]. In this case it is possible to identify the E_2-term of the E-theory Adams-Novikov spectral sequence (which converges, under appropriate assumptions, to $\pi_*(S^0)$ or rather $\pi_*(S^0_E)$, where X_E denotes the E-theory localisation of X) with $\mathrm{Ext}_E^{*,*} = \mathrm{Ext}^{*,*}_{E_*(E)}(E_*, E_*)$. These Ext groups are to be calculated in the category of comodules over $E_*(E)$. For example if $E = H\mathbf{F}_p$, the mod p Eilenberg-Mac Lane spectrum then $E_*(E)$ is Milnor's dual of the Steenrod algebra [19].

We will consider *complex oriented* ring spectra [3]. Thus we have a morphism of ring spectra $\varphi : MU \to E$, so that the induced map $\varphi_* : MU_* \to E_*$ determines a formal group law over E_*. More particularly we

---

1991 *Mathematics Subject Classification.* Primary 55S25; Secondary 11F33, 19L20, 33E05, 55N20, 55N22.

Key words and phrases. Elliptic cohomology, Hopf algebroids, modular forms.

The authors would like to acknowledge the financial support of the S. E. R. C. (grant number GR/F/61288), the N. S. E. R. C. (grant number A8829), and M. S. R. I., Berkeley. In particular Francis Clarke would like to thank M. S. R. I. for providing a superb working environment.

restrict attention to those theories such that the MU_*-module structure on E_* determined by φ_* induces a Conner-Floyd type isomorphism

$$E_*(X) \cong E_* \otimes_{MU_*} MU_*(X). \tag{1}$$

The original examples of theories of this type were rational homology, and K-theory [7]. Elliptic homology provides another example, as do the theories $E(n)$; see [18]. Recall that the Landweber exact functor theorem [15] provides necessary conditions on φ_* for this to hold. The first of these conditions is that multiplication by p must be a monomorphism on E_*. By considering the degree p map on S^0, it is clear that this condition is necessary for the isomorphism (1) to hold. Hence E_* must be torsion-free. In fact Rudjak showed in [22] that all the conditions of the Landweber exact functor theorem are necessary.

In the case where the isomorphism (1) holds we have

$$E_*(E) \cong E_* \otimes_{MU_*} MU_*(MU) \otimes_{MU_*} E_*$$

and the flatness of $E_*(E)$ over E_* follows from that of $MU_*(MU)$ over MU_*. This is an observation of Miller and Ravenel; see [18] Remark 3.7. Since E_* is torsion-free so is $E_*(E)$.

As part of the structure alluded to above, there are left and right unit maps $\eta_L, \eta_R : E_* \to E_*(E)$, which are induced from the inclusion of the left and right factors in $E \wedge E$. This gives us inclusions of rings

$$E_* \otimes E_* \hookrightarrow E_*(E) \hookrightarrow E_* \otimes E_* \otimes \mathbf{Q}.$$

How can we identify this extension? Abstractly this is straight forward. The homomorphisms η_L and η_R induce two formal group laws over $E_*(E)$ whose logarithm and exponential series we will denote by $\log_L^E(x)$, $\log_R^E(x)$, $\exp_L^E(x)$ and $\exp_R^E(x)$. In the case $E = MU$ it is known that

$$MU_*(MU) = MU_*[m_1, m_2, \dots],$$

where the elements m_i are the coefficients of the power series

$$\exp_L^{MU}(\log_R^{MU}(x)) = \sum_{i\geq 0} m_i x^{i+1}.$$

This follows from Landweber's description of $MU_*(MU)$ in terms of isomorphisms between formal group laws [14]. Since MU_* is the universal ring for formal group laws, it follows that for any E, the coefficients of the series

$$\exp_L^E(\log_R^E(x)) \in E_* \otimes E_* \otimes \mathbf{Q},$$

which we will also denote m_i, generate $E_*(E)$. We thus need to have some hold on these elements, with an eye to characterising the ring which they generate. Note that the same approach works for $E_*(F)$ where E and F are both complex oriented ring spectra, with just one of them satisfying the Landweber exact functor conditions. For example one can compute $H_*(K)$ or $H_*(\mathrm{Ell})$ this way.

2. K-THEORY AND $KO[\frac{1}{2}]$-THEORY

We begin this section by sketching the description of the ring $K_*(K)$ of cooperations in complex, periodic K-theory in terms of the elements m_i. This ring was first determined by Adams, Harris and Switzer in [4]; see §1 of [5] for a third way of computing $K_*(K)$.

We have $K_* = \mathbf{Z}[t, t^{-1}]$ with $t \in K_2$. The formal group law is multiplicative with

$$\exp^K(x) = t^{-1}(e^{tx} - 1), \qquad \log^K(x) = t^{-1}\ln(1 + tx).$$

Let $u = \eta_L(t)$ and $v = \eta_R(t)$, then

$$\begin{aligned}\exp_L^K(\log_R^K(x)) &= u^{-1}(e^{uv^{-1}\ln(1+vx)} - 1)\\ &= u^{-1}((1 + vx)^{uv^{-1}} - 1)\\ &= \sum_{j\geq 1} u^{-1}v^j \binom{uv^{-1}}{j} x^j.\end{aligned}$$

Thus

$$m_i = m_i(u, v) = u^{-1}v^{i+1}\binom{uv^{-1}}{i+1} = \frac{(u - v)(u - 2v)\ldots(u - iv)}{(i+1)!}.$$

Now we notice that $m_i(k, l) \in \mathbf{Z}[\frac{1}{kl}]$ for all non-zero integers k and l. Since the polynomials $m_i(u, v)$ are generators this is true for all elements of $K_*(K)$.

But the converse of this statement is true: *every* homogeneous Laurent polynomial $f(u, v) \in \mathbf{Q}[u, v, u^{-1}, v^{-1}]$ such that $f(k, l) \in \mathbf{Z}[\frac{1}{kl}]$ for $kl \neq 0$ can be expressed in terms of the $m_i(u, v)$. The details may be found in [4]; the key technique is to use Newton interpolation.

Thus we have a characterisation of $K_*(K)$ in terms of an integrality condition. It is this that we seek to generalise to elliptic homology.

The elements in

$$\begin{aligned}\mathrm{Ext}_K^1 &= \mathrm{Ext}_{K_*(K)}^{1,*}(K_*, K_*)\\ &\cong \{\,\alpha \in K_* \otimes \mathbf{Q} : \eta_L(\alpha) - \eta_R(\alpha) \in K_*(K)\,\}/K_*,\end{aligned}$$

which detect the image of J are as follows [1]. To compute these groups we need to decide how divisible is $(\eta_L - \eta_R)(t^n) = u^n - v^n$ in $K_{2n}(K)$. The map $\eta_L - \eta_R$ is the first differential in the cobar resolution; see A1.2.11 of [21].

We have $k^{\varphi(m)} \equiv l^{\varphi(m)} \mod m$ if k and l are prime to m, since both sides are congruent to 1 modulo m. Therefore $k^n \equiv l^n \mod m(n)$ where $m(n)$ is twice the least common multiple of those natural numbers m such that $\varphi(m)$ divides n, and k and l are prime to $m(n)$. It follows that $(u^n - v^n)/m(n) \in K_{2n}(K)$, and it is not hard to show that this element is indivisible. Hence $\mathrm{Ext}_K^{1,2n} \cong \mathbf{Z}/m(n)\mathbf{Z}$.

For future reference we note that, if n is even, $m(n)$ is the denominator of the divided Bernoulli number $B_n/(2n)$; see [24]. We index the Bernoulli numbers so that

$$\frac{x}{e^x - 1} = \sum_{n \geq 0} B_n \frac{x^n}{n!}.$$

It was pointed out by Adams [2] that if one localises the spectrum K at a prime p, then K splits as a wedge of $p-1$ suspensions of a spectrum now usually denoted by $E(1)$. The coefficient ring $E(1)_*$ is identified with the subring $\mathbf{Z}_{(p)}[t^{p-1}, t^{-p+1}]$ of $K_* \otimes \mathbf{Z}_{(p)}$. The spectrum inherits a complex orientation from K and so has an associated formal group law whose logarithm is given by the series

$$\log^{E(1)}(x) = \sum_{n \geq 0} \frac{t^{p^n - 1} x^{p^n}}{p^n}.$$

Let

$$\chi(x) = t^{-1} \left(\prod_{(m,p)=1} (1 - t^m x^m)^{-\mu(m)/m} - 1 \right) \in K_* \otimes \mathbf{Z}_{(p)}[[x]],$$

the Artin-Hasse series. We have (see [10] or (17.2.14) of [11])

$$\log^{E(1)}(x) = \log^K(\chi(x))$$

so that

$$\exp_L^{E(1)}(\log_R^{E(1)}(x)) = \chi_L^{-1}(\exp_L^K(\log_R^K(\chi_R(x)))) \in K_*(K) \otimes \mathbf{Z}_{(p)}[[x]].$$

Hence each of the coefficients of this series, and thus all the elements of $E(1)_*(E(1))$, satisfy the same integrality conditions as the elements of $K_*(K)$. Since we have localised at p the integrality conditions in this case are nontrivial only for integers k and l prime to p.

We conclude this section by considering the case of KO-theory localised away from 2. We denote the resulting spectrum by $KO[\frac{1}{2}]$. Note that the spectrum $K[\frac{1}{2}]$ is a wedge of two copies of $KO[\frac{1}{2}]$. We have the coefficient ring $KO[\frac{1}{2}]* = \mathbf{Z}[\frac{1}{2}][t^2, t^{-2}]$ and we use the orientation $MU \to KO[\frac{1}{2}]$ which induces the formal group law with logarithm and exponential

$$\log^{KO}(x) = \frac{2}{t} \sinh^{-1}(\frac{tx}{2}), \qquad \exp^{KO} = \frac{2}{t} \sinh(\frac{tx}{2}).$$

We need to understand the coefficients of

$$\exp_L^{KO}(\log_R^{KO}(x)) = \frac{2}{u}\sinh(uv^{-1}\sinh^{-1}(\frac{tx}{2})).$$

If we followed the same route as for K-theory we would be led to consider the coefficients of Chebyshev polynomials, for, just as the K-theory calculation depended on being able to express e^{kx} in terms of e^x, here we are simply seeking the general answer to the problem: find $\sinh(kx)$ in terms of $\sinh(x)$. Instead we take the same approach which we used for $E(1)$-theory and which we will adopt later for the elliptic case.

Note that

$$\exp^{KO}(\log^K(x)) = \frac{x}{\sqrt{1+tx}} \in \mathbf{Z}[\tfrac{1}{2}, t][[x]].$$

Let us denote this power series by $\Phi(x)$, then

$$\exp_L^{KO}(\log_R^{KO}(x)) = \Phi_L(\exp_L^K(\log_R^K(\Phi_R^{-1}(x)))) \in K_*(K)\otimes \mathbf{Z}[\tfrac{1}{2}][[x]],$$

showing that in $KO[\frac{1}{2}]$-theory the m_i satisfy the same integrality conditions, away from 2, as in K-theory.

3. Elliptic homology

We recall the definition of (one version of) elliptic homology due to Landweber, Ravenel and Stong; see [16] for details.

Form the graded ring $\mathrm{Ell}_* = \mathbf{Z}[\frac{1}{2}][\delta, \varepsilon, (\delta^2-\varepsilon)^{-1}]$, where $\delta \in \mathrm{Ell}_4$ and $\varepsilon \in \mathrm{Ell}_8$, over which there is a formal group law whose logarithm is the elliptic integral

$$\log^{\mathrm{Ell}}(x) = \int_0^x \frac{dt}{\sqrt{1-2\delta t^2+\varepsilon t^4}} = \sum_{n\geq 0} P_n(\delta/\sqrt{\varepsilon})\varepsilon^{n/2}\frac{t^{2n+1}}{2n+1},$$

where $P_n(\ )$ is the Legendre polynomial.

The fact that this formal group law is defined over Ell_* (in fact over $\mathbf{Z}[\frac{1}{2}][\delta,\varepsilon]$), which goes back to Euler [9], provides us with a homomorphism $MU_* \to \mathrm{Ell}_*$.

Congruences amongst the Legendre polynomials, and the inverted element, give us the conditions of Landweber's exact functor theorem so that

$$\mathrm{Ell}_*(X) = \mathrm{Ell}_* \otimes_{MU_*} MU_*(X)$$

defines a homology theory and a ring spectrum Ell. Our aim in this paper is to describe $\mathrm{Ell}_*(\mathrm{Ell})$.

The coefficient ring Ell_* can be interpreted as a ring of modular forms. Recall that a *modular form* of *weight* k, with respect to a subgroup Γ of $SL_2(\mathbf{Z})$ which contains the matrix $\begin{pmatrix} 1 & 1 \\ 0 & 1 \end{pmatrix}$, is a holomorphic function $f(\tau)$ of τ in the upper half plane, which (i) satisfies the transformation rule

$$f\left(\frac{a\tau+b}{c\tau+d}\right) = (c\tau+d)^{-k} f(\tau),$$

for all $\begin{pmatrix} a & b \\ c & d \end{pmatrix} \in \Gamma$, and (ii) is holomorphic at all cusps, in particular at ∞ it has a Fourier series (or *q-expansion*)

$$f(\tau) = \sum_{n\geq 0} a_n q^n, \qquad \text{where } q = e^{2\pi i\tau}.$$

Important examples of modular forms are the Eisenstein series:

$$\begin{aligned} G_k(\tau) &= \frac{(k-1)!}{2(2\pi i)^k} \sum_{(m,n)\neq(0,0)} \frac{1}{(m\tau+n)^k} \\ &= -\frac{B_k}{2k} + \sum_{n\geq 1} \sigma_{k-1}(n) q^n, \end{aligned}$$

which is a modular form of weight k for $\Gamma(1) = SL_2(\mathbf{Z})$ if $k \geq 4$; and the modified versions

$$\begin{aligned} G_k^*(\tau) &= G_k(\tau) - 2^{k-1} G_k(2\tau) \\ &= (2^{k-1}-1)\frac{B_k}{2k} + \sum_{n\geq 1} \sigma_{k-1}^*(n) q^n, \end{aligned}$$

and

$$\begin{aligned} \tilde{G}_k(\tau) &= -G_k(\tau) + 2G_k(2\tau) \\ &= -\frac{B_k}{2k} + \sum_{n\geq 1} \tilde{\sigma}_{k-1}(n) q^n, \end{aligned}$$

which are both modular forms of weight k for the group

$$\Gamma_0(2) = \{ \begin{pmatrix} a & b \\ c & d \end{pmatrix} \in SL_2(\mathbf{Z}) : c \text{ is even} \},$$

if $k \geq 2$. Here

$$\sigma_j(n) = \sum_{d|n} d^j,$$
$$\sigma_j^*(n) = \sum_{\substack{d|n \\ d \text{ odd}}} d^j,$$
$$\tilde{\sigma}_j(n) = \sum_{d|n} (-1)^{n/d} d^j.$$

Note that G_k, G_k^* and $\tilde{G}_k$ all have rational q-series; apart from the constant term they are integral. See [25] for these formulas.

If we let $\delta = -3G_2^*$ and $\varepsilon = -(G_4^* + 7\tilde{G}_4)/6$ then

$$\delta = -\tfrac{1}{8} - 3\sum_{n\geq 1} \sigma_1^*(n) q^n,$$
$$\varepsilon = \sum_{n\geq 1} \left(\sum_{\substack{d|n \\ n/d \text{ odd}}} d^3 \right) q^n.$$

In fact $\mathbf{Z}[\frac{1}{2}][\delta, \varepsilon]$ is equal to the ring of modular forms with respect to $\Gamma_0(2)$ whose q-series lie in $\mathbf{Z}[\frac{1}{2}][[q]]$. Note that $(\delta^2 - \varepsilon)^{-1} \in \mathbf{Z}[\frac{1}{2}][[q]]$ too.

We have, then, a ring homomorphism

$$\lambda : \mathrm{Ell}_* \to \mathbf{Z}[\tfrac{1}{2}][[q]].$$

We have just written λ as the identity, but now we need to distinguish it. Let us remark that λ is a monomorphism since a holomorphic function is uniquely determined by its Fourier series.

Following the programme outlined in §1, to understand $\mathrm{Ell}_*(\mathrm{Ell})$ we need to study the coefficients of the series $\exp_L^{\mathrm{Ell}}(\log_R^{\mathrm{Ell}}(x))$. They are complicated rational polynomials in δ_L, δ_R, ϵ_L and ϵ_R. (Here, and below, we adopt the general notation: $\alpha_L = \eta_L(\alpha)$ and $\alpha_L = \eta_L(\alpha)$.) The first three are

$$\frac{-\delta_L + \delta_R}{3},$$
$$\frac{\delta_L^2 + 10\delta_L\delta_R + 10\delta_R^2 + 3\epsilon_L - 3\epsilon_R}{30},$$
$$\frac{-\delta_L^3 + 35\delta_L^2\delta_R - 259\delta_L\delta_R^2 + 45\delta_R^3 - 33\delta_L\epsilon_L + 105\epsilon_L\delta_R + 63\delta_L\epsilon_R - 135\delta_R\epsilon_R}{630}.$$

However, when looked at in terms of q-series, we have the following result.

Theorem 1. *The q-series of these generators (in the variables q_L and q_R) are integral, i.e.,*

$$\lambda \otimes \lambda : \mathrm{Ell}_* \otimes \mathrm{Ell}_* \otimes \mathbf{Q} \to \mathbf{Q}[[q_L, q_R]],$$

where $q_L = q \otimes 1$ and $q_R = 1 \otimes q$, maps $\exp_L^{\mathrm{Ell}}(\log_R^{\mathrm{Ell}}(x))$ into the subring $\mathbf{Z}[[q_L, q_R]][[x]]$.

Corollary. *$\lambda \otimes \lambda$ maps $\mathrm{Ell}_*(\mathrm{Ell})$ into $\mathbf{Z}[\frac{1}{2}][[q_L, q_R]]$.*

Conjecture. *The converse is true, i.e., $\mathrm{Ell}_*(\mathrm{Ell})$ consists of all those elements of $\mathrm{Ell}_* \otimes \mathrm{Ell}_* \otimes \mathbf{Q}$ which $\lambda \otimes \lambda$ maps into $\mathbf{Z}[\frac{1}{2}][[q_L, q_R]]$.*

We will return to the evidence for the conjecture after outlining the proof of Theorem 1.

Proof. Let $\exp^{\lambda\mathrm{Ell}}(x), \log^{\lambda\mathrm{Ell}}(x) \in \mathbf{Q}[[q]][[x]]$ be the series obtained after applying λ. The notation here is meant to indicate that we should think of these expressions as power series in x whose coefficients are power series in q. We need to look at

$$\exp_L^{\lambda\mathrm{Ell}}(\log_R^{\lambda\mathrm{Ell}}(x)) \in \mathbf{Q}[[q_L, q_R]][[x]].$$

The key fact is the following result.

Theorem 2.

$$\exp^{\lambda\mathrm{Ell}}(2\sinh^{-1}(x/2)) = x \prod_{n\geq 1} \left(1 - \frac{q^n x^2}{(1-q^n)^2}\right)^{(-1)^n} = \Theta(x)$$

and $\Theta(x) \in \mathbf{Z}[[q]][[x]]$.

For this formula see [17] and [25]. It expresses the fact that, once one converts to q-series, the elliptic formal group law is multiplicative.

Now

$$\begin{aligned}\exp_L^{\lambda\mathrm{Ell}}(\log_R^{\lambda\mathrm{Ell}}(x)) &= \Theta_L(2\sinh(\sinh^{-1}(\Theta_R^{-1}(x)/2))) \\ &= \Theta_L(\Theta_R^{-1}(x)) \in \mathbf{Z}[[q_L, q_R]][[x]].\end{aligned}$$

The important point is that $2\sinh(x/2)$ contains no parameters so its left and right versions are the same.

We discuss now why the conjecture might be true. We have two pieces of evidence.

Firstly we show how, assuming the conjecture, we can compute $\mathrm{Ext}^{1,*}_{\mathrm{Ell}}$ and get the expected answer, consistent with the corresponding results for K-theory and MU-theory. Analogously with K-theory,

$$\mathrm{Ext}^{1,*}_{\mathrm{Ell}} = \{\,\alpha \in \mathrm{Ell}_* \otimes \mathbf{Q} : \alpha_L - \alpha_R \in \mathrm{Ell}_*(\mathrm{Ell})\,\}/\mathrm{Ell}_*.$$

If

$$\alpha = \sum_{n\geq 0} a_n q^n$$

then

$$\alpha_L - \alpha_R = \sum_{n\geq 1} a_n(q_L^n - q_R^n).$$

Note that the constant terms have cancelled.[1]

Thus, if the conjecture is correct, we need to look for modular forms whose q-series are in $\mathbf{Q} + q\mathbf{Z}[\frac{1}{2}][[q]]$. An example, if k is even, is

$$\tilde{G}_k(\tau) = -\frac{B_k}{2k} + \sum_{n\geq 1} \tilde{\sigma}_{k-1}(n)q^n,$$

which provides an element of $\mathrm{Ext}^{1,2k}_{\mathrm{Ell}}$ of order $m(k)_{\mathrm{odd}}$.

The map $\alpha \mapsto a_0$ provides a monomorphism

$$\mathrm{Ext}^{1,2k}_{\mathrm{Ell}} \to \mathbf{Q}/\mathbf{Z}[\tfrac{1}{2}].$$

This shows that the group is cyclic, but could the class of $\tilde{G}_k$ be divisible?

A theorem of Serre [23] shows that the answer is no. Using results of Swinnerton-Dyer, Serre shows that $m(k)_{\mathrm{odd}}$ is the largest denominator occurring in the constant term of any modular form of weight k which has its q-series in $\mathbf{Q} + q\mathbf{Z}[\frac{1}{2}][[q]]$. The result is in fact stated for modular forms with respect to the group $\Gamma(1)$ but it is not difficult to adapt to the case in hand. Thus we have $\mathrm{Ext}^{1,2k}_{\mathrm{Ell}} \cong \mathbf{Z}/m(k)_{\mathrm{odd}}$, as we expect.

The second reason to believe our conjecture is that a simpler, but analogous result is true. Let us consider the problem of characterising the ring $KO_*(\mathrm{Ell})$. As we remarked at the end of §1, the same approach works. The generators of $KO_0(\mathrm{Ell})$ are the coefficients of the series $\Theta(x)$.

Theorem 3. *Let D denote the ring of non-homogeneous sums of rational modular forms with respect to $\Gamma_0(2)$ such that the q-series of the whole sum is in $\mathbf{Z}[\frac{1}{2}][[q]]$. Then $KO_0(\mathrm{Ell}) \cong D[(\delta^2 - \varepsilon)^{-1}]$.*

The inclusion in one direction follows from Theorem 2, which shows that the coefficients of $\Theta(x)$ lie in D. The converse is a result of Katz [13].

[1] We are reverting to writing λ as the identity here.

Katz's proof is more general than we need and is phrased in the language of schemes. It would be nice to find a more down to earth proof which will generalise to the case of $\mathrm{Ell}_*(\mathrm{Ell})$.

Theorem 3 shows how Miller's elliptic character [17] arises from the map $\mathrm{Ell} \to KO \wedge \mathrm{Ell}$, by analogy with the Chern character map $K \to K \wedge H$.

Appendix. The elliptic exponential series

We include here some remarks on the computation of the exponential series for the elliptic formal group law. Since $\log^{\mathrm{Ell}}(x)$ is an elliptic integral (of the first kind), $\exp^{\mathrm{Ell}}(x)$ is, up to constants, a Jacobi sine function. In fact

$$\exp^{\mathrm{Ell}}(x) = a\,\mathrm{sn}(\frac{x}{a}, \delta + \sqrt{\delta^2 - \varepsilon}),$$

where

$$a = \sqrt{\frac{\delta + \sqrt{\delta^2 - \varepsilon}}{\varepsilon}}.$$

One ought then to be able to obtain the coefficients of $\exp^{\mathrm{Ell}}$ from the Taylor series coefficients of $\mathrm{sn}(u, k)$. Hermite worked on this expansion; see [12] and [6]. But the first fully satisfactory treatment was published only in 1981 by Dumont [8]. He shows that if $\mathrm{sn}(u; y, z)$ is defined by

$$u = \int_0^{\mathrm{sn}(u;y,z)} \frac{dt}{\sqrt{(1 + y^2t^2)(1 + z^2t^2)}},$$

so that $\mathrm{sn}(u, k) = \mathrm{sn}(u; i, ik)$, then

$$\mathrm{sn}(u; y, z) = \frac{1}{yz}\left(e^{u\mathcal{D}}x\right)\Big|_{x=0},$$

where $\mathcal{D}$ is the operator

$$\mathcal{D} = yz\frac{\partial}{\partial x} + zx\frac{\partial}{\partial y} + xy\frac{\partial}{\partial z}.$$

It is possible, though extremely messy, to substitute into the expansion for $\mathrm{sn}(u; y, z)$ for y and z in terms of δ and ε. It is, however, much better to write $\mathcal{D}$ in terms of δ and ε.

We have $\exp^{\mathrm{Ell}}(u) = \mathrm{sn}(u; y, z)$ with $-2\delta = y^2 + z^2$ and $\varepsilon = y^2z^2$, so that

$$\frac{\partial}{\partial y} = -y\frac{\partial}{\partial \delta} + 2yz^2\frac{\partial}{\partial \varepsilon},$$
$$\frac{\partial}{\partial z} = -z\frac{\partial}{\partial \delta} + 2y^2z\frac{\partial}{\partial \varepsilon}.$$

Hence

$$zx\frac{\partial}{\partial y} + xy\frac{\partial}{\partial z} = x\left(-2yz\frac{\partial}{\partial \delta} + 2yz(y^2+z^2)\frac{\partial}{\partial \varepsilon}\right),$$
$$= -2x\sqrt{\varepsilon}\frac{\partial}{\partial \delta} - 4x\sqrt{\varepsilon}\delta\frac{\partial}{\partial \varepsilon},$$

and so

$$\mathcal{D} = \sqrt{\varepsilon}\left(\frac{\partial}{\partial x} - 2x\frac{\partial}{\partial \delta} - 4x\delta\frac{\partial}{\partial \varepsilon}\right).$$

If we let $\eta = \sqrt{\varepsilon} = yz$, we have

$$\mathcal{D} = \eta\frac{\partial}{\partial x} - 2x\eta\frac{\partial}{\partial \delta} - 2x\delta\frac{\partial}{\partial \eta},$$

and the coefficient of $u^n/n!$ in $g(u)$ is

$$\frac{1}{\eta}(\mathcal{D}^n x)\bigg|_{x=0}.$$

Giving x degree 2 and δ and η degree 4, $\mathcal{D}^n x$ is homogeneous of degree $2n+2$. If we write

$$\mathcal{D}^n x = \begin{cases} \sum_{i,j} b_{n,i,j} x^{2i+1}\delta^j\eta^{m-i-j}, & \text{if } n \text{ is even, } n = 2m, \\ \sum_{i,j} b_{n,i,j} x^{2i}\delta^j\eta^{m-i-j}, & \text{if } n \text{ is odd, } n = 2m-1, \end{cases}$$

we have the recurrence

$$b_{n,i,j} = (2i+2)b_{n-1,i+1,j} - (2j+2)b_{n-1,i,j+1} - (n-2i-2j+2)b_{n-1,i,j-1},$$

if n is even, and

$$b_{n,i,j} = (2i+1)b_{n-1,i,j} - (2j+2)b_{n-1,i-1,j+1} - (n-2i-2j+3)b_{n-1,i-1,j-1},$$

if n is odd; compare (6.2) of [8]. Thus

$$\exp^{\mathrm{Ell}}(u) = \sum_{m\geq 1}\left(\sum_{k=0}^{[(m-1)/2]} b_{2m-1,0,m-2k-1}\delta^{m-2k-1}\eta^k\right)\frac{u^{2m-1}}{(2m-1)!}.$$

We find that

$$\begin{aligned}
\mathcal{D}x &= \eta,\\
\mathcal{D}^2x &= -2x\delta,\\
\mathcal{D}^3x &= -2\delta\eta + 4x^2\eta,\\
\mathcal{D}^4x &= 12x\eta^2 + 4x\delta^2 - 8x^3\delta,\\
\mathcal{D}^5x &= 12\eta^3 + 4\delta^2\eta - 88x^2\delta\eta + 16x^4\eta,\\
\mathcal{D}^6x &= -264x\delta\eta^2 - 8x\delta^3 + 240x^3\eta^2 + 176x^3\delta^2 - 32x^5\delta,\\
\mathcal{D}^7x &= -264\delta\eta^3 - 8\delta^3\eta + 1248x^2\eta^3 + 1632x^2\delta^2\eta - 1824x^4\delta\eta + 64x^6\eta,
\end{aligned}$$

and thus

$$\begin{aligned}
\exp^{\mathrm{Ell}}(u) = u \\
&- 2\delta\frac{u^3}{3!}\\
&+ 4(3\varepsilon + \delta^2)\frac{u^5}{5!}\\
&- 8(33\delta\varepsilon + \delta^3)\frac{u^7}{7!}\\
&+ 16(189\varepsilon^2 + 306\delta^2\varepsilon + \delta^4)\frac{u^9}{9!}\\
&- 32(8289\delta\varepsilon^2 + 2766\delta^3\varepsilon + \delta^5)\frac{u^{11}}{11!}\\
&+ 64(68607\varepsilon^3 + 255987\delta^2\varepsilon^2 + 24909\delta^4\varepsilon + \delta^6)\frac{u^{13}}{13!}\\
&- 128(7660737\delta\varepsilon^3 + 6988167\delta^3\varepsilon^2 + 224199\delta^5\varepsilon + \delta^7)\frac{u^{15}}{15!}\\
&+ \cdots .
\end{aligned}$$

References

1. J. F. Adams, *On the groups $J(X)$–IV*, Topology **5** (1966), 21–71.
2. ———, *Lectures on generalised cohomology*, Category theory, homology theory and their applications III, Lecture Notes in Math., vol. 99, Springer-Verlag, Berlin and New York, 1969, pp. 1–138.
3. ———, *Stable homotopy and generalised homology*, Chicago Lectures in Mathematics, University of Chicago Press, Chicago and London, 1974.
4. J. F. Adams, A. S. Harris and R. M. Switzer, *Hopf algebras of cooperations for real and complex K-theory*, Proc. London Math. Soc. (3) **23** (1971), 385–408.
5. A. Baker, F. Clarke, N. Ray and L. Schwartz, *On the Kummer congruences and the stable homotopy of BU*, Trans. Amer. Math. Soc. **316** (1989), 385–432.
6. C. A. Briot and J. C. Bouquet, *Théorie des fonctions elliptiques*, second edition, Gauthier-Villars, Paris, 1875.

7. P. E. Conner and E. E. Floyd, *The relation of cobordism to K-theories*, Lecture Notes in Math. vol. 28, Springer-Verlag, Berlin and New York, 1966.
8. D. Dumont, *Une approche combinatoire des fonctions elliptiques de Jacobi*, Adv. in Math. **41** (1981), 1–39.
9. L. Euler, *De integratione aequationis differentialis* $\frac{m\,dx}{\sqrt{1-x^4}} = \frac{n\,dy}{\sqrt{1-y^4}}$, Novi Comm. Acad. Sci. Petropolitanae **6** (1756/7), 1761, 37–57; Collected Works, first series, Vol. 20, B. G. Teuber, Leipzig and Berlin, 1912.
10. H. Hasse, *Die Gruppe der p^n-primären Zahlen für einen Primteiler $\mathfrak{p}$ von p*, J. Reine Angew. Math. **176** (1936), 174–183.
11. M. Hazewinkel, *Formal Groups and Applications*, Academic Press, New York, 1978.
12. C. Hermite, *Remarques sur le développement de* $\cos \operatorname{am} x$, C. R. Acad. Sci. (Paris) **57** (1863), 613–618; J. Math. Pure Appl. **9** (1864), 289-295; Collected Works, Gauthier-Villars, Paris, 1908.
13. N. M. Katz, *Higher congruences between modular forms*, Ann. of Math. (2) **101** (1975), 332–367.
14. P. S. Landweber, *$BP_*(BP)$ and typical formal groups*, Osaka J. Math. **12** (1975), 357–363.
15. ———, *Homological properties of comodules over $MU_*(MU)$ and $BP_*(BP)$*, Amer. J. Math. **98** (1976), 591–610.
16. ———, *Elliptic cohomology and modular forms*, Elliptic curves and modular forms in topology, Princeton 1986, Lecture Notes in Math., vol. 1326, Springer-Verlag, Berlin Heidelberg and New York, 1988, pp. 55–68.
17. H. Miller, *The elliptic character and the Witten genus*, Algebraic Topology (M. Mahowald and S. Priddy, eds.), Contemp. Math. vol. 96, 1989, pp. 281–289.
18. H. R. Miller and D. C. Ravenel, *Morava stabilizer algebras and the localisation of Novikov's E_2-term*, Duke Math. J. **44** (1977), 433–447.
19. J. Milnor, *The Steenrod algebra and its dual*, Ann. of Math. (2) **67** (1958), 150–171.
20. J. Morava, *Noetherian localizations of categories of cobordism comodules*, Ann. of Math. (2) **121** (1985), 1–39.
21. D. C. Ravenel, *Complex cobordism and stable homotopy groups of spheres*, Academic Press, Orlando, Florida, 1986.
22. Yu. B. Rudjak, *Exactness theorems in the cohomology theories MU, BP and $P(n)$*, Mat. Zametki **40** (1986), 115–126 (Russian); English transl. in Math. Notes **40** (1986), 562–569.
23. J. P. Serre, *Congruences et formes modulaires (d'après H. P. F. Swinnerton-Dyer)*, Exposé n° 416, Séminaire Bourbaki 1971/72, Lecture Notes in Math., vol. 317, Springer-Verlag, Berlin Heidelberg and New York, 1973, pp. 319–338.
24. K. G. C. von Staudt, *De numeris Bernoullianis, commentatio altera*, Erlangen, 1845.
25. D. Zagier, *Note on the Landweber-Stong elliptic genus*, Elliptic curves and modular forms in topology, Princeton 1986, Lecture Notes in Math., vol. 1326, Springer-Verlag, Berlin Heidelberg and New York, 1988, pp. 216–224.

Department of Mathematics and Computer Science, University College of Swansea, Swansea SA2 8PP, Wales
E-mail address: mafred@uk.ac.swan.pyr

Department of Mathematics and Computer Science, Dalhousie University, Halifax, Nova Scotia, Canada B3H 4H8
E-mail address: johnson@ca.dal.cs

Completions of G-spectra at ideals of the Burnside ring

by J. P. C. Greenlees and J. P. May

On May 3, 1988, Frank Adams wrote us a letter suggesting the topic of the present paper. Referring to the first author's paper [**10**], he wrote as follows:

> You interest the reader in a problem which makes sense for all finite groups G; but you only give the answer when G is a p-group. Perhaps one should try to give the answer in a form valid for all finite groups G? I hope the answer is
>
> $$D(EG_+) \simeq (S^0)^{\wedge}_I.$$
>
> Here EG_+ is the universal G-space and $D(EG_+)$ is its equivariant functional dual, just as you have it; and the operation $(\cdot)^{\wedge}_I$ takes the equivariant completion of a G-spectrum at an ideal of the Burnside ring $A(G)$—in this case the augmentation ideal. This would be pleasingly reminiscent of Graeme's original statement.
>
> The first point to settle is whether the necessary completion already exists in the literature. I don't know off hand. [He then points out that references [**17**], [**19**], and [**20**] fail to construct such completions.] Did someone pass up a chance to set up a useful general theory?

We took up Frank's challenge, and we offer the results in his memory. We think that he would have enjoyed the ideas.

Expanding on Frank's letter, let G be a finite group, let EG_+ be the union of EG and a disjoint basepoint and let $\pi : EG_+ \to S^0$ be the projection. Then the Segal conjecture asserts that π induces an isomorphism from the I-adic completion of the stable G-cohomotopy groups of S^0 to the stable G-cohomotopy groups of EG_+, where I is the augmentation ideal of $A(G)$. This is a statement about the behavior on homotopy groups of the map of dual G-spectra

$$\pi^* : S = D(S^0) \to D(EG_+),$$

where S is the sphere G-spectrum and $D(X) = F(X, S)$.

Generalizations of the Segal conjecture, such as those in [**3**, **7**, **23**], admit similar descriptions in terms of more general G-spectra than S_G and more general ideals in $A(G)$. It is natural to ask whether such results can be interpreted as saying that the relevant maps of G-spectra are completions at the appropriate ideals. If we define the notion of completion of a G-spectrum in terms of behavior on homotopy groups, then of course the answer is obviously yes. We shall give a more interesting and powerful notion of completion, one specified in terms of an appropriate universal property, and we shall prove that, on suitably restricted kinds of G-spectra, our completions of G-spectra do indeed induce completions of homotopy groups and are characterized by that property.

In fact, we shall construct $X_I^\wedge$ when G is an arbitrary compact Lie group, I is a finitely generated ideal in $A(G)$, and X is an arbitrary G-spectrum. The G-spectrum $X_I^\wedge$ will be a suitable Bousfield localization of X, and the homotopy groups of $X_I^\wedge$ will be determined algebraically by the homotopy groups of X via the left derived functors $L_i^I(M)$ of the I-adic completion functor $M_I^\wedge$ on $A(G)$-modules M. These derived functors are identically zero for $i \geq 2$. When $A(G)$ is Noetherian and M is finitely generated, $L_0^I(M) \cong M_I^\wedge$ and $L_1^I(M) = 0$.

For finite groups G, the new construction leads to generalizations of both the Segal conjecture and the Atiyah-Segal completion theorem. Previous versions of these results start with the restrictions of the appropriate cohomology theory to, say, finite G-CW complexes or to pro-group valued theories defined in terms of such restricted theories. Our constructions allow us to extend these results to the represented theories defined on arbitrary G-spaces or G spectra. In a later paper [**13**], which gives a number of applications of the present theory, we shall prove the very surprising result that, for the periodic K-theory spectrum K_G, the projection $K_G \wedge EG_+ \to K_G$ induces an equivalence upon completion at the augmentation ideal I of $A(G)$; equivalently, the Tate spectrum associated to K_G, which turns out to be a rational G-spectrum, has trivial completion.

In section 1, we give an elementary cohomological construction of completions that applies only when $A(G)$ is Noetherian and is restricted to G-spectra that are bounded below and of finite type. However, this version of completion has the important virtue of being characterized by its behavior on homotopy groups or on certain ordinary homology or cohomology groups. We give our general construction of completions in section 2, and we describe its behavior on homotopy groups in section 3. We give the promised application to the Segal conjecture and Atiyah-Segal completion theorem in section 4. This requires both constructions. The homotopical characterization of our cohomological construction allows us to use existing completion theorems to identify $Y_I^\wedge$ for certain G-spectra Y and ideals I. Our general construction allows us to deduce an identification of $F(X,Y)_I^\wedge$ for arbitrary G-spectra X.

The proofs of our calculational results (as opposed to the construction and characterization of completions) involve an algebraic study of the relationship between derived functors of completions and Grothendieck's local cohomology groups. In deference to Adams' dictum about the writing of algebraic topology "... one writes algebra only as required" [**22**, p.48], we have separated out the general algebraic study for publication elsewhere [**12**]. However, the key piece of algebra is the proof that $L_i^I(M) = 0$ for $i \geq 2$. This follows from a general theorem of Grothendieck when $A(G)$ is Noetherian, but we give a self-contained proof which works for all G in section 6; several other algebraic proofs are also deferred to that final section. Topological proofs, in particular the determination of the homotopy groups of completed G-spectra, are given in section 5. For completeness, we quickly run through the simpler theory of localization of G-spectra in an appendix.

At the referee's request, we have added a brief introduction that explains our parallel definitions of homological and cohomological completions. In his words, "the connection between corresponding completions is interesting even in the nonequivariant case". The material of section 1 originally appeared in the short preprint [**14**], with Mike Hopkins as a joint author. Also at the referee's request, we have incorporated that elementary part of the theory into the more sophisticated general theory, at the price of jettisoning Hopkins' very pretty proof of Theorem 1.4 below. It is a pleasure to thank Hopkins for his contribution to this project. We also thank Bill Dwyer, Dick Swan, and Gennady Lyubeznik for enjoyable and helpful conversations.

§0. General definitions of localizations and completions.

We begin by describing a general conceptual approach to the theory of localizations and completions that has been part of the second author's way of thinking about the subject since the mid 1970's. Like Bousfield's much more substantive study [4], these ideas were inspired by Frank Adams' marvelous 1973 lectures at the University of Chicago.

Our general definitions make sense in any category with suitable properties, but we shall use the language of spectra for specificity. The reader is warned that the words "localization" and "completion" are used with a certain whimsical interchangeability, both here and in the literature. The following notion is all inclusive.

DEFINITION 0.1. *Let $\mathcal{W}$ be any class of spectra, to be thought of as a class of acyclic spectra.*

(i) *A map $f : X \to Y$ is a $\mathcal{W}$-equivalence if its cofiber is in $\mathcal{W}$.*

(ii) *A spectrum Z is $\mathcal{W}$-complete if $[W, Z] = 0$ for all $W \in \mathcal{W}$.*

(iii) *A $\mathcal{W}$-completion of a spectrum X is a $\mathcal{W}$-equivalence from X to a $\mathcal{W}$-complete spectrum.*

There are Eckmann-Hilton dual ways of giving content to these general definitions, one homological and the other cohomological.

DEFINITION 0.2. *Let $\mathcal{T}$ be any class of spectra, to be thought of as a class of test spectra.*

(i) *A spectrum W is $\mathcal{T}$-acyclic if $W \wedge T$ is contractible for all $T \in \mathcal{T}$.*

(ii) *A spectrum W is $\mathcal{T}^*$-acyclic if $F(W, T)$ is contractible for all $T \in \mathcal{T}$.*

When $\mathcal{W}$ is the class of $\mathcal{T}$-acyclic spectra, we refer to the concepts of Definition 0.1 as "$\mathcal{T}$-equivalence", "$\mathcal{T}$-local", and "$\mathcal{T}$-localization". When $\mathcal{W}$ is the class of $\mathcal{T}^*$-acyclic spectra, we refer to these concepts as "$\mathcal{T}^*$-equivalence", "$\mathcal{T}^*$-complete", and "$\mathcal{T}^*$-completion".

We think of (i) as specifying homological acyclicity and (ii) as specifying cohomological acyclicity. If $\mathcal{T}$ consists of a single spectrum E, then $\mathcal{T}$-localization is exactly Bousfield's notion of E-localization [4], and he proved that E-localizations always exist. His proof generalizes readily to the equivariant setting. (He works in Adams' version [1] of the stable category, but the argument is similar and a bit simpler in the equivariant stable category constructed in [17].) We omit the proof since we shall give explicit constructions of the localizations that we need.

Various elementary properties of E-localizations are listed in [4, §1], and they remain valid in the equivariant context. In particular, two spectra E and E' are said to be Bousfield equivalent if they determine the same class of homologically acyclic spectra. They then also determine the same class of equivalences, the same local spectra, and the same localization functor. It is an important observation that the function spectrum $F(E', X)$ is E-local for any spectrum E' that is Bousfield equivalent to E, by an immediate application of the adjunction

$$[W, F(E', X)] \cong [W \wedge E', X].$$

In fact, our completions will turn out to be just such function spectra.

More generally, we say that two classes $\mathcal{T}$ and $\mathcal{T}'$ are Bousfield equivalent if they determine the same homologically acyclic spectra. Again, they then determine the same equivalences, the same local spectra, and the same localizations. In practice, $\mathcal{T}$ is usually Bousfield equivalent to $\{E\}$ for some spectrum E.

Dually, we say that two classes $\mathcal{T}$ and $\mathcal{T}'$ are cohomologically equivalent if they determine the same class of cohomologically acyclic spectra. Of course, this is a different notion than Bousfield equivalence. The relationship between acyclic and complete spectra has a different flavor in the cohomological and homological contexts. In the former, we have the following tautological observation, which shows that taking the complete spectra is a kind of closure operation.

LEMMA 0.3. *Let $C\mathcal{T}$ be the class of $\mathcal{T}^*$-complete spectra. Then $\mathcal{T}$ and $C\mathcal{T}$ are cohomologically equivalent, and $C\mathcal{T}$ is the largest class of spectra that is cohomologically equivalent to $\mathcal{T}$.*

We will first define I-completions cohomologically, using a certain class $\mathcal{K}$ of I-torsion Eilenberg-MacLane spectra as test spectra. That class will be cohomologically equivalent to a larger class $\mathcal{T}$ of I-torsion spectra, and we will take $\mathcal{T}$-localization as our primary definition of I-completion. In turn, $\mathcal{T}$ will be Bousfield equivalent to $\{E\}$ for a certain spectrum E. We do not claim that $\mathcal{K}$ and $\{E\}$ are either cohomologically or Bousfield equivalent. However, when restricted to bounded below spectra, which seems to be the limit of utility of the cohomological notion, $\mathcal{K}^*$-completion and E-localization will automatically agree because the class of bounded below $\mathcal{K}^*$-acyclic spectra will turn out to be the same as the class of bounded below $\mathcal{T}$-acyclic spectra.

§1. A cohomological construction of completions at I.

Let G be a compact Lie group. We work in the stable category $\overline{h}\mathcal{GS}$ of G-spectra constructed in [17], and spectra and maps are understood to be G-spectra and G-maps throughout. Let $\mathcal{O}$ be the full subcategory of $\overline{h}\mathcal{GS}$ whose objects are the suspension spectra of orbits G/H_+ (where H is a closed subgroup of G and the plus denotes addition of a disjoint basepoint). A Mackey functor M is an additive contravariant functor $\mathcal{O} \to \mathcal{A}b$, written $M(G/H)$ on objects. When G is finite, this is equivalent to the standard definition of Dress [9], by [17, V.9.9]. Since $A(G) = [S,S]_G$ and S is the suspension spectrum of G/G_+, it follows formally from the definition that each $M(G/H)$ is an $A(G)$-module. For any spectrum X and integer n, we have the nth homotopy group Mackey functor $\underline{\pi}_n(X)$. Its value on G/H is

$$\pi_n(X^H) = [G/H_+ \wedge S^n, X]_G,$$

and its contravariant functoriality on $\mathcal{O}$ is obvious.

Let I be an ideal of $A(G)$. For an $A(G)$-module M, we define $M_I^\wedge = \lim M/I^r M$. We define the I-adic completion of Mackey functors termwise, so that

$$M_I^\wedge(G/H) = M(G/H)_I^\wedge.$$

We extend any other functor on $A(G)$-modules to Mackey functors in the same termwise fashion. There results a map of Mackey functors $\gamma : M \to M_I^\wedge$. When M is of finite type, in the sense that it takes values in finitely generated Abelian groups and thus in finitely generated $A(G)$-modules, we call γ the I-adic completion of M. We say that a Mackey functor is "I-adic" if it takes values in finitely generated $A(G)_I^\wedge$-modules. Observe that $\gamma : M \to M_I^\wedge$ is an isomorphism if M is I-adic. Experience with the case $G = e$ and $I = (p)$ teaches us that, due to lack of exactness, I-adic completion cannot be the appropriate algebraic completion functor on general Mackey functors M. We shall define the appropriate general functor in section 3.

A Mackey functor M determines an Eilenberg-MacLane spectrum $HM = K(M,0)$ which is uniquely characterized up to isomorphism in $\overline{h}\mathcal{GS}$ by $\underline{\pi}_0(HM) = M$ and $\underline{\pi}_n(HM) = 0$ if $n \neq 0$. Maps $M \to M'$ of Mackey functors determine and are determined by maps $HM \to HM'$ of Eilenberg-MacLane spectra; we shall denote corresponding maps by the same letter. See [16] and [17, V§9] for background. We let $H_G^*(X;M)$ denote the cohomology theory on G-spectra X which is represented by HM. All of our cohomology theories will be taken to be Z-graded. However, since our theories are represented by G-spectra, they are $RO(G)$-gradable. When X is a based G-space, $H_G^*(X;M)$ is just the classical reduced Bredon cohomology of X [**6**].

Now return to the context of Definitions 0.1 and 0.2.

DEFINITION 1.1. *Let $\mathcal{K}_I$ be the class of Eilenberg-MacLane spectra $K(N,q)$, where N runs over the Mackey functors N such that $IN = 0$ and q runs over the integers. Note that a spectrum W is $\mathcal{K}_I^*$-acyclic if and only if $H_G^*(W;N) = 0$ for all such Mackey functors N.*

LEMMA 1.2. *Let W be $\mathcal{K}_I^*$-acyclic. Then $H_G^*(W;N) = 0$ if $I^rN = 0$ for some $r \geq 1$ or if $N = M_I^\wedge$ for some Mackey functor M. Therefore $H(M_I^\wedge)$ is $\mathcal{K}_I^*$-complete for any Mackey functor M.*

PROOF: Inductive use of Bockstein exact sequences gives the result when $I^rN = 0$. Since cohomology commutes with products in the coefficient variable, the conclusion for $M_I^\wedge$ follows from the Bockstein exact sequence associated to the evident short exact sequence

$$0 \to M_I^\wedge \to \prod M/I^rM \to \prod M/I^rM \to 0.$$

A spectrum X is said to be bounded below if the groups $\pi_n(X^H)$ are zero for all H and all n less than some fixed n_0.

PROPOSITION 1.3. *If Z is bounded below and if each $H\underline{\pi}_n(Z)$ is $\mathcal{K}_I^*$-complete, then Z is $\mathcal{K}_I^*$-complete.*

PROOF: Since Z is bounded below, it has a Postnikov decomposition. The conclusion follows by induction up the tower and passage to limits. In fact, the dualization of the proof of the Whitehead theorem for CW-complexes that is explained on the space level in [**18**, Theorems 5*, 5#] works equally well on the spectrum level. The essential technical point is that, in the good category of spectra of [**17**], we can work with fibration sequences of spectra exactly as we work with fibration sequences of spaces. (Even nonequivariantly, such an argument would not work quite so simply in traditional categories of CW-spectra, which bear essentially the same relationship to our category of spectra as the category of CW-complexes and cellular maps bears to the category of general spaces.)

Henceforward in this section, we assume that $A(G)$ is Noetherian. This holds if and only if $A(G)$ is a finitely generated Abelian group. More concretely, it holds if and only if the Weyl group of G acts trivially on the maximal torus or, equivalently, if G is a central toral extension over a finite group; see [**8**, §5.10]. By the latter description, $A(H)$ is Noetherian for all (closed) subgroups H of G. These are precisely the compact Lie groups G for which the most naive version of the Segal conjecture is valid [**24**]. With this assumption, the Artin-Rees lemma implies that I-adic completion is an exact functor on the category of Mackey functors of finite type.

Of course, provided that it exists, the $\mathcal{K}_I^*$-completion of a spectrum X is unique up to canonical equivalence since it is specified by a universal property. We shall construct $\mathcal{K}_I^*$-completions by induction up Postnikov towers, and the following cohomological result will allow us to start the induction. It will be given a best possible generalization in Theorem 3.2, so we omit the short direct proof that was found by Hopkins.

THEOREM 1.4. *If M is a Mackey functor of finite type, then the map $\gamma : HM \to H(M_I^\wedge)$ is a $\mathcal{K}_I^*$-equivalence and therefore a $\mathcal{K}_I^*$-completion.*

The homology theory represented by the Burnside ring Mackey functor $\underline{A}$, $\underline{A}(G/H) = A(H)$, plays the same role equivariantly that integral homology plays nonequivariantly, and we have a Mackey functor version of this theory with $\underline{H}_n(X;\underline{A}) = \underline{\pi}_n(X \wedge H\underline{A})$. Since $H\underline{A}$ can be constructed by killing the higher homotopy groups of S, we see that there is a Hurewicz isomorphism $\underline{\pi}_n(X) \cong \underline{H}_n(X;\underline{A})$ when X is (n-1)-connected.

We can now state and prove the main results of this section.

PROPOSITION 1.5. *Let Z be bounded below. The following conditions on Z are equivalent, and they imply that Z is $\mathcal{K}_I^*$-complete.*

(i) *Each Mackey functor $\underline{\pi}_n(Z)$ is I-adic.*

(ii) *$[K, Z]_G$ is I-adic for all finite G-CW spectra K.*

If G is finite, the following condition is also equivalent to (i).

(iii) *Each Mackey functor $\underline{H}_n(Z;\underline{A})$ is I-adic.*

PROOF: By Lemma 1.2 and Proposition 1.3, (i) implies that Z is I-complete.

(i) if and only if (ii). Since I-adic completion is exact, (i) implies (ii) by induction on the number of cells of K; (ii) implies (i) by taking $K = (G/H)_+ \wedge S^n$ for any subgroup H and integer n.

(ii) if and only if (iii). Given (ii), Spanier-Whitehead duality implies that $\underline{\pi}_*(Z \wedge K)$ is I-adic for any finite G-spectrum K. The skeleta of $H\underline{A}$ are finite if G is finite, and (iii) follows. Conversely, let Z_n be the (n-1)-connected cover of Z and consider the fibration $Z_{n+1} \to Z_n \to K(\underline{\pi}_n(Z), n)$. Assume inductively that $\underline{H}_*(Z_n; \underline{A})$ is I-adic and that $\underline{\pi}_q(Z)$ is I-adic for $q \leq n$. By the implication (ii) implies (iii), $\underline{H}_*(K(\underline{\pi}_n(Z), n); \underline{A})$ is I-adic, and it follows that $\underline{H}_*(Z_{n+1}; \underline{A})$ is I-adic. By the Hurewicz theorem, $\underline{\pi}_{n+1}(Z) \cong \underline{\pi}_{n+1}(Z_{n+1})$ is isomorphic to $\underline{H}_{n+1}(Z_{n+1}; \underline{A})$ and is therefore I-adic.

A spectrum X is said to be of finite type if each $\pi_n(X^H)$ is finitely generated.

THEOREM 1.6. *Let X be bounded below and of finite type. Then the following conditions on a map $\gamma : X \to X_I^\wedge$ from X to a $\mathcal{K}_I^*$-complete spectrum are equivalent. Moreover, there exists one and, up to homotopy, only one such map γ.*

(i) *γ is a $\mathcal{K}_I^*$-equivalence; that is, γ is a $\mathcal{K}_I^*$-completion of X.*

(ii) *Each $\gamma_* : \underline{\pi}_n(X) \to \underline{\pi}_n(X_I^\wedge)$ is I-adic completion.*

iii) *$\gamma_* : [K, X]_G \to [K, X_I^\wedge]_G$ is I-adic completion for all finite K.*

If G is finite, the following condition is also equivalent to (i)–(iii).

iv) *$X_I^\wedge$ is bounded below and $\gamma_* : \underline{H}_*(X; \underline{A}) \to \underline{H}_*(X_I^\wedge; \underline{A})$ is I-adic completion.*

PROOF: Conditions (ii), (iii), and (iv) are equivalent by arguments exactly like those in the previous proof. We shall construct γ satisfying (i) through (iv). Since (i) implies the uniqueness of γ, it will follow that (i) implies (ii). We first give the promised construction and then prove that (ii) implies (i). We may replace X by a Postnikov tower $\lim X_n$, and we construct $(X_n)_I^\wedge$ inductively; X_n is the trivial G-spectrum for sufficiently small n, and we take $(X_n)_I^\wedge$ to be trivial for such n. Assume given a map $\gamma_n : X_n \to (X_n)_I^\wedge$ which satisfies (i) and (ii) and consider the following diagram:

$$\begin{array}{ccccccc}
K(\underline{\pi}_{n+1}(X), n+1) & \to & X_{n+1} & \to & X_n & \xrightarrow{k} & K(\underline{\pi}_{n+1}(X), n+2) \\
\gamma \downarrow & & \gamma_{n+1} \downarrow & & \gamma_n \downarrow & & \downarrow \gamma \\
K(\underline{\pi}_{n+1}(X)_I^\wedge, n+1) & \to & (X_{n+1})_I^\wedge & \to & (X_n)_I^\wedge & \xrightarrow{k_I^\wedge} & K(\underline{\pi}_{n+1}(X)_I^\wedge, n+2).
\end{array}$$

The top row is the fiber sequence induced by the k-invariant $k = k^{n+2}$. Since γ_n is an I-completion and $K(\underline{\pi}_{n+1}(X)_I^\wedge, n+2)$ is I-complete, there is a map $k_I^\wedge : (X_n)_I^\wedge \to K(\underline{\pi}_{n+1}(X)_I^\wedge, n+2)$, unique up to homotopy, such that the right square commutes up to homotopy. The bottom row is defined to be the fiber sequence induced by $k_I^\wedge$. A standard fiber sequence argument gives a map γ_{n+1} which makes the left square commute up to homotopy and the middle square commute on the nose. By the five lemma, γ_{n+1} is a $\mathcal{K}_I^*$-equivalence, and γ_{n+1} clearly satisfies (ii). By the previous proposition, $(X_{n+1})_I^\wedge$ is $\mathcal{K}_I^*$-complete. The commutativity of the middle squares for all n allows us to define $X_I^\wedge = \lim(X_n)_I^\wedge$ and obtain a well-defined inverse limit map $\gamma : X \to X_I^\wedge$. Clearly $H_G^*(X; M)$ is the colimit of the $H_G^*(X_n; M)$, the colimit being achieved in each degree, and similarly for $X_I^\wedge$. Therefore γ is a $\mathcal{K}_I^*$-equivalence, and γ clearly satisfies (ii). Again by the previous proposition, $X_I^\wedge$ is $\mathcal{K}_I^*$-complete. Returning to the remaining implication, assume (ii). By "cocellular approximation", we may assume that X and $X_I^\wedge$ are given as Postnikov towers and that γ is cocellular, that is, the inverse limit of a tower of compatible maps $X_n \to (X_I^\wedge)_n$. We then have diagrams like those of the construction and can deduce (i).

REMARK 1.7: Even for nonequivariant p-adic completion, the integral homological conditions in the above two results do not appear in the literature. It is critical to observe that these characterizations only work stably, because of our use of duality in their proofs. Bousfield and Kan [**5**, 5.7] point out that the integral homology of the (unstable) p-adic completion of an n-sphere, n odd, contains large $\mathbb{Q}$-modules in dimensions kn for $k \geq 2$.

§2. The general construction of completions at I.

Return again to the context of Definitions 0.1 and 0.2. We must specify the appropriate class $\mathcal{T}_I$ of test spectra.

DEFINITION 2.1. *Let I be an ideal in $A(G)$. A spectrum T is an I-torsion spectrum if, for each $\alpha \in I$, there exists a positive integer k such that $\alpha^k : T \to T$ is null homotopic. Observe that, if I is finitely generated, we can choose a k such that $\alpha^k : T \to T$ is null homotopic for all $\alpha \in I$. let $\mathcal{T}_I$ be the collection of all I-torsion spectra. Define the I-completion of X (or the completion of X at I), denoted $X_I^\wedge$, to be its $\mathcal{T}_I$-localization.*

When I is finitely generated, we shall prove shortly that $X_I^\wedge$ exists for any X. In general, an ideal I may be countably generated. We are confident that I-completions still exist but, as explained at the end of the section, we do not have a construction starting from the given definitions (which may not be quite the right ones for infinitely generated ideals). We assume that I is finitely generated except where otherwise specified.

For $\alpha \in A(G)$, let S/α be the cofiber of $\alpha : S \to S$, let $S[\alpha^{-1}]$ be the telescope of countably many instances of $\alpha : S \to S$, and let $M(\alpha)$ be the fiber of the evident inclusion $S \to S[\alpha^{-1}]$. For $r \geq 1$, there are maps of cofiber sequences

$$\begin{array}{ccccccc} S & \xrightarrow{\alpha^r} & S & \longrightarrow & S/\alpha^r & \longrightarrow & S^1 \\ \| & & \downarrow{\scriptstyle \alpha} & & \downarrow{\scriptstyle \overline{\alpha}} & & \| \\ S & \xrightarrow{\alpha^{r+1}} & S & \longrightarrow & S/\alpha^{r+1} & \longrightarrow & S^1. \end{array}$$

Passing to telescopes, we obtain the cofiber sequence

$$S \to S[\alpha^{-1}] \to S/\alpha^{\infty} \to S^1.$$

Thus $M(\alpha) \simeq \Sigma^{-1} S/\alpha^{\infty}$. For a finite set $\underline{\alpha} = \{\alpha_1, \ldots, \alpha_n\}$ of elements of $A(G)$, let $S/\underline{\alpha} = S/\alpha_1 \wedge \cdots \wedge S/\alpha_n$ and $M(\underline{\alpha}) = M(\alpha_1) \wedge \cdots \wedge M(\alpha_n)$. By Lemma 5.5 below, $M(\underline{\alpha})$ is equivalent to a G-CW spectrum whose cellular chain complex admits a convenient algebraic description.

The following result implies that I-completion, alias $\mathcal{T}_I$-localization, is the same thing as Bousfield localization at either $S/\underline{\alpha}$ or $M(\underline{\alpha})$.

THEOREM 2.2. *Let $\underline{\alpha} = \{\alpha_1, \ldots, \alpha_n\}$ be a finite set of generators for the ideal I. Then the following properties of a spectrum W are equivalent.*
(i) W is I-acyclic: $W \wedge T$ is contractible for all I-torsion spectra T.
(ii) W is $S/\underline{\alpha}$-acyclic: $W \wedge S/\underline{\alpha}$ is contractible.
(iii) W is $M(\underline{\alpha})$-acyclic: $W \wedge M(\underline{\alpha})$ is contractible.
That is, $\mathcal{T}_I$ is Bousfield equivalent to $\{S/\underline{\alpha}\}$ and $S/\underline{\alpha}$ is Bousfield equivalent to $M(\underline{\alpha})$.

PROOF: Let $\underline{\beta} = \{\alpha_1, \ldots, \alpha_{n-1}\}$, $J = (\alpha_1, \ldots, \alpha_{n-1}) \subset I$, and $\alpha = \alpha_n$. We proceed by induction on n.

Step 1: the case $n = 1$: Clearly $\alpha^2 : S/\alpha \to S/\alpha$ is null homotopic, hence so is $(b\alpha)^2$ for any $b \in A(G)$. Thus S/α is an I-torsion spectrum and (i) implies (ii). Assume (ii). By Verdier's axiom in the stable category, there are cofibrations $S/\alpha \to S/\alpha^r \to S/\alpha^{r-1}$. Inductively, $W \wedge S/\alpha^r$ is contractible for all r. Since telescopes commute with smash products, $W \wedge M(\alpha)$ is contractible. Now assume (iii). Then $S \wedge W \to S[\alpha^{-1}] \wedge W$ is an equivalence and therefore so is $\alpha : W \to W$. If T is an I-torsion spectrum, then $W \wedge T$ is contractible since $\alpha : W \wedge T \to W \wedge T$ is a nilpotent equivalence. Therefore (i)–(iii) are equivalent when $I = (\alpha)$.

Step 2: the inductive step. Assume that (i)–(iii) are equivalent for $\underline{\beta}$ and observe that $S/\underline{\alpha} = S/\underline{\beta} \wedge S/\alpha$ and $M(\underline{\alpha}) = M(\underline{\beta}) \wedge M(\alpha)$. If $\lambda = \sum b_i\alpha_i \in I$, then $\lambda^{2n} : S/\underline{\alpha} \to S/\underline{\alpha}$ is null homotopic. Thus $S/\underline{\alpha}$ is an I-torsion spectrum and (i) implies (ii). Assume (ii). Then $W \wedge S/\alpha \wedge M(\underline{\beta})$ is contractible by the induction hypothesis and $W \wedge M(\underline{\alpha})$ is contractible by Step 1. Finally, assume (iii). Since an I-torsion spectrum T is also a J-torsion spectrum, $W \wedge M(\alpha) \wedge T$ is contractible by the induction hypothesis. As in Step 1, $W \wedge T$ is then contractible since the map $\alpha : W \wedge T \to W \wedge T$ is a nilpotent equivalence.

We now change notation and write $M(\underline{\alpha}) = M(I)$. Up to equivalence, $M(I)$ is in fact independent of the choice of generating set $\underline{\alpha}$ (as would be false for $S/\underline{\alpha}$). Observe that there is a canonical map $M(I) \to S$ and thus a natural map $X = F(S, X) \to F(M(I), X)$. We have already noted (in §0) that its target is $M(I)$-local and thus I-complete.

THEOREM 2.3. *The natural map $X \to F(M(I), X)$ is an I-completion.*

PROOF: We use the notations of the previous theorem and its proof. By that result, it suffices to show that $X \to F(M(\underline{\alpha}), X)$ is an $S/\underline{\alpha}$-equivalence. We proceed by induction on n.

Step 1: the case $n = 1$. We must show that $X \wedge S/\alpha \to F(M(\alpha), X) \wedge S/\alpha$ is an equivalence or, equivalently, that $F(S[\alpha^{-1}], X) \wedge S/\alpha$ is contractible. Since S/α is finite and is the dual of S^{-1}/α, standard equivalences give

$$F(S[\alpha^{-1}], X) \wedge S/\alpha \simeq F(S[\alpha^{-1}], X \wedge S/\alpha) \simeq F(S[\alpha^{-1}] \wedge S^{-1}/\alpha, X).$$

Since $\alpha : S[\alpha^{-1}] \to S[\alpha^{-1}]$ is an equivalence and $\alpha^2 : S^{-1}/\alpha \to S^{-1}/\alpha$ is null homotopic, $S[\alpha^{-1}] \wedge S^{-1}/\alpha$ is contractible.

Step 2: the inductive step. The map $X \to F(M(\underline{\alpha}), X)$ can be written as the composite

$$X \to F(M(\underline{\beta}), X) \to F(M(\alpha), F(M(\underline{\beta}), X)) \simeq F(M(\underline{\alpha}), X).$$

Smashed with $S/\underline{\alpha} = S/\underline{\beta} \wedge S/\alpha$, the first map becomes an equivalence by the induction hypothesis and the second map becomes an equivalence by the case $n = 1$ applied with X replaced by $F(M(\underline{\beta}), X)$.

This explicit construction of I-completions has the following important immediate consequence.

COROLLARY 2.4. *For any spectra X and Y,*

$$F(X, Y_I^\wedge) = F(X, F(M(I), Y)) \cong F(M(I), F(X, Y)) = F(X, Y)_I^\wedge.$$

Therefore, if $\gamma : Y \to Y_I^\wedge$ is an I completion, then so is

$$\gamma_* : F(X, Y) \to F(X, Y_I^\wedge).$$

We must still connect up our two constructions of completions. The following parenthetical observation was suggested by the referee.

PROPOSITION 2.5. *Let $\mathcal{T}_I^!$ be the class of bounded below I-torsion spectra.*
(i) *$\mathcal{T}_I$ is Bousfield equivalent to $\mathcal{T}_I^!$.*
(ii) *$\mathcal{T}_I^!$ is cohomologically equivalent to $\mathcal{K}_I$.*

PROOF: Part (i) is immediate from Theorem 2.2. Part (ii) is implied by Lemma 1.2 and Proposition 1.3 since, if T is an I-torsion spectrum, then some power of I annihilates each $\underline{\pi}_n(T)$.

Although we do not know how to make mathematical use of this fact, it may lend plausibility to the following comparison of our two notions of completion at I. For clarity, we use the language of Definition 0.2.

THEOREM 2.6. *Let W, X, Y, and Z be spectra and let $f : X \to Y$ be a map.*
(i) *If W is $\mathcal{T}_I$-acyclic, then W is $\mathcal{K}_I^*$-acyclic.*
(ii) *If W is bounded below and $\mathcal{K}_I^*$-acyclic, then W is $\mathcal{T}_I$-acyclic.*
(iii) *If Z is $\mathcal{K}_I^*$-complete, then Z is $\mathcal{T}_I$-local.*
(iv) *If Z is bounded below and $\mathcal{T}_I$-local, then Z is $\mathcal{K}_I^*$-complete.*
(v) *If X is bounded below and $\gamma : X \to X_I^\wedge$ is a $\mathcal{T}_I$-localization, then γ is a $\mathcal{K}_I^*$-completion.*
(vi) *If X is bounded below and $\gamma : X \to X_I^\wedge$ is a $\mathcal{K}_I^*$-completion, then γ is a $\mathcal{T}_I$-localization.*

Obviously (i) implies (iii), and (i) and (iv) imply (v). Moreover, (vi) implies (ii) since, if W is $\mathcal{K}_I^*$-acyclic, then $W \to *$ is a $\mathcal{K}_I^*$-completion. Thus it suffices to prove (i), (iv), and (vi).

PROOF OF (i). Let $I = (J, \alpha)$. The result is trivial if $I = 0$ and we assume inductively that it holds for J. Assume that $IN = 0$ and thus $JN = 0$ and $\alpha N = 0$. Suppose that W is $\mathcal{T}_I$-acyclic and note that $W \wedge S^{-1}/\alpha$ is then

$\mathcal{T}_J$-acyclic. Since $S^{-1}/\alpha \simeq D(S/\alpha)$, we have isomorphisms

$$[W,(HN)/\alpha]^*_G \cong [W, HN\wedge S/\alpha]^*_G \cong [W\wedge S^{-1}/\alpha, HN]^*_G = H^*_G(W\wedge S^{-1}/\alpha; N)$$

These groups are zero because $W\wedge S^{-1}/\alpha \simeq *$ if $J = 0$, and by the induction hypothesis if $J \neq 0$. Since $\alpha N = 0$, the evident cofibration $HN \to HN \to (HN)/\alpha$ gives rise to short exact sequences

$$0 \to H^q_G(W;N) \to [W,(HN)/\alpha]^q_G \to H^{q+1}_G(W;N) \to 0.$$

Therefore $H^*_G(W;N) = 0$.

PROOF OF (iv). In the next section, we will give a new algebraic definition of an "I-complete Mackey functor". If Z is $\mathcal{T}_I$-local, the homotopy group Mackey functors $\underline{\pi}_*(Z)$ are I-complete by Theorems 3.3 and 3.5 and the spectra $H\underline{\pi}_*(Z)$ are therefore $\mathcal{K}^*_I$-complete by Theorem 3.1. If, further, Z is bounded below, then Z is $\mathcal{K}^*_I$-complete by Proposition 1.3.

PROOF OF (vi). By (v), the $\mathcal{T}_I$-localization $\gamma : HM \to (HM)^\wedge_I$ is a $\mathcal{K}^*_I$-completion for any M. If X is bounded below, then its $\mathcal{K}^*_I$-completion $\gamma : X \to X^\wedge_I$ can and, up to equivalence, must be constructed by induction up the Postnikov tower, exactly as in the proof of Theorem 1.6 but with $K(M^\wedge_I, n)$'s replaced by $K(M,N)^\wedge_I$'s. Since the functor $F(M(I), ?)$ preserves fibrations and limits, the $\mathcal{T}_I$-localizations $F(M(I), X_n)$ and $F(M(I), X)$ give a precise realization of this inductive construction of the $\mathcal{K}^*_I$-completion of X.

We complete this section by considering the behavior of completions with respect to inclusions of ideals. The following observations are easily verified from the definitions.

PROPOSITION 2.7. *Let $J \subset I$, where I is any ideal, not necessarily finitely generated. Then the following conclusions hold.*

(i) *If T is an I-torsion spectrum, then T is a J-torsion spectrum.*

(ii) *T is an I-torsion spectrum if and only if T is a J-torsion spectrum for all finitely generated ideals $J \subset I$.*

(iii) *If W is a J-acyclic spectrum, then W is an I-acyclic spectrum.*

(iv) *If $f : X \to Y$ is a J-equivalence, then f is an I-equivalence.*

(v) *If Z is I-complete, then Z is J-complete.*

(vi) *If it exists, the completion $X \to X_I^\wedge$ of X at I factors uniquely through the completion $X \to X_J^\wedge$ of X at J, and the resulting map $X_J^\wedge \to X_I^\wedge$ is the completion of $X_J^\wedge$ at I.*

(vii) *Let $Y = \operatorname{hocolim} X_J^\wedge$, where the homotopy colimit runs over the set of finitely generated subideals $J \subset I$. Then any map from X to an I-complete spectrum factors uniquely through the canonical map $X \to Y$, and $\underline{\pi}_*(Y) \cong \operatorname{colim} \underline{\pi}_*(X_J^\wedge)$.*

PROOF OF (vii): If Z is I-complete, then Z is J-complete for all J and, for $J \subset J'$, the map $[X_{J'}^\wedge, Z]_G \to [X_J^\wedge, Z]_G$ induced by $X_J^\wedge \to X_{J'}^\wedge$ is an isomorphism. Therefore the relevant $\lim^1$ term vanishes and the canonical map $[X_I^\wedge, Z]_G \to \lim [X_J^\wedge, Z]_G$ is an isomorphism. The rest follows.

Assuming that I-completions exist, we conclude that $X_I^\wedge \cong Y_I^\wedge$. Of course, Y itself would be $X_I^\wedge$ if it were I-complete, but presumably this fails.

§3. Statements of results about the homotopy groups of completions at I.

We wish to describe the homotopy groups $X_I^\wedge$ in terms of the homotopy groups of X. In general, the I-adic completion functor is neither left nor even right exact. Its left derived functors L_i^I are studied in [**12**]. They are constructed on an $A(G)$-module M simply by taking the homology of the complex obtained by applying I-adic completion to a free resolution of M. There is a natural epimorphism $\varepsilon : L_0^I M \to M_I^\wedge$, which is an isomorphism when $A(G)$ is Noetherian and M is finitely generated. In general, the natural map $\gamma : M \to M_I^\wedge$ factors as $\varepsilon \circ \eta$, $\eta : M \to L_0^I M$. Setting aside any preconceived algebraic ideas, let us say that a Mackey functor M is I-complete if the natural map $\eta : M \to L_0^I M$ is an isomorphism. The following definitive generalization of Lemma 1.2 indicates the relevance of this notion to cohomology. It will be proven in section 5.

THEOREM 3.1. *The following conditions on a Mackey functor M are equivalent.*

(i) *HM is $\mathcal{K}_I^*$-complete: if W is $\mathcal{K}_I^*$-acyclic, then $H_G^*(W; M) = 0$.*

(ii) *HM is I-complete: if W is $\mathcal{T}_I$-acyclic, then $[W, HM]_G = 0$.*

(iii) *M is I-complete: $\eta : M \to L_0^I M$ is an isomorphism.*

Similarly, the following result is the definitive generalization of Theorem 1.4 (compare Theorem 3.11 below). Parts (ii) and (iii) are immediate consequences of part (i) and Theorem 2.6.

THEOREM 3.2. *Let M be a Mackey functor.*

(i) *The I-completion of HM is a two-stage Postnikov system with*

$$\underline{\pi}_0((HM)_I^\wedge) = L_0^I(M) \quad \text{and} \quad \underline{\pi}_1((HM)_I^\wedge) = L_1^I(M).$$

(ii) *If $L_I^1 M = 0$, then $\gamma : HM \to (HM)_I^\wedge = H(L_0^I M)$ is a $\mathcal{K}_I^*$-equivalence.*

(iii) *If $L_I^1 M \neq 0$, then there does not exist a $\mathcal{K}_I^*$-equivalence $HM \to HM'$ where HM' is $\mathcal{K}_I^*$-complete.*

Theorem 3.2(i) will also be proven in section 5, where it will be shown to imply the first part of the following theorem. The second part follows from the first by Corollary 2.4. Let Y_G^* denote the cohomology theory represented by a spectrum Y.

THEOREM 3.3. *Let X and Y be any spectra.*

(i) *There are natural short exact sequences*

$$0 \to L_1^I(\underline{\pi}_{q-1}(X)) \to \underline{\pi}_q(X_I^\wedge) \to L_0^I(\underline{\pi}_q(X)) \to 0.$$

(ii) *There are natural short exact sequences*

$$0 \to L_1^I(Y_G^{q+1}(X)) \to (Y_I^\wedge)_G^q(X) \to L_0^I(Y_G^q(X)) \to 0.$$

As a matter of algebra, the $L_i^I(M)$ admit the following descriptions. We abbreviate $A = A(G)$ in the rest of this section.

THEOREM 3.4. *Let M be either an A-module or a Mackey functor.*

(i) *There is a natural short exact sequence*

$$0 \to \lim{}^1 \operatorname{Tor}_1^A(A/I^r, M) \to L_0^I(M) \to M_I^\wedge \to 0.$$

(ii) *There is a natural short exact sequence*

$$0 \to \lim{}^1 \operatorname{Tor}_2^A(A/I^r, M) \to L_1^I(M) \to \lim \operatorname{Tor}_1^A(A/I^r, M) \to 0;$$

if I is a principal ideal, then the $\lim^1$ *kernel term is zero.*

(iii) $L_i^I(M) = 0$ *for* $i \geq 2$.

Except for (iii), this is proven in [**12**, 1.1 and 1.5]. The notion of an I-complete Mackey functor is clarified by the following result, which is proven in [**12**, 4.1].

THEOREM 3.5. *For any Mackey functor M, the Mackey functors $M_I^{\wedge}$, $L_0^I M$, and $L_1^I M$ are I-complete. If M is I-complete, then $L_1^I M = 0$.*

So far we have stated our results in a language that should be reasonably familiar to topologists. However, Grothendieck's local cohomology groups $H_I^n(M)$ and certain analogous local homology groups $H_n^I(M)$ of A-modules M play an essential intermediate role in the passage from the topology to the algebra. We shall give the relevant definitions in section 5. As the remarks after the next theorem make clear, local cohomology occurred implicitly in the Bousfield-Kan treatment [**5**] of p-adic completion. The following result is proven in [**12**, 2.5 and 3.5].

THEOREM 3.6. *There are natural isomorphisms $L_i^I(M) \cong H_i^I(M)$, and*
(i) *there is a natural short exact sequence*

$$0 \to \mathrm{Ext}_A^1(H_I^1(A), M) \to L_0^I(M) \to \mathrm{Hom}_A(H_I^0(A), M) \to \mathrm{Ext}_A^2(H_I^1(A), M) \to 0;$$

(ii) *there is a natural isomorphism*

$$L_1^I(M) \cong \mathrm{Hom}_A(H_I^1(A), M).$$

REMARK 3.7: Working nonequivariantly, with $A = Z$, suppose that $I = (p)$ for a prime p. Then $H_I^0(A) = 0$ and $H_I^1(A) = Z/p^\infty$. Therefore

$$L_0^{(p)}(M) = \mathrm{Ext}(Z/p^\infty, M) \quad \text{and} \quad L_1^{(p)}(M) = \mathrm{Hom}(Z/p^\infty, M),$$

as was first observed by Bousfield and Kan [**5**, VI.2.1]. Since $\mathrm{Tor}(Z/qZ, M)$ is the kernel $\Gamma(q; M)$ of $q : M \to M$, Theorem 3.4 gives a short exact sequence

$$0 \to \lim{}^1 \Gamma(p^r; M) \to L_0^{(p)}(M) \to M_I^{\wedge} \to 0$$

and reinterprets $L_1^{(p)}(M)$ as $\lim \Gamma(p^r; M)$. Therefore, $L_0^{(p)}(M) = M_p^{\wedge}$ and $L_1^{(p)}(M) = 0$ if the p-power torsion of M is of bounded order.

There is an analog of the last statement in our general context.

DEFINITION 3.8. *For an A-module or Mackey functor M and an element $\alpha \in A$, let $\Gamma(\alpha; M)$ be the kernel of $\alpha : M \to M$. Observe that $\Gamma(\alpha^r; M)$ is contained in $\Gamma(\alpha^{r+1}; M)$ and say that M has bounded α-torsion if this* *ascending chain stabilizes. For A itself,* $\Gamma(\alpha; A) = \Gamma(\alpha^2; M)$ *since A is a subring of a product of copies of Z. Say that a sequence* $\underline{\alpha} = \{\alpha_1, \dots, \alpha_n\}$ *of elements of A is pro-regular for M if* $M/(\alpha_1, \dots, \alpha_{i-1})^r M$ *has bounded* α_i*-torsion for* $r \geq 0$ *and* $1 \leq i \leq n$.

Clearly any $\underline{\alpha}$ is pro-regular for M if A is Noetherian and M is finitely generated. The following result, which will be proven in section 6, implies that any $\underline{\alpha}$ is pro-regular for M when M is A itself and therefore when M is any free A-module.

THEOREM 3.9. *For any finitely generated ideal J and any element* $\alpha \in A$, *the A-module* A/J *has bounded* α*-torsion.*

Given Theorem 3.9, the following result is a special case of [**12**, 1.9]. Actually, its first conclusion holds under a somewhat weaker notion of a "pro-regular sequence" than that specified in Definition 3.8; see [**12**, 1.8]

THEOREM 3.10. *Let* $I = (\alpha_1, \dots, \alpha_n)$ *and write* $J = (\alpha_1, \dots, \alpha_{n-1})$ *and* $\alpha = \alpha_n$. *If* $\underline{\alpha}$ *is a pro-regular sequence for M, then* $L_0^I(M) \cong M_I^\wedge$ *and* $L_i^I(M) = 0$ *for* $i > 0$. *Moreover, the following conclusions hold for any M.*

(i) $L_0^I(M) \cong L_0^\alpha(L_0^J(M))$.

(ii) *There is a short exact sequence*

$$0 \to L_0^\alpha(L_1^J(M)) \to L_1^I(M) \to L_1^\alpha(L_0^J(M)) \to 0.$$

Theorems 3.3 and 3.10 have the following implication.

THEOREM 3.11. *If* $\underline{\alpha}$ *is a pro-regular sequence for each* $\underline{\pi}_q(X)$, *then* $\gamma_* : \underline{\pi}_*(X) \to \underline{\pi}_*(X_I^\wedge)$ *is completion at I. In particular, if* $\underline{\alpha}$ *is a pro-regular sequence for M, then* $(HM)_I^\wedge = H(M_I^\wedge)$.

§4. The Segal conjecture and the Atiyah-Segal completion theorem.

Let G be finite in this section. Recall that a family $\mathcal{F}$ in G is a collection of subgroups of G which is closed under subconjugacy. A G-space X is an $\mathcal{F}$-space if X^H is empty for $H \notin \mathcal{F}$. There is a universal $\mathcal{F}$-space $E\mathcal{F}$ characterized by the property that $(E\mathcal{F})^H$ is non-empty and contractible for $H \in \mathcal{F}$. A family determines an ideal $I\mathcal{F}$ in $A(G)$, namely the intersection over $H \in \mathcal{F}$ of the kernels I_H^G of the restriction homomorphisms $A(G) \to A(H)$.

The cohomology groups of a spectrum X with coefficients in a spectrum Y are the homotopy groups of $F(X,Y)^G$. In particular, the stable cohomotopy groups of X are the homotopy groups of $D(X)^G$. If X is a finite G-CW spectrum and Y is a G-CW spectrum with finite skeleta, then $F(X,Y)$ is bounded below and of finite type, so that the elementary cohomological construction of completions of Theorem 1.6 applies and we can recognize completions by their behavior on homotopy groups. In any case, it is natural to ask if the $I\mathcal{F}$-completion of $F(X,Y)$ is equivalent to $F(E\mathcal{F}_+ \wedge X, Y)$. The latter kind of geometric completion has long played an important role in the study of equivariant cohomology theories; see for example [**8**, Ch.7; **17**, Ch.V].

Our definitions imply that, for any ideal $I \subset A(G)$ and any $H \subset G$, $M(I)$ regarded as an H-spectrum is $M(r^G_H(I)A(H))$, where r^G_H is the restriction $A(G) \to A(H)$; moreover, $M(0) \simeq S$. For a family $\mathcal{F}$, the canonical map $M(I\mathcal{F}) \to S$ is an $\mathcal{F}$-equivalence, in the sense that it is an H-equivalence for all $H \in \mathcal{F}$. Therefore, by the $\mathcal{F}$-Whitehead theorem [**17**, II.2.2], there is a unique map $\xi : \Sigma^\infty E\mathcal{F}_+ \to M(I\mathcal{F})$ over S. Thus, for any spectra X and Y, we have a commutative diagram of canonical maps

$$\begin{array}{ccccc} & & F(M(I\mathcal{F}) \wedge X, Y) & \cong & F(X, F(M(I\mathcal{F}), Y)) \\ & \overset{\gamma}{\nearrow} & \big\downarrow \xi^* & & \big\downarrow \xi^* \\ F(X,Y) & \underset{\pi^*}{\searrow} & & & \\ & & F(E\mathcal{F}_+ \wedge X, Y) & \cong & F(X, F(E\mathcal{F}_+, Y)). \end{array}$$

Via Corollary 22.4, this implies a striking reinterpretation of completion theorems in terms of the cohomological behavior of the map ξ; via Theorem 3.3(ii), it also implies a calculational generalization of such theorems.

THEOREM 4.1. *The map* $\xi^* : F(M(I\mathcal{F}),Y) \to F(E\mathcal{F}_+,Y)$ *is an equivalence if and only if the map* $\pi^* : Y \to F(E\mathcal{F}_+,Y)$ *is an* $I\mathcal{F}$*-completion. When this holds,*

$$\pi^* : F(X,Y) \to F(E\mathcal{F}_+ \wedge X, Y)$$

is an $I\mathcal{F}$*-completion for any* X*, and there are short exact sequences*

$$0 \to L_1^{I\mathcal{F}}(Y_G^{q+1}(X)) \to Y_G^q(E\mathcal{F}_+ \wedge X) \to L_0^{I\mathcal{F}}(Y_G^q(X)) \to 0.$$

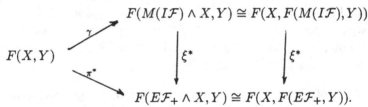

In particular, if Y has finite skeleta and $\pi^* : Y \to F(E\mathcal{F}_+,Y)$ induces I-adic completion on homotopy groups or satisfies one of the other equivalent conditions of Theorem 1.6, then we can deduce the conclusion of the theorem for arbitrary X. This allows us to generalize all of the existing completion theorems to the corresponding represented theories without doing any further work. For example, the generalized Segal conjecture proved in [**3**, 1.6] asserts that $S \to D(E\mathcal{F}_+)$ is an $I\mathcal{F}$-completion for any family $\mathcal{F}$. The arguments of [**10**] [**11**] interpret the conclusion in terms of ordinary p-adic completion.

There is no general criterion for determining whether or not the conclusion of Theorem 4.1 holds for a given spectrum Y. It fails if Y represents connective equivariant K-theory, and it also fails for the suspension spectra of general G-spaces. The following results, which are immediate consequences of [**23**, 1.1] and [**7**, 3.1], respectively, give the most general positive conclusions presently available.

THEOREM 4.2. *Let $G = \Gamma/\Pi$, where Γ is a compact Lie group and Π is a normal subgroup, let $E(\Pi,\Gamma)$ be the universal Π-free Γ-space, and let $Y = E(\Pi,\Gamma)_+ \wedge_\Pi Z$ for a finite Γ-CW spectrum Z. Then $\pi^* : Y \to F(E\mathcal{F}_+, Y)$ is an $I\mathcal{F}$-completion for any family $\mathcal{F}$.*

THEOREM 4.3. *Let Y be the suspension G-spectrum of the function G-space $F(K,L)$, where K is a G-space of finite type and L is a finite G-CW complex such that, for $H \subset G$, L^H is a simple space and $\pi_i(L^H) = 0$ if $i \le \dim(K^H)$. Then $\pi^* : Y \to F(EG_+, Y)$ is an IG-completion, where IG is the augmentation ideal of $A(G)$.*

Both of these results generalize the original Segal conjecture, which is the case $Y = S$ and $I = IG$. In this case, we have

$$\pi^q_G(EG_+ \wedge X) \cong \pi^q(EG_+ \wedge_G X),$$

by [**17**, II.8.4], and we can now compute the nonequivariant stable cohomotopy groups of the Borel construction $EG_+ \wedge_G X$ from the equivariant stable cohomotopy groups of X for any X.

COROLLARY 4.4. *For any spectrum X, there are short exact sequences*

$$0 \to L_1^{IG}(\pi_G^{q+1}(X)) \to \pi^q(EG_+ \wedge_G X) \to L_0^{IG}(\pi_G^q(X)) \to 0.$$

By the generalization of the Atiyah-Segal completion theorem given in [**2**] together with the following algebraic result, which will be proven at the end of section 6, Theorem 4.1 also applies when Y represents real or complex periodic equivariant K-theory. Recall that passage from finite G-sets to their permutation representations induces a ring homomorphism $\rho : A(G) \to R(G)$ which factors through the real representation ring $RO(G)$. Regard $R(G)$-modules and $RO(G)$-modules as $A(G)$-modules by pullback along ρ, and remember that G is finite here.

THEOREM 4.5. *Let $J\mathcal{F} \subset R(G)$ be the intersection over $H \in \mathcal{F}$ of the kernels J^G_H of the restrictions $R(G) \to R(H)$. Then, for any $R(G)$-module M, the $A(G)$-modules $M^\wedge_{J\mathcal{F}}$ and $M^\wedge_{I\mathcal{F}}$ are naturally isomorphic. The analogous result holds for $RO(G)$.*

Just as for cohomotopy, we have natural isomorphisms

$$K^*_G(EG_+ \wedge X) \cong K^*(EG_+ \wedge_G X) \text{ and } KO^*_G(EG_+ \wedge X) \cong KO^*(EG_+ \wedge_G X),$$

by [**17**, II.8.4], and we can now compute the nonequivariant K-groups of $EG_+ \wedge_G X$ from the equivariant K-groups of X.

COROLLARY 4.6. *For any spectrum X, there are short exact sequences*

$$0 \to L^{IG}_1(K^{q+1}_G(X)) \to K^q(EG_+ \wedge_G X) \to L^{IG}_0(K^q_G(X)) \to 0,$$

and similarly for real K-theory.

As a bizarre application, we have the following purely algebraic conclusion about completions of representation rings of finite groups.

COROLLARY 4.7. *Let Π be a normal subgroup of a finite group Γ with quotient group G and let $I(\Pi;\Gamma) = I\mathcal{F}$, where $\mathcal{F}$ is the family of subgroups Λ of Γ such that $\Lambda \cap \Pi = e$. Then*

$$R(\Gamma)^\wedge_{I\Gamma} \cong L^{IG}_0 R(\Gamma)^\wedge_{I(\Pi;\Gamma)} ,$$

where $A(G)$ acts on $R(\Gamma)^\wedge_{I(\Pi;\Gamma)}$ through $A(G) \to R(G) \to R(\Gamma)$. The same conclusion holds for real representation rings.

PROOF: Let $X = B(\Pi;\Gamma) = E(\Pi;\Gamma)/\Pi$, where $E(\Pi;\Gamma)$ is the universal Π-free Γ-space. Then $EG \times E(\Pi;\Gamma)$ is a free contractible Γ-space, so that $B\Gamma \simeq (EG \times E(\Pi;\Gamma))/\Gamma = EG \times_G X$. By [**2**, 2.1 and 2.2], $K^1_G(X) = 0$ and $K^0_G(X) \cong R(\Gamma)^\wedge_{I(\Pi;\Gamma)}$. The previous corollary gives the conclusion.

The reader may object that it is unnatural to appeal to the Burnside ring when studying K-theory. Ideally, we should instead study completions of K-module spectra at ideals of $R(G)$. By doing so, we could expect to generalize the K-theory case of Theorem 4.1 to arbitrary compact Lie groups. We believe that this can be done, but the arguments would entail a discussion of highly structured module spectra over highly structured ring spectra, which would take us far afield.

As a final observation, the following parenthetical remark on duality generalizes [**21**], to which we refer for more discussion. Note that, since completions are given by Bousfield localizations, $(X \wedge Y)^{\wedge}_I \simeq (X^{\wedge}_I \wedge Y)^{\wedge}_I$.

REMARK 4.8: Let $\nu : D(X) \wedge Y \to F(X,Y)$ be the canonical duality map. If X is finite and Y satisfies the conclusion of Theorem 4.1, then both horizontal arrows and the left vertical arrow, hence also the right vertical arrow, are equivalences in the commutative diagram

$$\begin{array}{ccc} (DX \wedge Y)^{\wedge}_{I\mathcal{F}} & \longrightarrow & (D(E\mathcal{F}_+ \wedge X) \wedge Y)^{\wedge}_{I\mathcal{F}} \\ \nu\downarrow & & \nu\downarrow \\ F(X,Y)^{\wedge}_{I\mathcal{F}} & \longrightarrow & F(E\mathcal{F}_+ \wedge X, Y). \end{array}$$

Thus, up to completion, duality holds for the infinite spectrum $E\mathcal{F}_+ \wedge X$ in the theory represented by Y. The intuitive explanation is that we can replace $E\mathcal{F}_+$ by the finite dimensional spectrum $M(I\mathcal{F})$.

§5. Algebraic definitions and topological proofs.

We begin by summarizing some basic definitions from [**12**]. For the moment, A can be any commutative ring.

Define the cofiber Ck of a map $k : X \to Y$ of chain complexes by $(Ck)_i = Y_i \oplus X_{i-1}$, with differential $d_i(y,x) = (d_i(y) + k_{i-1}(x), -d_{i-1}(x))$. Define the suspension ΣX by $(\Sigma X)_i = X_{i-1}$, with differential $-d$. We have a short exact sequence $0 \to Y \to Ck \to \Sigma X \to 0$, and the connecting homomorphism of the derived long exact sequence in homology is k_*. Given a sequence of chain maps $f^r : X^r \to X^{r+1}$, $r \geq 0$, define a map $\iota : \oplus X^r \to \oplus X^r$ by $\iota(x) = x - f^r(x)$ for $x \in X^r$. Define the homotopy colimit, or telescope, of the sequence $\{f^r\}$ to be $C\iota$ and denote it $\mathrm{Tel}(X^r)$. Then $H_i(\mathrm{Tel}(X^r)) = \mathrm{Colim}\, H_i(X^r)$. The composite of the projection from $C\iota$ to its second variable and the canonical map $\oplus X^r \to \mathrm{Colim}(X^r)$ is a homology isomorphism $\mathrm{Tel}(X^r) \to \mathrm{Colim}(X^r)$.

DEFINITIONS 5.1. *For $\alpha \in A$, let $K_*(\alpha)$ denote the map $\alpha : A \to A$ regarded as a chain complex $d_0 : K_0(\alpha) \to K_{-1}(\alpha)$. The identity map in degree 0 and multiplication by α in degree -1 specify a chain map $K_*(\alpha^r) \to K_*(\alpha^{r+1})$. Define $K_*(\alpha^\infty) = \operatorname{Colim} K_*(\alpha^r)$, so that $K_*(\alpha^\infty)$ is the cochain complex $A \to A[1/\alpha]$. For a sequence $\underline{\alpha} = \{\alpha_1, \dots, \alpha_n\}$, let $K_*(\underline{\alpha}) = K_*(\alpha_1) \otimes \cdots \otimes K_*(\alpha_n)$. Taking tensor products, we obtain a chain map $K_*(\underline{\alpha}^r) \to K_*(\underline{\alpha}^{r+1})$. Define $K_*(\underline{\alpha}^\infty) = \operatorname{Colim} K_*(\underline{\alpha}^r)$. Observe that the homology isomorphism $\operatorname{Tel} K_*(\underline{\alpha}^r) \to K_*(\underline{\alpha}^\infty)$ gives a projective approximation of the flat cochain complex $K_*(\underline{\alpha}^\infty)$. For any of these complexes $K_*(?)$, let $K^{\cdot}(?)$ denote the complex regraded as a cochain complex, $K^i(?) = K_{-i}(?)$.*

DEFINITIONS 5.2. *Let M be an A-module and define the local cohomology groups of M at $I = (\alpha_1, \dots, \alpha_n)$ to be*

$$H_I^*(M) \cong H^*(K^{\cdot}(\underline{\alpha}^\infty) \otimes M) \cong H^*(\operatorname{Tel} K^{\cdot}(\underline{\alpha}^r) \otimes M).$$

Define the local homology groups of M at I by

$$H_*^I(M) \cong H_*(\operatorname{Hom}(\operatorname{Tel} K^{\cdot}(\underline{\alpha}^r), M)).$$

Note that both $H_I^*(M)$ and $H_*^I(M)$ are covariant functors of M. It is not obvious from the definitions just given that these functors depend only on I and not on the choice of its generators. We refer the reader to [**15**] and [**12**, §2] for discussion of this point.

Now let $A = A(G)$. We must prove Theorems 3.1, 3.2(i), and 3.3(i). We begin with the last two. By Theorem 3.6 and the algebraic fact that $H_q^I(M) = 0$ for $q \geq 2$, as will be proven in the next section, these results can be restated as follows in terms of local homology.

THEOREM 5.3. *For any Mackey functor M, $(HM)_I^\wedge$ is (-1)-connected and*

$$\underline{\pi}_q((HM)_I^\wedge) = H_q^I(M) \text{ for } q \geq 0.$$

THEOREM 5.4. *For any spectrum X, there is a natural short exact sequence*

$$0 \to H_1^I(\underline{\pi}_{q-1}(X)) \to \underline{\pi}_q(X_I^\wedge) \to H_0^I(\underline{\pi}_q(x)) \to 0.$$

The proof of Theorem 5.3 depends on a chain level understanding of the cohomology theory represented by HM, as set out briefly in [**16**] where these $RO(G)$ gradable cohomology theories were introduced. Details on G-CW spectra are given in [**17**, II§5].

If X is a G-CW spectrum, we have a cellular chain complex $\underline{C}_*(X)$ in the Abelian category of Mackey functors with $\underline{C}_q(X) = \underline{\pi}_q(X^q/X^{q-1})$. For a Mackey functor M, we obtain a cochain complex of Abelian groups by taking homomorphisms of Mackey functors:

$$C^*(X;M) = \mathrm{Hom}(\underline{C}_*(X), M).$$

Its cohomology groups $H^q_G(X;M)$ agree with the cohomology groups represented by HM, namely

$$[X, HM]^q_G = \pi^G_{-q}F(X, HM).$$

Let us say that a G-CW spectrum X is special if all of its cells have trivial orbit type, so that X^q/X^{q-1} is a wedge of copies of S^q. We then write $C_*(X) = \underline{C}_*(X)(G/G)$. This is a chain complex of $A(G)$-modules, and we have an isomorphism

$$(*) \qquad C^*(X;M) = \mathrm{Hom}(\underline{C}_*(X), M) \cong \mathrm{Hom}_{A(G)}(C_*(X), M(G/G))$$

since a map of Mackey functors with represented domain functor $[?, S]_G$ is determined by the image of the identity map of S.

The cellular theory of special G-CW spectra is exactly like the cellular theory of nonequivariant spectra, as set out in [**17**, VIII§2]. The chain complex $C_*(X \wedge Y)$ of a smash product of special G-CW spectra is isomorphic to $C_*(X) \otimes C_*(Y)$. If $f : X \to Y$ is a cellular map, then the chain complex $C_*(Cf)$ of its cofiber is isomorphic to the algebraic cofiber $C(C_*(f))$ of its map of cellular chains. If $f_r : X_r \to X_{r+1}$ is a sequence of cellular maps, we may take the telescope of the X_r to be the cofiber of the map $\bigvee X_r \to \bigvee X_r$ whose restriction to X_r is $1 - f_r$. With this definition, the chain complex $C_*(\mathrm{Tel}\, X_r)$ is isomorphic to the algebraic telescope $\mathrm{Tel}(C_*(X_r))$. Moreover, if we regard a special G-CW spectrum as an H-spectrum for $H \subset G$, then it inherits a structure of special H-CW spectrum with the same cells.

Returning to the matter at hand, recall that $X^\wedge_I = F(M(\underline{\alpha}), X)$. We have the following application of the discussion just given.

LEMMA 5.5. *The spectrum* $M(\underline{\alpha})$ *is equivalent to a special* G*-CW spectrum whose cellular chain complex is isomorphic to* $\mathrm{Tel}\, K_*(\underline{\alpha}^r)$.

PROOF: Up to equivalence, telescopes commute with smash products, hence $M(\underline{\alpha})$ is equivalent to the telescope of the spectra $S^{-1}/\alpha_1^r \wedge \cdots \wedge S^{-1}/\alpha_n^r$. This telescope is clearly a special G-CW spectrum. In view of the remarks above and the description of $K_*(\underline{\alpha}^r)$ given in Definitions 5.1, its cellular chain complex is isomorphic to $\mathrm{Tel}\, K_*(\underline{\alpha}^r)$.

PROOF OF THEOREM 5.3: We have

$$\pi_q^G(HM)_I^\wedge = \pi_q^G F(M(\underline{\alpha}), HM) = H_G^{-q}(M(\underline{\alpha}); M).$$

The regrading in the last equality matches that in the last sentence of Definitions 5.1, and $(*)$ together with the previous lemma imply that this group is $H_q^I(M(G/G))$. The argument passes to subgroups H of G via the chain of isomorphisms

$$\pi_q^H(HM)_I^\wedge = \pi_q^H(HN)_J^\wedge \cong H_q^J(N(H/H)) = H_q^I(M(G/H)).$$

Here $J = r(I)A(H)$, $r : A(G) \to A(H)$, and N is the Mackey functor $M|\, H$, so that $N(H/K) = M(G/K)$. The first equality holds since, when regarded as H-spectra, $M(\underline{\alpha})$ and HM are $M(r(\underline{\alpha}))$ and HN. The last equality holds since the chain complexes from which the two groups are computed are the same.

PROOF OF THEOREM 5.4: The proof is essentially the same as the folklore argument for the special case of nonequivariant p-adic completion. Like any other Bousfield localization, I-completion preserves cofiber sequences. If X is bounded below, it has a Postnikov tower $\{X_i\}$ with cofiber sequences $K(\underline{\pi}_i(X), i) \to X_i \to X_{i-1}$. By Theorem 5.3 and inductive use of the long exact homotopy sequences of the completions of these cofiber sequences, Theorem 5.4 holds for each X_i and thus for X. If X is bounded above, say $\underline{\pi}_r(X) = 0$ for $r > m$, then $X_I^\wedge$ is also bounded above. In fact, $\pi_r^H(X_I^\wedge) \cong [G/H_+ \wedge S^r \wedge M(\underline{\alpha}), X]_G$, and it follows from the dimensions of the cells occurring in the CW decomposition of $M(\underline{\alpha})$ that $\underline{\pi}_r(X_I^\wedge) = 0$ for $r > m+n$, where n is the number of generators of I. To determine $\underline{\pi}_q(X_I^\wedge)$ for a general X, we form the cofibration

$$X[q-n, \infty) \to X \to X(-\infty, q-n),$$

where the second arrow is obtained by killing the homotopy groups $\underline{\pi}_r(X)$ for $r \geq q-n$. When we pass to completions, the first arrow still induces an isomorphism on $\underline{\pi}_q$, and the calculation of $\underline{\pi}_q(X)$ follows.

PROOF OF THEOREM 3.1: This can be proven by a detailed study of the algebraic structure of I-complete Mackey functors, starting from Theorem 3.4(i), but it is easier to use topology. Statements (ii) and (iii) are equivalent by Theorems 3.2(i) and 3.5, which imply that each says that the identity map $HM \to HM$ is an I-completion. Theorem 2.6(iii) gives that (i) implies (ii). It remains to prove that (ii) implies (i). Thus assume that HM is I-complete and W is $\mathcal{K}_I^*$-acyclic. We must show that $H_G^*(W;M) = 0$. Since HM is equivalent to $F(M(I), HM)$ and $M(I)$ is the telescope of a sequence of finite I-torsion spectra T, HM is a homotopy inverse limit of spectra $F(T, HM) \simeq HM \wedge D(T)$. Since $D(T)$ is a finite I-torsion spectrum, the homotopy group systems of $HM \wedge D(T)$ are I-torsion. This implies that $[W, HM \wedge D(T)]_G^* = 0$, by Lemma 1.2 and Proposition 1.3, and therefore $H_G^*(W;M) = 0$ by the $\lim^1$ exact sequence.

§6. Algebraic proofs.

Let $A = A(G)$. We proved Theorems 3.4, 3.5, 3.6, and 3.10 in [**12**], modulo Theorem 3.9 and the assertion that $L_i^I(M) = 0$ for $i \geq 2$. We prove these facts here, starting with the latter. Since $L_i^I(M) \cong H_i^I(M)$, it suffices to show that $H_i^I(M) = 0$ for $i \geq 2$. By [**12**, **3.1**], there is a spectral sequence which converges to $H_*^I(M)$ and has E_2-term $\mathrm{Ext}^*(H_I^*(A), M)$. It is immediate from the form of the spectral sequence that the following result implies the desired vanishing.

THEOREM 6.1. *Let $I = (\alpha_1, \ldots, \alpha_n)$ be a finitely generated ideal in the Burnside ring $A = A(G)$. Then $H_I^i(A) = 0$ for $i \geq 2$.*

PROOF: Let ΦG be the set of conjugacy classes of those subgroups H of G which have finite index in their normalizers. Additively, A is the free abelian group on generators $[G/H]$ with $(H) \in \Phi G$. Let NH be the normalizer of H in G and $WH = NH/H$. The set of orders of the groups WH has a least common multiple w. Give ΦG the Hausdorff metric. Then ΦG is a totally disconnected compact Hausdorff space. Let C be the ring of continuous functions $\Phi G \to \mathbb{Z}$. There is an embedding $\varphi : A \to C$ and we let $Q = C/A$. Then $wQ = 0$. See [8] or [**17**, Ch.V] for the proofs of the given statements. We have a long exact sequence

$$\cdots \to H_I^{i-1}(Q) \to H_I^i(A) \to H_I^i(C) \to H_I^i(Q) \to \cdots.$$

It suffices to prove that $H_I^i(C) = 0$ for $i \geq 2$ and $H_I^i(Q) = 0$ for $i \geq 1$.

LEMMA 6.2. *Let D be the product of finitely many copies of $\mathbb{Z}$. If $i \geq 2$, then $H_I^i(M) = 0$ for any ideal I in D and any D-module M.*

PROOF: If $D = \mathbb{Z}$, then I is principal and the conclusion is immediate from the definitions. The constructions all commute with finite direct products, and the conclusion follows.

LEMMA 6.3. *Under the hypotheses of Theorem 6.1, $H_I^i(C) = 0$ for $i \geq 2$.*

PROOF: Let $J = IC \subset C$. Inspection of definitions shows that the groups $H_I^i(C)$ and $H_J^i(C)$ are isomorphic since they are computable from isomorphic cochain complexes. In C, the $\varphi(\alpha_i)$ take finitely many values. We can choose finitely many disjoint open and closed subsets Γ_k of ΦG with characteristic functions δ_k, $\delta_k(H) = 1$ for $(H) \in \Gamma_k$ and $\delta_k(H) = 0$ for $(H) \notin \Gamma_k$, such that each $\varphi(\alpha_i)$ is a linear combination of the δ_k. If D is the subring of C generated by the δ_k, then D is isomorphic to a finite product of copies of $\mathbb{Z}$. If K is the ideal in D generated by the $\varphi(\alpha_i)$, then $J = KC$ and the groups $H_J^i(C)$ and $H_K^i(C)$ are isomorphic since they are computable from isomorphic cochain complexes. Now the previous lemma gives the conclusion.

LEMMA 6.4. *Under the hypotheses of Theorem 6.1, if R is any sub A-module of Q, then $H_I^i(R) = 0$ for $i \geq 1$.*

PROOF: Observe that $H_I^0(R) = \{r \mid I^k \cdot r = 0 \text{ for some } k\} \subset R$. Write $I = (J, \alpha)$, where $J = (\alpha_1, \dots, \alpha_{n-1})$ and $\alpha = \alpha_n$. The result is trivial when $I = 0$, and we assume it for J. If $\underline{\beta} = \{\alpha_1, \dots, \alpha_{n-1}\}$, then $K^\cdot(\underline{\alpha}^\infty) = K^\cdot(\underline{\beta}^\infty) \otimes K^\cdot(\alpha^\infty)$ and a standard double complex argument gives a spectral sequence which converges from $E_2^{p,q} = H_\alpha^p(H_J^q(R))$ to $H_I^{p+q}(R)$. We have $E_2^{p,q} = 0$ for $q \geq 1$ by the induction hypothesis, and it is obvious that $E_2^{p,q} = 0$ for $p \geq 2$. Since $H_J^0(R) \subset R$, it only remains to prove that $H_\alpha^1(R) = 0$ for $R \subset Q$. Since $H_\alpha^1(R)$ is the cokernel of the natural map $\iota : R \to R[1/\alpha]$, it suffices to prove that ι is an epimorphism. The kernel of ι is the group $H_\alpha^0(R)$ of α-power torsion elements in R. Let $\overline{R} = R/H_\alpha^0(R)$. It suffices to prove that $\alpha : \overline{R} \to \overline{R}$ is an isomorphism, and it is clearly a monomorphism. We shall prove that, for any $r \in R$, the subgroup $\langle \alpha, r \rangle$ of R generated by $\{\alpha^k r \mid k \geq 0\}$ is finite. The same will then hold for the image $\langle \alpha, \overline{r} \rangle$ of this group in $\overline{R}$. Therefore $\alpha : \langle \alpha, \overline{r} \rangle \to \langle \alpha, \overline{r} \rangle$ will be a bijection and the conclusion will follow. Let $\rho \in C$ map to $r \in Q$. We can choose finitely many open and closed subsets Γ_j of ΦG such that $\varphi(\alpha)$ and ρ are both linear combinations of the characteristic functions δ_j of the Γ_j.

Then each $\alpha^k \rho = \varphi(\alpha)^k \rho$ is also a linear combination of the δ_j. If d_j is the image of δ_j in Q, then $wd_j = 0$. Thus the d_j generate a finite subgroup of Q which contains $\langle \alpha, r \rangle$.

We must still prove Theorem 3.9, which can be restated as follows.

THEOREM 6.5. *Let $J = (\beta_1, \ldots, \beta_q)$ be a finitely generated ideal in the Burnside ring $A = A(G)$ and let α be any element of A. There is an $n > 0$ such that if $\gamma \in A$ and $\alpha^s \gamma \in J$ for some $s \geq n$, then $\alpha^n \gamma \in J$. That is, α^n annihilates any element of A/J that α^s annihilates.*

We shall derive this from the following result.

THEOREM 6.6. *Let B' be a finitely generated subring of A that contains the β_i among its generators. Then there is a Noetherian ring B such that $B' \subset B \subset A$ and $P^n \subset Q$ for each P-primary ideal Q in a well chosen reduced primary decomposition of the ideal K of B generated by $\{\beta_1, \ldots, \beta_q\}$, where $n > 0$ is an integer depending only on the β_i.*

PROOF OF THEOREM 6.5, ASSUMING THEOREM 6.6: Assume that $\alpha^s \gamma \in J$. Write $\alpha^s \gamma = \sum a_i \beta_i$ and let B' be generated by α, γ, the β_i and the a_i. Let B and K be as in Theorem 6.6. Then $\alpha^s \gamma \in K$, hence $\alpha^s \gamma$ is in each primary ideal Q of any reduced primary decomposition of K. Either $\gamma \in Q$ or $\alpha \in P$, where Q is P-primary. In the latter case, Theorem 6.6 gives that $\alpha^n \in Q$ if our reduced primary decomposition is well chosen. Then $\alpha^n \gamma \in Q$, hence $\alpha^n \gamma \in K$ in B and thus $\alpha^n \gamma \in J$ in A.

We can specify n as follows. Let ΦG, C, φ, and w be as in the proof of Theorem 6.1 and let ν be the maximum over all primes p of the exponents $\nu_p(w)$. Each $\varphi(\beta_i)$ takes finitely many values $m_{i,r}$. For each prime p and each sequence $m_R = \{m_{1,r_1}, \ldots, m_{q,r_q}\}$, let $n(p, R)$ be maximal such that $p^{n(p,R)}$ divides each m_{i,r_i}. Then let n' be the maximum over p and R of the $n(p, R)$. Define $n = n' + \nu$.

PROOF OF THEOREM 6.6: Let $\{\xi_j\} \supset \{\beta_j\}$ be a finite set of generators for B'. Each $\varphi(\xi_j)$ takes finitely many values. For each sequence $S = \{s_j\}$ of such values, the subset Γ_S of ΦG consisting of those (H) such that $\varphi(\xi_j)(H) = s_j$ for each j is open and closed. Let $\Pi = \{\Gamma_S\}$. Then Π is a finite partition of ΦG into a disjoint union of open and closed subsets. Let $D \subset C$ be the ring of those continuous functions on ΦG which are constant on each $\Gamma \in \Pi$. Then D is the product of copies of $\mathbb{Z}$ indexed on the $\Gamma \in \Pi$. Let $B = A \cap D$ in C. Since $wC \subset A$, $wD \subset B$. Moreover, for any ideal L

of B, $wLD \subset L(wD) \subset LB = L$, so that w annihilates LD/L. Clearly D is Noetherian, and it follows that B is Noetherian.

The analysis of prime ideals and localizations in B is precisely parallel to the analysis in A that is given in [8] and [17, V§§3,5]. Of course, $\mathrm{Spec}(D) = \Pi \times \mathrm{Spec}(\mathbb{Z})$; the prime ideals of D are of the form

$$\overline{q}(\Gamma, p) = \{d \mid d(\Gamma) \equiv 0 \mod p\},$$

where $\Gamma \in \Pi$ and p is a prime or zero. The prime ideals of B are of the form $q(\Gamma, p) = \overline{q}(\Gamma, p) \cap B$. Here $q(\Gamma, 0) = q(\Gamma', 0)$ if and only if $\Gamma = \Gamma'$, and $B \to B/q(\Gamma, 0) \otimes \mathbb{Q}$ is the localization of B at $q(\Gamma, 0)$. Let $\Pi(\Gamma, p)$ be the set of Γ' such that $q(\Gamma', p) = q(\Gamma, p)$ and let $I(\Gamma, p)$ be the intersection over $\Gamma' \in \Pi(\Gamma, p)$ of the prime ideals $q(\Gamma', 0)$. Then $B \to B/I(\Gamma, p) \otimes \mathbb{Z}_{(p)}$ is the localization of B at $q(\Gamma, p)$. Moreover, the localization of D at $q(\Gamma, p)$ is the product of copies of $\mathbb{Z}_{(p)}$ indexed on the $\Gamma' \in \Pi(\Gamma, p)$, and the map $B_{q(\Gamma,p)} \to D_{q(\Gamma,p)}$ is a monomorphism.

Let K be the ideal of B generated by $\{\beta_1, \ldots, \beta_q\}$. Let $K = \bigcap Q_i$ be a reduced primary decomposition, let Q be any of the Q_i, and let $P = q(\Gamma, p)$ be the radical of Q. We want to show that $P^n \subset Q$. Let $\lambda : B \to B_P$ be the localization of B at P and let M be the maximal ideal of B_P. Recall (e.g. [26, pp.223–228]) that Q is P-primary if and only if $Q = \lambda^{-1}(R)$ where R is M-primary and that R is M-primary if and only if $M \supset R \supset M^t$ for some t. If $p = 0$, then P is the only P-primary ideal and there is nothing to prove. Thus assume that p is a non-zero prime. With $Q = \lambda^{-1}(R)$, let $R' = R + M^n$ (with n defined as above) and $Q' = \lambda^{-1}(R')$. We claim that if we replace Q by Q', then we still have a reduced primary decomposition of K. Since $P^n \subset Q'$, this claim will immediately imply Theorem 6.6.

To prove the claim, let L be the intersection of the prime ideals $q(\Gamma', 0)_P$ in B_P, where Γ' runs over those elements of $\Pi(\Gamma, p)$ such that $K \subset q(\Gamma', 0)$. Obviously these $q(\Gamma', 0)$ are minimal prime ideals of K; since they are themselves the only $q(\Gamma', 0)$-primary ideals, they must appear in any reduced primary decomposition of K [26, p.211]. Thus it suffices to show that $R \cap L = R' \cap L$. This means that $R' \cap L \subset R$, and this will hold if $M^n \cap (R + L) \subset R$. Regarded as a B-module via φ, the localization D_P of D is the sum of a copy of $Z_{(p)}$ for each $\Gamma' \in \Pi(\Gamma, p)$. Its submodule $K_P D_P$ is the sum of (0) for $\Gamma' \in \Pi(\Gamma, p)$ such that $K \subset q(\Gamma', 0)$ and $(p^{m(\Gamma')})$ for $\Gamma' \in \Pi(\Gamma, p)$ such that $K \not\subset q(\Gamma', 0)$, where $m(\Gamma')$ is maximal such that $p^{m(\Gamma')}$ divides each $\beta_i(\Gamma')$. By the definition of n', $n' \geq m(\Gamma')$ for each Γ'.

Let RD_P be the sum of $(p^{r(\Gamma')})$. Since $K_P \subset R$, $m(\Gamma') \geq r(\Gamma')$. Clearly LD_P is the sum of (0) for Γ' such that $K \subset q(\Gamma',0)$ and $Z_{(p)}$ for Γ' such that $K \not\subset q(\Gamma',0)$. It follows that $(R+L)D_P$ is the sum of $(p^{r(\Gamma')})$ for Γ' such that $K \subset q(\Gamma',0)$ and $Z_{(p)}$ for Γ' such that $K \not\subset q(\Gamma',0)$. Therefore

$$M^{n'}D_P \cap (R+L)D_P \subset RD_P.$$

Also, $M^{\nu_p(w)} \subset (p^{\nu_p(w)})$ and $p^{\nu_p(w)}D_P \subset B_P$. Since $n \geq n' + \nu_p(w)$,

$$\begin{aligned} M^n \cap (R+L) &\subset (M^{\nu_p(w)}M^{n'})D_P \cap (R+L)D_P \\ &\subset (p^{\nu_p(w)}M^{n'})D_P \cap (R+L)D_P \\ &= p^{\nu_p(w)}(M^{n'}D_P \cap (R+L)D_P) \\ &\subset p^{\nu_p(w)}RD_P \subset RB_P = R. \end{aligned}$$

The middle equality follows from the inequalities $n' \geq r(\Gamma')$ and a check of submodules in the summands $\mathbb{Z}_{(p)}$ of D_P.

We left the following unfinished piece of business in section 4.

PROOF OF THEOREM 4.5: Remember that G is finite in this result. Clearly $M/(I\mathcal{F}\cdot M)^r = M/(I'\mathcal{F}\cdot M)^r$, where $I'\mathcal{F}$ is the ideal of $R(G)$ generated by $\rho(I\mathcal{F})$, and $I'\mathcal{F} \subset J\mathcal{F}$ since ρ commutes with restriction. It suffices to show that some power of $J\mathcal{F}$ is contained in $I'\mathcal{F}$ or, equivalently, that $J\mathcal{F}$ is nilpotent in $R(G)/I'\mathcal{F}$. This will hold if any prime ideal P of $R(G)$ which contains $I'\mathcal{F}$ also contains $J\mathcal{F}$. Clearly $Q = \rho^*P$ contains $I\mathcal{F}$, hence Q contains I^G_H for some $H \in \mathcal{F}$. It suffices to prove that P contains J^G_H, and this will certainly hold if the support C of P is contained in H (up to conjugacy). Recall that the support of P is a subgroup C which is minimal such that P comes from a prime ideal of $R(C)$. Since the support of Q is contained in H (see e.g. [**3**, p.17]), it suffices to prove that C is also the support of Q.

Let $\mathcal{H}$ and $\mathcal{E}$ denote the sets of conjugacy classes of subgroups of G and of elements of G, respectively. Passage from elements g to cyclic groups $\langle g\rangle$ gives a function $\mathcal{E} \to \mathcal{H}$. Let $\mathbb{Z}^{\mathcal{H}}$ and $\mathbb{C}^{\mathcal{E}}$ be the rings of functions $\mathcal{H} \to \mathbb{Z}$ and $\mathcal{E} \to \mathbb{C}$, and let $\pi : \mathbb{Z}^{\mathcal{H}} \to \mathbb{C}^{\mathcal{E}}$ be the evident map of rings. Then the following diagram commutes, where $\varphi(S)(H)$ is the cardinality of S^H for a finite G-set S and $\chi(\alpha)(g)$ is the trace of $\alpha(g)$ for a representation α:

$$\begin{array}{ccc} A(G) & \xrightarrow{\rho} & R(G) \\ {\scriptstyle\varphi}\downarrow & & {\scriptstyle\chi}\downarrow \\ Z^{\mathcal{H}} & \xrightarrow{\pi} & C^{\mathcal{E}}. \end{array}$$

Incidentally, this implies that ρ is injective if and only if G is cyclic. According to Segal [**25**], the support C is a cyclic group $\langle g\rangle$. If the prime P is minimal, then P consists of those α such that $\chi(\alpha)(g) = 0$, and the diagram implies that $Q = q(C, 0)$, whose support is also C. If P is maximal (the only other case), then P has residual characteristic a prime p, and Segal shows [**25**, p.122] that p does not divide $|C|$. Since ρ commutes with the restrictions $A(G) \to A(C)$ and $R(G) \to R(C)$, it is clear that the support of Q is a subgroup D of C, so that $Q = q(D,p)$. Since P contains the minimal prime with support C, Q contains $q(C,0)$ and therefore Q also equals $q(C,p)$. This implies that C/D is a p-group and thus that $C = D \subset H$.

Appendix: localizations of G-spectra

Let C be a multiplicatively closed subset of $A(G)$. Since we are interested in localizations, we may as well assume that $0 \notin C$ and that C is saturated, in the sense that $xy \in C$ implies $x \in C$ and $y \in C$. Equivalently, C is the complement of the (set theoretical) union of a nonempty set of prime ideals. An $A(G)$-module N is C-local if $\gamma : N \to N$ is an isomorphism for all $\gamma \in C$. A map $\lambda : M \to C^{-1}M$ is the C-localization of an $A(G)$-module M if $C^{-1}M$ is C-local and any $A(G)$-map from M to another C-local $A(G)$-module N factors uniquely through $C^{-1}M$. The kernel of λ is the set of elements annihilated by some $\gamma \in C$. When $C = A(G) - P$, it is usual to write $C^{-1}M = M_P$ and to call it the localization of M at P.

We show how to mimic this algebraic construction on spectra.

DEFINITIONS A.1. *A spectrum Y is said to be C-local if $\gamma : Y \to Y$ is an equivalence for all $\gamma \in C$. A map $\lambda : X \to C^{-1}X$ is said to be a C-localization if $C^{-1}X$ is C-local and if any map from X to a C-local spectrum Y factors uniquely (up to homotopy) through $C^{-1}X$.*

Say that T is a C-torsion spectrum if $\gamma : T \to T$ is null homotopic for some $\gamma \in C$.

LEMMA A.2. *The following conditions on a spectrum Y are equivalent.*

(i) *Y is C-local.*

(ii) *$Y \wedge T$ is contractible for all C-torsion spectra T.*

(iii) *$Y \wedge T$ is contractible for all finite C-torsion spectra T.*

(iv) *Each Mackey functor $\underline{\pi}_n(Y)$ is C-local.*

PROOF: If Y is C-local and $\gamma : T \to T$ is null homotopic, then the map $\gamma : Y \wedge T \to Y \wedge T$ is a null homotopic equivalence and $Y \wedge T$ is contractible. If $Y \wedge T$ is contractible for all finite C-torsion spectra T, then $\gamma : Y \to Y$ is an equivalence for $\gamma \in C$ since $Y \wedge S/\gamma$ is contractible. This shows the equivalence of (i)–(iii). Since $\gamma_* : \underline{\pi}_*(X) \to \underline{\pi}_*(X)$ is multiplication by γ, (i) and (iv) are equivalent by the Whitehead theorem in the stable category.

We can choose a sequence $\{\gamma_i\}$ of elements of C such that, for any $\gamma \in C$, there exist $i \geq 1$ and $\delta \in C$ such that $\gamma_1 \cdots \gamma_i = \gamma\delta$.

PROPOSITION A.3. *Let $C^{-1}X$ be the telescope of the sequence of maps $\gamma_i : X \to X$ and let $\lambda : X \to C^{-1}X$ be the canonical map. Then λ is a C-localization of X. Moreover, $\lambda_* : \underline{\pi}_*(X) \to \underline{\pi}_*(C^{-1}X)$ is C-localization, and λ is characterized by this property.*

PROOF: For $\gamma \in C$, $\gamma : C^{-1}X \to C^{-1}X$ is the telescope of the maps $\gamma : X \to X$ on the terms of the telescope. Its cofiber is the telescope of the sequence of maps $\gamma_i : X/\gamma \to X/\gamma$. Applying the cofinality assumed of the sequence $\{\gamma_i\}$ to the powers of γ, we see that γ divides a cofinal sequence of the γ_i. Since $\gamma^2 : X/\gamma \to X/\gamma$ is null homotopic, the cofiber is contractible and $\gamma : C^{-1}X \to C^{-1}X$ is an equivalence. Thus $C^{-1}X$ is C-local. If Y is C-local, the maps $(\gamma_i)^* : [X,Y]_G \to [X,Y]_G$ are isomorphisms for any X. Applying the $\lim^1$ exact sequence, we find immediately that $\lambda^* : [C^{-1}X, Y]_G \to [X,Y]_G$ is an isomorphism. The first clause of the last statement holds since our construction mimics a standard algebraic construction of C-localization, and the second clause follows from the universal property and the Whitehead theorem.

Observe that $C^{-1}X$ is naturally equivalent to $X \wedge C^{-1}S$ and that $C^{-1}(X \wedge Y)$ is naturally equivalent to $C^{-1}X \wedge C^{-1}Y$.

PROPOSITION A.4. *Y is C-local if and only if it is $C^{-1}S$-local, and C-localization coincides with Bousfield localization at $C^{-1}S$.*

PROOF: Write $E = C^{-1}S$ and let F be the fiber of the canonical map $S \to E$. Observe that F is the telescope of the sequence $S^{-1}/\gamma_1 \cdots \gamma_n$. If Y is C-local, then $[W \wedge S^{-1}/\gamma, Y]^G_* = [W, Y \wedge S/\gamma]^G_* = 0$ for all $\gamma \in C$ since

S^{-1}/γ is dual to S/γ and $Y \wedge S/\gamma$ is contractible. If W is E-acyclic, then $W \simeq W \wedge F$ and $[W, Y]^G_* = 0$ by the $\lim^1$ exact sequence. Conversely, let Y be E-local. If T is a finite C-torsion spectrum then so is $D(T)$, hence $X \wedge D(T)$ is E-acyclic for any X. Thus $[X, Y \wedge T]^G_* = [X \wedge D(T), Y]^G_* = 0$ and $Y \wedge T$ is contractible. It is clear from the construction that the C-localization $X \to C^{-1}X$ is an E-equivalence.

BIBLIOGRAPHY

1 J. F. Adams. Stable homotopy and generalized homology. Chicago Lecture Notes in Mathematics. Chicago University Press, 1974.

2 J. F. Adams, J.-P. Haeberly, S. Jackowski, and J. P. May. A generalization of the Atiyah-Segal completion theorem. Topology 27(1988), 1–6.

3 J. F. Adams, J.-P. Haeberly, S. Jackowski, and J. P. May. A generalization of the Segal conjecture. Topology 27(1988), 7–21.

4 A. K. Bousfield. The localization of spectra with respect to homology. Topology 18(1979), 257–281.

5 A.K. Bousfield and D. M. Kan. Homotopy limits, completions, and localizations. Springer Lecture Notes in Mathematics, Vol. 304, 1972.

6 G.E. Bredon. Equivariant cohomology theories. Springer Lecture Notes in Mathematics, Vol. 34, 1967.

7 G. Carlsson. On the homotopy fixed point problem for free loop spaces and other function complexes. K-theory 4(1991), 339–361.

8 T. tom Dieck. Transformation groups and representation theory. Springer Lecture Notes in Mathematics, Vol. 766. 1979.

9 A. Dress. Contributions to the theory of induced representations. Springer Lecture Notes in mathematics, Vol. 342, 1973. 183–240.

10 J. P. C. Greenlees. Equivariant functional duals and universal spaces. Journal London Math. Soc. 40 (1989), 347–354.

11 J. P. C. Greenlees. Equivariant functional duals and completions. Bull. London Math. Soc. To appear.

12 J. P. C. Greenlees and J. P. May. Derived functors of I-adic completion and local homology. Journal of Algebra. To appear.

13 J.P.C. Greenlees and J. P. May. Generalized Tate, Borel, and coBorel cohomology. To appear.

14 J. P. C. Greenlees, M. J. Hopkins, and J. P. May. Completions of G-

spectra at ideals of the Burnside ring, I. (Preprint, 1990.)

15 A. Grothendieck (notes by R. Hartshorne). Local Cohomology. Lecture Notes in Math. Vol 41. Springer-Verlag. 1967.

16 L. G. Lewis, J. P. May, and J. E. McClure. Ordinary $RO(G)$-graded cohomology. Bull. Amer. Math. Soc. 4(1981), 208–212.

17 L. G. Lewis, J. P. May, and M. Steinberger. Equivariant stable homotopy theory. Springer Lecture Notes in Mathematics. Vol. 1213, 1986.

18 J. P. May. The dual Whitehead theorems. London Math. Soc. Lecture Note Series vol. 86, 1983, 46–54.

19 J. P. May. Equivariant completion. Bull. London Math. Soc. 14(1982), 231–237.

20 J. P. May, J. E. McClure, and G. Triantafillou. Equivariant localization. Bull. London Math. Soc. 14(1982), 223–230.

21 J. P. May. A remark on duality and the Segal conjecture. Springer Lecture Notes in Mathematics. Vol 1217, 1986, pp. 303–305.

22 J. P. May. Memorial address for J. Frank Adams and Reminiscences on the life and mathematics of J. Frank Adams. The Mathematical Intelligencer 12 #1 (1990), 40–48.

23 J. P. May, V. P. Snaith, and P. Zelewski. A further generalization of the Segal conjecture. Quart. J. Math. Oxford (2) 40(1989), 457–473.

24 N. Minami. On the $I(G)$-adic topology of the Burnside ring of compact Lie groups. Publ. RIMS Kyoto Univ. 20(1984), 447–460.

25 G. B. Segal. The representation ring of a compact Lie group. Inst. des Hautes Etudes Sci. Publ. Math. 34(1968), 113–128.

26 O. Zariski and P. Samuel. Commutative Algebra, Vol 1. D. Van Nostrand Co. 1958.

Theorems of Poisson, Euler, and Bernoulli on the Adams spectral sequence

Martin C. Tangora

University of Illinois at Chicago

Introduction

Let Ext denote the E_2 term of the classical Adams spectral sequence for the stable sphere mod 2 [A1]. Then Ext is a bi-graded algebra over the field of 2 elements. Write $e(s,t)$ for the rank of this vector space in bi-grading (s,t).

It would be very helpful if we could find a method to compute these ranks, with substantially less work than it now requires to obtain Ext itself, or Ext with much of its algebra structure. Perhaps this is not a well-posed problem; we have certainly not "solved" it. What we do have to present at this time is some interesting observations about these numbers.

First, the numbers $e(s,t)$ are not randomly distributed in the plane (or in any part of the plane). We call this "Poisson's theorem."

Next, the alternating sums of the $e(s,t)$ can easily be computed, in the form of a simple generating function. We call this "Euler's Theorem."

If these alternating sums are defined in the right way, they are all 0's and 1's, and the gaps between the 1's can be described in terms of the Bernoulli numbers. This gives us "Bernoulli's theorem."

Topologists will already have noticed that our theorems are not really about the Adams spectral sequence, but rather about its E_2 term.

Our title may seem facetious, but it signals to the reader that our results have a classical (and elementary) flavor, and that this note is meant to be enjoyed.

1. Testing the Poisson hypothesis

Using an old calculation of Ext in bi-gradings $t-s \leq 70$ [T], we can display the numbers $e(s,t)$ in that range; see Figure 1.

As usual, we plot s on the vertical axis and $t-s$ on the horizontal axis; we leave a blank where the rank is zero, and note that non-zero ranks occur only in the first quadrant. Recall that $e(s,s)=1$ for every $s \geq 0$, *i.e.*, the upper half of the s-axis is all 1's; these are suppressed from Figure 1, but it will be important to take them into account later on.

What strikes the eye immediately — though we are unaware that anyone has ever made mention of this fact before — is that a great many of the non-zero entries are 1's. We claim, in fact, that there are *unnaturally many ones.* To make this precise, we imagine that the numbers in the range of Figure 1, *i.e.*, $t-s \leq 70$, are summed up — the sum is 782 — and then test the hypothesis that the numbers are distributed at random.

```
 0    v    1    v    2    v    3    v    4    v    5    v    6    v    7
35                                                                  1
34                                                                 11  1
33                                                                1 1  1
32                                                              1 1    1
31                                                          1   111
30                                                         11  1111  1
29                                                        1 1  12 11 11
28                                                      1 1    11 11 11
27                                                  1   111    11  1  1
26                                                 11  1111  1 12  1  1
25                                                1 1  12 11 11121 11 11
24                                              1 1    11 11 11121 11 21
23                                          1   111     1  1  11111 1 12
22                                         11  1111  1  1  1   1111  111
21                                        1 1  12 11 11 11 11  12  1  21
20                                      1 1    11 11 11 11 11 1111  1 12
19                                  1   111    11  1  1 11  1  21 1  112
18                                 11  1111  1 12  1  1  11  1 12  1  21
17                                1 1  12 11 11121 11 111  1  1112  1  2
16                              1 1    11 11 11121 11 21 1  1 2211  11 1
15                          1   111     1  1  11111 1 12  1  1 221111111
14                         11  1111  1  1  1   1111  1111  1 111121 21 3
13                        1 1  12 11 11 1111    2  1  21 2  11121 2  21
12                      1 1    11 11 11 11 11 1  1  1 12111 12121 2212 1
11                  1   111    11  1  1  11 1  1  1  112111 12223 22222
10                 11  1111  1 12  1  1  11  1  1  1  12 21 111322121211
 9                1 1  12 11 11121 11 111  1  1  2  1 11 12  211312 2121
 8              1 1    11 11 11121 11 21 1 1   121  11 1  21 11323111221
 7          1   111     1  1  11111 1 211 11  1112  1 1 1 11 12132221212
 6         11  1111  1  1  1   111 1 112 2 1 111111111 1   1  2221112121
 5        1 1  12 11 11 11     12 1 1 11121  12 111 111        3111 2112
 4      1 1    11 12 1 11      11111   2121  11  1 1          111111 112
 3  1   111    11 111 1        12 11  111 1   1                12 21  11
 2 11  111     111 1           111 1   1                       111 1   1
 11 1   1       1               1                               1
 0    v    1    v    2    v    3    v    4    v    5    v    6    v    7
```

$(t-s)$

Figure 1

$e(s,t)$ = rank of Ext at bi-grading (s,t)

For the convenience of homotopy theorists, $e(s,t)$ is plotted with $t-s$ and s on the horizontal and vertical axes, respectively. Thus lines of constant t are downward diagonals.

On the horizontal axis n means $t-s=10n$ and the v's mark odd multiples of 5.

The simplest kind of randomness test that comes to mind is perhaps to compare our numbers with numbers from a Poisson distribution. Suppose 782 darts had been thrown at the plane — or rather at the part of the first quadrant where the non-zero numbers occur — so that the average number of darts per "box" (or bi-grading) was that which we observe, namely, 782 / 1287 = 0.6076146, approximately. The number 1287 is roughly $70^2/4$, but it takes exact account of the "Adams edge" above which all entries are 0.

The well-known Poisson distribution predicts that the resulting numbers will look like the first column in the table below, but the observed numbers are those in the second column.

	Poisson frequency	Observed frequency
0	701	622
1	426	556
2	129	101
3	26	8
4	4	0
≥5	0 or 1	0

This is clearly a very bad fit. In the observed numbers there are simply too many ones. (We leave the reader to perform a goodness-of-fit test if he has any doubts.)

Of course if we included the 1's in column $t - s = 0$ it would only make matters worse, and similarly if we made a more precise use of the Adams vanishing and periodicity theorems, so as to more carefully exclude the 0's established by those theorems.

If one reflects on the definition of Ext as the cohomology of the Steenrod algebra, one may or may not believe that these excess ones are "natural." In fact the sudden emergence of large numbers of 2's and 3's for $t - s \geq 60$ muddies the waters. In any case, the discussion of Euler characteristics that follows will suggest that a "flat" distribution is required, or in other words, a more uniform distribution than the Poisson.

2. The Euler characteristic

If we cannot easily predict the numbers $e(s,t)$, the next best thing would be to state some conditions that those numbers must satisfy. Such conditions are given by Euler characteristics.

In fact Ext is obtainable in various ways as the homology of differential graded algebras. The differential or boundary operator in such a complex preserves the Euler characteristic of the complex.

Consider the May spectral sequence [M] which has as its E_∞ term an algebra with the same vector-space ranks as Ext. Each differential in the spectral sequence

$$d_r : E_r^{s,t} \to E_r^{s+1,t}$$

preserves the internal degree t, and hence preserves the Euler characteristics

$$\chi_r(t) = \sum_s (-1)^s e_r(s,t)$$

where by $e_r(s,t)$ we mean the E_r-analogue of $e(s,t)$, *i.e.*, the vector-space rank of E_r at (s,t). The convergence is such that for any given t we have $e(s,t) = e_\infty(s,t) = e_r(s,t)$ for some finite stage r.

This would reduce the problem to enumerating some initial stage of the May spectral sequence. Recall that the E_2 term is the homology of a polynomial algebra with generators R^i_j in bi-gradings $s = 1, t = 2^i(2^j - 1)$ for all $i \geq 0$ and all $j \geq 1$.

Fact: The Euler characteristics $\chi(t) = \sum_s (-1)^s e(s,t)$ are as follows:

$$(1, -1, 0, -1, 1, 0, 0, -1, 1, 0, 1, -1, 0, 0, 0, -1, 1, \ldots)$$

These numbers were obtained experimentally, and then the following generating function was derived:

Theorem: The Euler characteristics $\chi(t) = \sum_s (-1)^s e(s,t)$ are the coefficients of $u^t v^t$ in

$$(1+u)^{-1} \prod (1 - uv^{t_k})^{-1}$$

where the t_k run through the numbers $2^i(2^j - 1)$.

That's what I called Euler's theorem. It is more interesting to replace the above Euler characteristic by

$$E(t) = (-1)^t \chi(t) = \sum_s (-1)^{t-s} e(s,t)$$

to obtain what I called Bernoulli's theorem:

Theorem: The modified Euler characteristics $E(t)$ have the following properties:

(1) $E(t)$ is always 0 or 1;

(2) The sequence $(E(t))$ has the form

$$vz(b_1)vz(b_2)vz(b_3)\ldots$$

where $v = 11011$ and $z(b_i)$ denotes a string of b_i 0's;

(3) $b_i = \nu(4i) =$ the "length" of the image of J in the $(4i-1)$–stem, and hence can be expressed in terms of the Bernoulli numbers.

Here "length" means the difference in s between top and bottom elements representing Im J in the Adams spectral sequence [A2].

Thus the $E(t)$ sequence, for $t \geq 0$, runs

11011 00 11011 000 11011 00 11011 0000
11011 00 11011 000 11011 00 11011 00000 . . . .

There is a certain amount of deliberate mystification in the above. Part of it goes back to Frank Adams, who introduced the Bernoulli numbers into homotopy theory, where a more elementary description would have served.

(Mark Mahowald tells me that he had this discussion with Frank Adams many years ago.) Part of it, however, is here because at the time of the Adams Memorial Symposium I had not yet seen through the above theorems with complete clarity. The right way to see it is the following.

Theorem: Let A be a bi-graded polynomial algebra over the field of two elements, with generators $\{g_k\}$ in bi-gradings $(1, d_k)$, with each d_k a positive integer. Let $e(s,t)$ denote the vector-space rank of A at bi-grading (s,t). Let $E(t) = \sum_s (-1)^{t-s} e(s,t)$. Then these Euler characteristics have the ordinary generating function

$$X(w) = \prod_k (1 + (-w)^{d_k})^{-1}.$$

Corollary: In the particular case where $A = \text{Ext} =$ the cohomology of the mod 2 Steenrod algebra, the generating function becomes

$$\begin{aligned} X(w) &= \prod_n (1 + w^{2^n - 1}) \\ &= (1+w)(1+w^3)(1+w^7)(1+w^{15})\ldots \end{aligned}$$

This is one of those situations where once you have the right formulation, the proof is easy. However, I should tip my hat to Herb Wilf for showing me how easy it really is.

The proof of the last theorem above (which explains all the previous theorems in this section) is just a direct calculation with the generating functions: the bi-variate generating function

$$f(u,v) = \sum \sum e(s,t) u^s v^t = \prod (1 - u v^{d_k})^{-1}$$

becomes, on putting $u = -1$ and $v = -w$,

$$\begin{aligned} f(-1,-w) &= \sum \sum e(s,t)(-1)^s(-w)^t \\ &= \sum_t \sum_s (-1)^{t-s} e(s,t) w^t \\ &= \prod (1 + (-w)^{d_k})^{-1} \end{aligned}$$

and that's the theorem.

The corollary follows from the observation that, for generators in bi-gradings

$$\{(1,d), (1,2d), (1,4d), (1,8d), \ldots\}$$

where d is odd, the product reduces, in Eulerian fashion:

$$\prod_{k=0}^{\infty} \frac{1}{1+(-w)^{2^k d}} = 1 + w^d$$

The theorem is really a theorem about partitions of integers: Given a family $\{d_i\}$ of odd integers, consider partitions of n by parts of the form $2^k d_i$. We have observed that *the number of such partitions of n in which the number of parts has the same parity as n, reduced by the number in which the number of parts has the opposite parity, equals the number of partitions of n by distinct parts d_i.*

For the Steenrod algebra, May's theory shows that we are in this situation with $d_i = 2^i - 1$, and that explains our results.

3. Applications

Frank Adams generally read with a skeptical and critical intelligence and we would expect him to ask whether the above observations, elementary as they are, would be of any use. In fact they arose from computations in the May spectral sequence, computations which are proving very difficult, and the Euler characteristics have actually detected several errors.

Of course certain types of errors are not detectable this way. If you are looking at a pair of elements in the May spectral sequence, and trying to decide whether or not d_r of one element equals the other, you cannot get any advice this way, since the two elements are paired off with opposite signs in the Euler characteristic.

The following are the three situations that have occurred so far in which the above combinatorial arguments have helped.

First, at bi-gradings (72,8) and (72,13) there are survivors in the May spectral sequence of a unexpected and original character, and one wonders in such a situation whether some relation has been overlooked. However, the survivors are needed to make the Euler characteristics in $t = 80$ and $t = 85$ agree with the other computations in those degrees; there are no other doubtful cases; so the plausibility of these survivors is increased.

Second, at (75,8), my unpublished calculations show two elements, while Nakamura's published chart shows only one [N]. The Euler characteristic is consistent with the even number. It turns out that Nakamura's tables give two elements, only one of which was copied to his chart. Thus the Euler characteristic has detected a copying error.

Finally, at the time of the Adams Symposium, there was a discrepancy at $t = 90$ between my tentative computations and the Euler characteristic. Further work showed several related discrepancies in the range $90 < t < 100$. These focussed suspicion on the element $h_0(1,3)h_0(1,2)$ and I concluded that there must have been a missed relation involving this product in E_2 of the May spectral sequence. In fact there is such a relation, and it is not given in May's thesis, but it does appear correctly in May's unpublished revision.

In that case, the mistake had been corrected long ago; but it seems reassuring that we caught it independently by such an elementary numerical check.

References

[A1] Adams, J.F., On the structure and applications of the Steenrod algebra, Comm. Math. Helv. 52 (1958), 180-214.

[A2] Adams, J.F., On the groups J(X)—IV, Topology 5 (1967), 21-71.

[M] May, J.P., The cohomology of the Steenrod algebra; stable homotopy groups of spheres, Bull. Amer. Math. Soc. 71 (1965), 377-380.

—, Thesis, Princeton, 1964.

—, Typescript revision, c. 1966.

[N] Nakamura, O., On the cohomology of the mod 2 Steenrod algebra, Bull. Sci. & Engg. Div., U. Ryukyus, No. 16, March 1973.

[T] Tangora, M.C., On the cohomology of the Steenrod algebra, Math. Z. 116 (1970), 18-64.

Author's address

Martin C. Tangora
Department of Mathematics, Statistics, and Computer Science
University of Illinois at Chicago (mail code 249)
Chicago, Illinois 60680-4348 U.S.A.

Algebras over the Steenrod Algebra and Finite H-Spaces

Alain Jeanneret

0. Let p be a prime and $P = \mathbb{Z}/p[x_1,..., x_n]$ be a graded polynomial algebra over the mod p Steenrod algebra $\mathcal{A}(p)$. A fundamental question in algebraic topology is to decide whether P can be realized as the mod p cohomology of a topological space. Ideas from the Galois and invariant theories provided a complete answer to the realization problem for the non - modular case (i.e. when $\deg(x_i) \not\equiv 0 \pmod p$, $i = 1,..., n$) see [A-W]. In contrast the modular case is still not settled.

An interesting family of modular polynomial algebras at the prime 2 is provided by the so called Dickson algebras defined as follows :
The canonical $GL(q;\mathbb{Z}/2)$ - action on $(\mathbb{Z}/2)^q$ induces a $GL(q;\mathbb{Z}/2)$ - action on $H^*(B(\mathbb{Z}/2)^q;\mathbb{Z}/2) \cong \mathbb{Z}/2[t_1,..., t_q]$, ($\deg(t_i) = 1$, $i = 1,..., n$). As the Steenrod algebra acts via the squaring map (which is linear in characteristic 2), the actions of $W = GL(q;\mathbb{Z}/2)$ and $\mathcal{A}(2)$ commute. The Dickson algebra is the invariant algebra

$$D(q) = H^*(B(\mathbb{Z}/2)^q;\mathbb{Z}/2)^W$$

which is again an algebra over $\mathcal{A}(2)$. Dickson has shown that :

$$D(q) \cong \mathbb{Z}/2[w_1,..., w_q], \ (\deg(w_i) = 2^q-2^{q-i}).$$

The $\mathcal{A}(2)$ - action on D(q) can be described as follows : $Sq^{2^{q-k-1}} w_k = w_{k+1}$ ($k = 1,..., q-1$). A natural question is therefore :

Does there exist a topological space Y(q) such that $H^*(Y(q);\mathbb{Z}/2) \cong D(q)$?

The well known classical examples are : $D(1) \cong H^*(B\mathbb{Z}/2;\mathbb{Z}/2)$, $D(2) \cong H^*(BSO(3);\mathbb{Z}/2)$ and $D(3) \cong H^*(BG_2;\mathbb{Z}/2)$ where G_2 denotes the exceptional

Lie group. L. Smith and R. Switzer have proved in [S-S] that Y(q) does not exist for $q \geq 6$.

Let $E(x_1, \ldots, x_n)$ denote the exterior algebra on $x_1, \ldots, x_n$ and let

$$A(q) = \mathbb{Z}/2\,[v_1]/(v_1^4) \otimes E(v_2, v_3, \ldots, v_{q-1}) \quad (\deg(v_i) = 2^q - 2^{q-i} - 1)$$

be an algebra over $\mathcal{A}(2)$. The action of the Steenrod algebra is given by

$$Sq^{2^{q-k-1}} v_k = v_{k+1} \ (k = 1, \ldots, q-2), \ Sq^1 v_{q-1} = v_1^2.$$

Standard methods show that the cohomology ring of $X(q) = \Omega Y(q)$ satisfies $H^*(X(q);\mathbb{Z}/2) \cong A(q)$. For n = 5, U. Suter (unpublished), J. Lin and F. Williams [L-W] have shown that X(5) cannot support an H - structure. This implies the non realizability of D(5) as the cohomology algebra of a topological space. In [J-S], U. Suter and the author proved a stronger version of the result of Lin - Williams and Smith - Switzer, namely :

For $q \geq 5$ the algebra A(q) cannot be realized as the mod 2 cohomology algebra of a topological space.

The purpose of this note is to give a proof of a further generalization of the last result. Set

$$A_i = A(q) \quad (i = 1, \ldots, m)$$

$$B_j = E(w_j) \quad (\deg(w_j) = 2^{n_j} - 1, \ n_j \geq q-1, \ j = 1, \ldots, n)$$

$$K(q, m, n) = \bigotimes_{i=1}^{m} A_i \otimes \bigotimes_{j=1}^{n} B_j \quad (m \geq 1, n \geq 0).$$

We assume that $\bigotimes_{i=1}^{m} A_i$ is an $\mathcal{A}(2)$ - subalgebra of K(q, m, n) and that $Sq^1 w_j = Sq^2 w_j = 0$ ($j = 1, \ldots, n$).

Theorem. *For* $q \geq 5$ *the* $\mathcal{A}(2)$ *- algebras* $K(q, m, n)$ *cannot be realized as the mod 2 cohomology of topological spaces.*

Remark. The hypothesis on the $\mathcal{A}(2)$ - action on $K(q, m, n)$ is not essential (see the final remark at the end of this note).

This result is related to the question of connectivity of finite H - spaces in the following way. Let us first recall a result of J. Lin :

Theorem.(see [L]) *Let* X *be a* 14 *- connected finite* H *- space with* $H_*(X;\mathbb{Z}/2)$ *associative, then* $\xi(H^{15}(X;\mathbb{Z}/2)) \neq 0$ *where* $\xi : H^n(X;\mathbb{Z}/2) \longrightarrow H^{2n}(X;\mathbb{Z}/2)$ *is the squaring map.*

The algebras $K(5, m, n)$ satisfy the condition $\xi(K(5, m, n)^{15}) \neq 0$, so they are not ruled out by Lin's theorem. Our methods show however that they cannot even be the cohomology algebras of topological spaces.

We shall give the proof of the theorem for $q = 5$ only, the other cases can be treated exactly in the same way.

I am grateful to U. Suter for his guidance and his help. I would also like to take the opportunity to thank J. Lin for the many useful discussions I had with him.

Notations. Throughout the paper we set

$$A_i = \mathbb{Z}/2[x_{15}^{(i)}]/((x_{15}^{(i)})^4) \otimes E(x_{23}^{(i)}, x_{27}^{(i)}, x_{29}^{(i)}) \quad (\deg(x_k^{(i)}) = k,\ i = 1,\dots, m)$$

$$B_j = E(w_j) \quad (\deg(w_j) = 2^{n_j}-1,\ n_j \geq 4,\ j = 1,\dots, n)$$

$$K = \bigotimes_{i=1}^{m} A_i \otimes \bigotimes_{j=1}^{n} B_j$$

We assume that there exists a simply connected CW - complex X such that $H^*(X;\mathbb{Z}/2) \cong K$. We shall show that the Adams operations and the action of the

Steenrod algebra are not compatible on ΩX. This will rule out the existence of such a space.

1. In this section we shall describe the cohomology of ΩX, the loop space on X, and give some results on the $\mathbb{Z}_{(2)}$ - cohomology of X where $\mathbb{Z}_{(2)}$ stands for the ring of integers localized at the prime 2.

Our main device to obtain the mod 2 cohomology of ΩX is the Eilenberg-Moore spectral sequence (E-Mss for short). Though we adopt the notation of [K] let us recall briefly some properties of the E-Mss. It is a second quadrant spectral sequence of $\mathcal{A}(2)$ - modules with

$$E_2 \cong \mathrm{Tor}_{H^*(X;\mathbb{Z}/2)}(\mathbb{Z}/2,\mathbb{Z}/2) \text{ and } E_\infty \cong E^0(H^*(\Omega X;\mathbb{Z}/2))$$

($E^0(H)$ is the graded module associated to a filtration of H). Let H denote the algebra E(x) or $\mathbb{Z}/2\,[x]/(x^4)$, and let $\overline{H}$ denote the augmentation ideal of H. A straightforward computation, using the bar construction, shows that :

$$\mathrm{Tor}_{E(x)}(\mathbb{Z}/2,\mathbb{Z}/2) \cong \Gamma(sx) \text{ and } \mathrm{Tor}_{\mathbb{Z}/2\,[x]/(x^4)}(\mathbb{Z}/2,\mathbb{Z}/2) \cong E(sx) \otimes \Gamma(tx)$$

The element sx, called suspension element has bidegree (-1, deg(x)) and is represented (via the bar construction) by $[x] \in \overline{H}$, whereas tx, called transpotence element has bidegree (-2, deg(x)) and is represented by $[x^2|x^2]$ or by $[x|x^3] \in \overline{H} \otimes \overline{H}$ (these two cycles differ from a boundary and therefore represent the same element in the E_2 term). As usual $\Gamma(u)$ is the divided polynomial algebra on u. We also mention the following property of Tor :

$$\mathrm{Tor}_{N \otimes M}(\mathbb{Z}/2,\mathbb{Z}/2) \cong \mathrm{Tor}_N(\mathbb{Z}/2,\mathbb{Z}/2) \otimes \mathrm{Tor}_M(\mathbb{Z}/2,\mathbb{Z}/2).$$

The E-Mss for X can then be computed. We obtain easily :

$$E_2 \cong \mathrm{Tor}_K(\mathbb{Z}/2,\mathbb{Z}/2) \cong \bigotimes_{i=1}^{m} C_i \otimes \bigotimes_{j=1}^{n} D_j$$

where

$$C_i \cong E(u_{14}^{(i)}) \otimes \Gamma(u_{22}^{(i)}, u_{26}^{(i)}, u_{28}^{(i)}, u_{58}^{(i)}) \quad (i = 1,..., m)$$

$$D_j \cong \Gamma(v_j) \qquad (j = 1,..., n)$$

with $u_{2n}^{(i)} = sx_{2n+1}^{(i)}$ $(n = 7, 11, 13, 14)$, $u_{58}^{(i)} = tx_{15}^{(i)}$ and $v_j = sw_j$.

The differentials of the E-Mss raise the total degree by one. As E_2 is concentrated in even total degree the spectral sequence is trivial, i.e. $E_2 \cong E_\infty$, and so we get the additive isomorphisms :

$$H^*(\Omega X;\mathbb{Z}/2) \cong E^0(H^*(\Omega X;\mathbb{Z}/2)) \cong \bigotimes_{i=1}^{m} C_i \otimes \bigotimes_{j=1}^{n} D_j.$$

For each $i = 1,..., m$ there is an element $u_{56}^{(i)} \in H^{56}(\Omega X;\mathbb{Z}/2)$ represented by $[x_{29}^{(i)}|x_{29}^{(i)}] = \gamma_2(u_{28}^{(i)})$ in $E^0(H^*(\Omega X;\mathbb{Z}/2)) \cong E_2$. As $\mathcal{Q}(2)$ acts on E_2 via the Cartan formula, one gets (with some abuse of notations) that $Sq^2u_{56} = Sq^2[x_{29}|x_{29}] = \sum_{i+j=2} [Sq^ix_{29}|Sq^jx_{29}] = [x_{15}^2|x_{15}^2] = [x_{15}|x_{15}^3] = u_{58} \neq 0$. So we have proved the following result :

Lemma 1. *The element* $u_{56}^{(i)} \in H^{56}(\Omega X;\mathbb{Z}/2)$, *represented by* $[x_{29}^{(i)}|x_{29}^{(i)}]$ *in* $E^0(H^*(\Omega X;\mathbb{Z}/2))$ $(i = 1,..., m)$ *satisfies :*

$$Sq^2u_{56}^{(i)} \neq 0.$$

An immediate consequence of the fact that $H^{2n+1}(\Omega X;\mathbb{Z}/2) = 0$ is the next lemma. Let $\rho_* : H^*(X;\mathbb{Z}_{(2)}) \longrightarrow H^*(X;\mathbb{Z}/2)$ denote the modulo 2 reduction homomorphism.

Lemma 2. *The algebra* $H^*(\Omega X;\mathbb{Z}_{(2)})$ *is torsion free and there are generators* $s_k^{(i)} \in H^k(\Omega X;\mathbb{Z}_{(2)})$ $(i = 1,\ldots, m$ *and* $k = 14, 22, 26, 28, 56)$ *with*

$$\rho_*(s_k^{(i)}) = u_k^{(i)} \text{ and } (s_{14}^{(i)})^2 = s_{28}^{(i)},\ (s_{14}^{(i)})^4 = (s_{28}^{(i)})^2 = 2s_{56}^{(i)},$$

and there exist $t_j \in H^{2^{n_j}-i}(\Omega X;\mathbb{Z}_{(2)})$ $(j = 1,\ldots, n)$ *with* $\rho_*(t_j) = v_j$.

Proof. The only point to be checked is the assertion concerning the multiplicative structure. As the E-Mss is a specral sequence of $\mathcal{Q}(2)$ - modules, we obtain from the $\mathcal{Q}(2)$ - action on K :

$$Sq^{14}\, u_{14}^{(i)} = (u_{14}^{(i)})^2 = u_{28}^{(i)} \text{ and } Sq^{28}\, u_{28}^{(i)} = (u_{28}^{(i)})^2 = 0.$$

Therefore we can choose $s_{14}^{(i)}, s_{28}^{(i)}$ $(i = 1,\ldots, m)$ such that $(s_{14}^{(i)})^2 = s_{28}^{(i)}$ and $(s_{14}^{(i)})^4 = (s_{28}^{(i)})^2 \equiv 0 \bmod 2$. Using the Hopf algebra structure on $H^*(\Omega X;\mathbb{Z}_{(2)})$ and the fact that $s_{14}^{(i)}$ is primitive, one can show there exists $s_{56}^{(i)}$ such that $(s_{14}^{(i)})^4 = (s_{28}^{(i)})^2 = 2s_{56}^{(i)}$. Moreover $\rho_*(s_{56}^{(i)}) = u_{56}^{(i)}$ $(i = 1,\ldots, m)$.

We end this section with a description of the $\mathbb{Z}_{(2)}$ - cohomology of X. Using the universal coefficient formula and the fact that Sq^1 detects $\mathbb{Z}/2$ - summands, one shows the existence of free generators $y_k^{(i)} \in H^k(X;\mathbb{Z}_{(2)})$ $(i = 1,\ldots, m;\ k = 15, 23, 27, 59)$ and $z_j \in H^{2^{n_j}-i}(X;\mathbb{Z}_{(2)})$ $(j = 1,\ldots, n)$ such that :

$$\rho_*(y_k^{(i)}) = x_k^{(i)} \quad (k = 15, 23, 27),\ \rho_*(y_{59}^{(i)}) = x_{29}^{(i)} \cdot (x_{15}^{(i)})^2 \text{ and } \rho_*(z_j) = w_j;$$

moreover we have the isomorphism of $\mathbb{Z}_{(2)}$ - modules :

$$H^*(X;\mathbb{Z}_{(2)}) \cong \bigotimes_{i=1}^{n} \{P(y_{15}^{(i)}) \otimes E_{\mathbb{Z}/2}(y_{23}^{(i)}, y_{27}^{(i)}, y_{59}^{(i)}) / ((y_{15}^{(i)})^2 \otimes y_{59}^{(i)})\}$$

$$\otimes \bigotimes_{j=1}^{n} E_{\mathbb{Z}_{(2)}}(z_j)$$

where $P(y_{15}^{(i)}) = \mathbb{Z}_{(2)}[y_{15}^{(i)}]/((y_{15}^{(i)})^4, 2(y_{15}^{(i)})^2) \cong \mathbb{Z}_{(2)} \oplus \mathbb{Z}_{(2)} \oplus \mathbb{Z}/2 \oplus \mathbb{Z}/2$ is additively generated by $1, y_{15}^{(i)}, (y_{15}^{(i)})^2, (y_{15}^{(i)})^3$. Note in particular that the two torsion of $H^*(X;\mathbb{Z}_{(2)})$ has only order two.

2. This section contains the computation of the K - theory of X. Let us first fix the notation and recall some basic facts about the K - theory with coefficients. If Y is a finite CW - complex, $K^*(Y)$ denotes the $\mathbb{Z}/2$ - graded ordinary complex K - theory [A-H]. For $G = \mathbb{Z}_{(2)}$ or $\mathbb{Z}/2$ the K - theory with coefficients in G, denoted by $K^*(Y;G)$, is defined in [A-T]. $K^*(Y;G)$ is again a $\mathbb{Z}/2$ - graded algebra. Moreover it satisfies $\widetilde{K}^0(S^n;G) = 0$ if n is odd and $\widetilde{K}^0(S^n;G) \cong G$ if n is even.

The Atiyah-Hirzebruch spectral sequence (A-Hss for short) is a first and fourth quadrant spectral sequence of algebras with

$$E_2(Y;G) \cong H^*(Y;K^*(pte;G)) \text{ and } E_\infty(Y;G) \cong E^0(K^*(Y;G)).$$

The modulo 2 reduction homomorphism induces a morphism of spectral sequences $\rho_* : E_r(Y;\mathbb{Z}_{(2)}) \longrightarrow E_r(Y;\mathbb{Z}/2)$. Finally, for $G = \mathbb{Z}/2$ the differential $d_3^{\mathbb{Z}/2}$ is given by

$$d_3^{\mathbb{Z}/2} = Sq^2Sq^1 + Sq^1Sq^2.$$

Our first step towards the determination of $K^*(X;\mathbb{Z}_{(2)})$ is the computation of $K^*(X;\mathbb{Z}/2)$. Recall that in this case $E_2 \cong E_3 \cong H^*(X;\mathbb{Z}/2)$. Using the $\mathcal{Q}(2)$ - action on K (see section 0) we infer that :

(1)
$$d_3(x_{15}^{(i)}) = d_3(x_{23}^{(i)}) = d_3(x_{29}^{(i)}) = 0 \text{ and } d_3(x_{27}^{(i)}) = (x_{15}^{(i)})^2 \quad (i = 1,\ldots, m)$$
$$d_3(w_j) = 0 \quad (j = 1,\ldots, n).$$

As d_3 is a derivation we readily get:

$$E_4 \cong \bigotimes_{i=1}^{m} E(x_{15}^{(i)}, x_{23}^{(i)}, x_{29}^{(i)}, x_{27}^{(i)}(x_{15}^{(i)})^2) \otimes \bigotimes_{j=1}^{n} E(w_j)$$

where x_k still denotes (for simplicity of notation) the element of E_4 represented by $x_k \in \ker d_3 \subset E_3$. For dimensional reasons ($\dim_{\mathbb{Z}/2} E_4(X;\mathbb{Z}/2) = \dim_{\mathbb{Q}} H^*(X;\mathbb{Q})$) the spectral sequence collapses after E_4, i.e. $E_4 \cong E_\infty$ and hence

$$K^*(X;\mathbb{Z}/2) \cong \bigotimes_{i=1}^{m} E(\theta_{15}^{(i)}, \theta_{23}^{(i)}, \theta_{29}^{(i)}, \theta_{57}^{(i)}) \otimes \bigotimes_{j=1}^{n} E(\tau_j)$$

where τ_j and $\theta_k^{(i)}$ are represented by the obvious classes in $E^0(K^*(X;\mathbb{Z}/2)) \cong E_4$.

From the universal coefficient formula for K -theory and the observation that $\dim_{\mathbb{Z}/2} K^*(X;\mathbb{Z}/2) = \dim_{\mathbb{Q}} H^*(X;\mathbb{Q})$ we conclude that $K^*(X;\mathbb{Z}_{(2)})$ is torsion free.

We are now ready to calculate the K-theory of X with $\mathbb{Z}_{(2)}$ - coefficients. The modulo 2 reduction homomorphism $\rho_* : H^*(X;\mathbb{Z}_{(2)}) \longrightarrow H^*(X;\mathbb{Z}/2)$ is an injection on the torsion subgroup (see the end of section 1) and since image$(d_r) \subset$ torsion(E_r) we obtain using (1) that

$$d_3(y_{15}^{(i)}) = d_3(y_{23}^{(i)}) = d_3(y_{59}^{(i)}) = 0,\ d_3(y_{27}^{(i)}) = (y_{15}^{(i)})^2 \text{ and } d_3(z_j) = 0.$$

We then compute (2):

$$E_4(X;\mathbb{Z}_{(2)}) \cong \bigotimes_{i=1}^{m} \{E_{\mathbb{Z}_{(2)}}(y_{15}^{(i)}, y_{23}^{(i)}, 2y_{27}^{(i)}) \oplus E_{\mathbb{Z}_{(2)}}(y_{15}^{(i)}, y_{23}^{(i)}, y_{27}^{(i)})\, y_{59}^{(i)}$$

$$\otimes (y_{15}^{(i)})^2 . y_{27}^{(i)}\, E_{\mathbb{Z}_{(2)}}(y_{15}^{(i)}, y_{23}^{(i)})\} \otimes \bigotimes_{j=1}^{n} E_{\mathbb{Z}_{(2)}}(z_j)$$

Observe that $\rho_* : E_4(X;\mathbb{Z}_{(2)}) \longrightarrow E_4(X;\mathbb{Z}/2)$ is again injective on the torsion subgroup. Since $d_r^{\mathbb{Z}/2} = 0$ for $r \geq 4$ we infer that $d_r = 0$ for $r \geq 4$, and therefore

(3) $$E_4(X;\mathbb{Z}_{(2)}) \cong E_\infty(X;\mathbb{Z}_{(2)}) \cong E^0(K^*(X;\mathbb{Z}_{(2)})).$$

Before giving the explicit description of $K^*(X;\mathbb{Z}_{(2)})$ we recall some relations between the K - theory and the rational cohomology.

Let $\xi \in K^*(Y;\mathbb{Z}_{(2)})$ and let $ch(\xi) = \sum_{q\geq 0} ch_q(\xi)$, $ch_q(\xi) \in H^q(X;\mathbb{Q})$, be its Chern character.

Definition. *The rational filtration of* ξ, rat.filt.(ξ), *equals* p *if* $ch_q(\xi) = 0$ *for* $q = 0,\ldots, p-1$ *and* $ch_p(\xi) \neq 0$. *If* ξ *is a torsion element, we set* rat.filt.(ξ) = ∞. *For* $\xi \in K^*(Y;\mathbb{Z}_{(2)})$ *we denote by* ch_ξ *the first non vanishing component of* $ch(\xi)$, *i.e.* $ch_\xi = ch_p(\xi)$ *with* p = rat.filt.(ξ).

The following result is not difficult to prove :

Lemma 3. *Let* Y *be a finite CW - complex. There exist free elements* $\xi_1,\ldots, \xi_n$ *providing a basis of* $K^*(Y;\mathbb{Z}_{(2)})$ / (torsion) *such that* $ch_{\xi_1},\ldots, ch_{\xi_n}$ *is a homogenous basis of* $H^*(X;\mathbb{Q})$.

From (2), (3) and lemma 3 we obtain the main proposition of this section.

Proposition. *There exist elements* $\xi_{15}^{(i)}, \xi_{23}^{(i)}, \xi_{27}^{(i)}, \xi_{59}^{(i)}$ (i = 1,..., m) *and* v_j (j = 1,.., n) *such that :*

$$K^*(X;\mathbb{Z}_{(2)}) \cong \bigotimes_{i=1}^{m} E(\xi_{15}^{(i)}, \xi_{23}^{(i)}, \xi_{27}^{(i)}, \xi_{59}^{(i)}) \otimes \bigotimes_{j=1}^{n} E(v_j)$$

where rat.filt.$(\xi_k^{(i)}) = k$ *and* rat.filt.$(v_j) = 2^{n_j}-1$. *Moreover* $\xi_{15}^{(i)}, \xi_{23}^{(i)}, \xi_{27}^{(i)}, \xi_{59}^{(i)}$, $2\xi_{59}^{(i)}$ *are represented by* $y_{15}^{(i)}, y_{23}^{(i)}, 2y_{27}^{(i)}, (y_{15}^{(i)})^2.y_{27}^{(i)}, y_{59}^{(i)}$ *whereas* v_j *is represented by* z_j *in* $E^0(K^*(X;\mathbb{Z}_{(2)}))$.

3. This section contains the description of the action of the Adams operations on some generators of $\widetilde{K}(\Sigma X;\mathbb{Z}_{(2)})$ and the proof of the main theorem. As we are going to use it many times, we first recall a generalisation of an important result of Atiyah [At].

Let Y be a finite 2 - torsion free CW - complex (i.e. $H^*(Y;\mathbb{Z}_{(2)})$ without 2 - torsion). The submodule of elements of filtration $\geq 2q$, $\mathcal{F}_{2q}\widetilde{K}(Y;\mathbb{Z}_{(2)}) \subseteq \widetilde{K}^0(Y;\mathbb{Z}_{(2)}) = \widetilde{K}(Y;\mathbb{Z}_{(2)})$ is defined by $\mathcal{F}_{2q}\widetilde{K}(Y;\mathbb{Z}_{(2)}) = \mathrm{Ker}\{\widetilde{K}(Y;\mathbb{Z}_{(2)}) \longrightarrow \widetilde{K}(Y^{(2n-1)};\mathbb{Z}_{(2)})\}$ where $Y^{(2n-1)}$ is the (2n-1) skeleton of Y. As Y is 2 - torsion free the A-Hss with $\mathbb{Z}_{(2)}$ - coefficients collapses and we may identify $H^*(Y;\mathbb{Z}/2)$ with $E^0(K^*(Y;\mathbb{Z}_{(2)})) \otimes \mathbb{Z}/2$; if $\alpha \in \mathcal{F}_{2q}\widetilde{K}(Y;\mathbb{Z}_{(2)})$ we denote the corresponding element of $H^{2q}(Y;\mathbb{Z}/2)$ by $\overline{\alpha}$. We are now ready to state the $\mathbb{Z}_{(2)}$ - localized version of Atiyah's result.

Theorem.(see [H]) *Let* Y *be a finite,* 2 - torsion *free CW-complex and* $\alpha \in \mathcal{F}_{2q}\widetilde{K}(Y;\mathbb{Z}_{(2)})$. *Then there exist elements* $\alpha_i \in \mathcal{F}_{2q+2i}\widetilde{K}(Y;\mathbb{Z}_{(2)})$ (i = 0,..., q) *such that*

$$\Psi^2(\alpha) = \sum_{i=0}^{q} 2^{q-i}\,\alpha_i.$$

Moreover $\overline{\alpha}_i = Sq^{2i}\,\overline{\alpha}$ *for* i = 0,..., q.

In order to state a first lemma about the Adams operations we define $\eta_k \in \widetilde{K}(\Sigma X;\mathbb{Z}_{(2)})$ by $\eta_k^{(i)} = \sigma(\xi_{k-1}^{(i)})$ $(i = 1,\dots, m,\ k = 16, 24, 28, 60)$ and $\zeta_j \in \widetilde{K}(\Sigma X;\mathbb{Z}_{(2)})$ by $\zeta_j = \sigma(v_j)$ $(j = 1,\dots, n)$, where $\sigma : \widetilde{K}^1(X;\mathbb{Z}_{(2)}) \cong \widetilde{K}^0(\Sigma X;\mathbb{Z}_{(2)})$ is the suspension isomorphism.

Lemma 4. *There exist a choice of generators* $\xi_{15}^{(i)}, \xi_{23}^{(i)}, \xi_{27}^{(i)}$ $(i = 1,\dots, m)$ *such that in* $\widetilde{K}(\Sigma X;\mathbb{Z}_{(2)}) / \mathcal{F}_{32}\widetilde{K}(\Sigma X;\mathbb{Z}_{(2)})$ *the following relation holds :*

$$\Psi^2(\eta_{16}^{(i)}) \equiv 2\eta_{28}^{(i)} \pmod 4.$$

Proof. Up to homotopy there is a subcomplex $Y = \times_{i=1}^{m} \{S^{16} \cup e^{24} \cup e^{28}\}_i$ of ΣX (and inclusion $i : Y \longrightarrow \Sigma X$) which carries the cohomology with $\mathbb{Z}_{(2)}$ - coefficients of ΣX up to dimension 28. The K - theory of Y is free on generators $\beta_{16}^{(i)} = i^*(\eta_{16}^{(i)})$, $\beta_{24}^{(i)} = i^*(\eta_{24}^{(i)})$ and $\beta_{28}^{(i)}$, where $2\beta_{28}^{(i)} = i^*(\eta_{28}^{(i)})$ $(i = 1,\dots, m)$.

The last equality comes from the fact that the A-Hss of Y collapses, whereas for ΣX we have $d_3(y_{27}^{(i)}) = (y_{15}^{(i)})^2$ $(i = 1,\dots, m)$. Atiyah's result now implies :

$$\Psi^2(\beta_{16}^{(i)}) = 2^8 \beta_{16}^{(i)} + 2^4 a^{(i)} \beta_{24}^{(i)} + 2^2 b^{(i)} \beta_{28}^{(i)} \qquad (i = 1,\dots, n),$$

where $a^{(i)}$ and $b^{(i)} \equiv 1 \pmod 2$. So we obtain :

$$\Psi^2(\eta_{16}^{(i)}) = 2^8 \eta_{16}^{(i)} + 2^4 \eta_{24}^{(i)} + 2\eta_{28}^{(i)} \pmod{\mathcal{F}_{32}\widetilde{K}(\Sigma X;\mathbb{Z}_{(2)})}. \tag{4}$$

It is clear that the lemma follows immediatly from (4).

The second lemma about the Adams operations concerns the K - theory of ΩX. Let us denote by $\varepsilon : \Sigma^2\Omega X \longrightarrow \Sigma X$ the suspension of the evaluation map and define $\mu_k^{(i)} \in \widetilde{K}(\Omega X;\mathbb{Z}_{(2)})$ by $\sigma^2(\mu_k^{(i)}) = \varepsilon^*(\eta_{k+2}^{(i)})$ (i = 1,..., m and k = 14, 22, 26) and $\lambda_j \in \widetilde{K}(\Omega X;\mathbb{Z}_{(2)})$ by $\sigma^2(\lambda_j) = \varepsilon^*(v_j)$ (j = 1,..., n).

Lemma 5. *In* $\widetilde{K}(\Omega X;\mathbb{Z}_{(2)}) / \mathcal{F}_{60}\widetilde{K}(\Omega X;\mathbb{Z}_{(2)})$ *the following relations hold :*

(i) $\Psi^2(\mu_{26}^{(i)}) \equiv 0 \pmod 2$ (i = 1,..., m)

(ii) $\Psi^2(\lambda_j) \equiv 0 \pmod 2$ (*for each* j *such that* $n_j = 5$).

Proof. Recall first that the suspension of decomposable elements is in the kernel of ε^*, therefore we obtain in $\widetilde{K}(\Sigma^2\Omega X;\mathbb{Z}_{(2)}) / \mathcal{F}_{62}\widetilde{K}(\Sigma^2\Omega X;\mathbb{Z}_{(2)})$ (for i = 1,..., m and j such that $n_j = 5$) :

$$\Psi^p(\varepsilon^*(\eta_{28}^{(i)})) = p^{14}\varepsilon^*(\eta_{28}^{(i)}) + \sum_{n_j=5} a_p^{(i,j)}\ \varepsilon^*(v_j) + \sum_{k=1}^{m} b_p^{(i,k)}\ \varepsilon^*(\eta_{60}^{(k)})$$

$$\Psi^p(\varepsilon^*(v_j)) = p^{16}\varepsilon^*(v_j) + \sum_{k=1}^{m} c_p^{(j,k)}\ \varepsilon^*(\eta_{60}^{(k)})$$

$$\Psi^p(\varepsilon^*(\eta_{60}^{(i)})) = p^{30}\varepsilon^*(\eta_{60}^{(i)})$$

Equating the coefficient of $\varepsilon^*(\eta_{60}^{(k)})$ in $\Psi^3\Psi^2(\varepsilon^*(v_j)) = \Psi^2\Psi^3(\varepsilon^*(v_j))$ gives :

(5) $c_2^{(j,k)} \equiv 0 \pmod{2^{12}}$ (k = 1,..., m and j such that $n_j = 5$).

The same method applied to the coefficient of $\varepsilon^*(v_j)$ in $\Psi^3\Psi^2(\varepsilon^*(\eta_{28}^{(i)})) = \Psi^2\Psi^3(\varepsilon^*(\eta_{28}^{(i)}))$ implies :

(6) $a_2^{(i,j)} \equiv 0 \pmod{2^{11}}$ (i = 1,..., m and j such that $n_j = 5$).

Equalities (5), (6) and the coefficient of $\varepsilon^*(\eta_{60}^{(k)})$ in $\Psi^3\Psi^2(\varepsilon^*(\eta_{28}^{(i)})) =$ $\Psi^2\Psi^3(\varepsilon^*(\eta_{28}^{(i)}))$ give :

(7) $$b_2^{(i,k)} \equiv 0 \pmod{2^5} \qquad (\text{i and k} = 1,\ldots, m).$$

In view of (5), (6) and (7) we have (for i = 1,..., m and j such that $n_j = 5$) in $\widetilde{K}(\Sigma^2\Omega X;\mathbb{Z}_{(2)}) / \mathcal{F}_{62}\widetilde{K}(\Sigma^2\Omega X;\mathbb{Z}_{(2)})$:

(8) $$\Psi^2(\varepsilon^*(\eta_{28}^{(i)})) \equiv \Psi^2(\varepsilon^*(\nu_j)) \equiv 0 \pmod 4.$$

Recall that the Adams operations do NOT commute with the Bott periodicity [Ad], the actual relation is :

$$\Psi^2(\sigma^2(\alpha)) = 2\sigma^2\Psi^2(\alpha).$$

As $K^*(\Omega X;\mathbb{Z}_{(2)})$ is torsion free (by lemma 2 $H^*(\Omega X;\mathbb{Z}_{(2)})$ is torsion free and thus the A-Hss is trivial), the relation (8) forces $\Psi^2(\mu_{26}^{(i)}) \equiv \Psi^2(\lambda_j) \equiv 0 \pmod 2$ in $\widetilde{K}(\Omega X;\mathbb{Z}_{(2)}) / \mathcal{F}_{60}\widetilde{K}(\Omega X;\mathbb{Z}_{(2)})$ and the lemma is proved.

The last part of this section is devoted to the proof of the main theorem. In $\widetilde{K}(\Omega X;\mathbb{Z}_{(2)}) / \mathcal{F}_{30}\widetilde{K}(\Omega X;\mathbb{Z}_{(2)})$ we have by lemma 4 :

(9) $$(\mu_{14}^{(i)})^2 \equiv \Psi^2(\mu_{14}^{(i)}) \equiv \mu_{26}^{(i)} \pmod 2 \quad (i = 1,\ldots, m).$$

Equality (4) can be writen more precisely as :

(10) $$\Psi^2(\varepsilon^*(\eta_{16}^{(i)})) = 2^8\varepsilon^*(\eta_{16}^{(i)}) + 2^4\varepsilon^*(\eta_{24}^{(i)}) + 2\varepsilon^*(\eta_{28}^{(i)}) + \sum_{n_j=5} d_j^{(i)}\varepsilon^*(\nu_j) + \sum_{k=1}^{m} e_k^{(i)}\varepsilon^*(\eta_{60}^{(k)}) \pmod{\mathcal{F}_{62}\widetilde{K}(\Sigma^2\Omega X;\mathbb{Z}_{(2)})}.$$

Therefore congruence (9), equality (10) and lemma 5 imply that :

$$(\mu_{14}^{(i)})^4 \equiv \Psi^2((\mu_{14}^{(i)})^2) \equiv \Psi^2(\mu_{26}^{(i)}) \equiv 0 \pmod 2 \qquad (i = 1,\dots, m)$$

in $\widetilde{K}(\Omega X;\mathbb{Z}_{(2)}) / \mathcal{F}_{60}\widetilde{K}(\Omega X;\mathbb{Z}_{(2)})$. So, there exists an element $\omega^{(i)} \in \widetilde{K}(\Omega X;\mathbb{Z}_{(2)})$ $(i = 1,\dots, m)$ such that $(\mu_{14}^{(i)})^4 = 2\omega^{(i)} \pmod{\mathcal{F}_{60}\widetilde{K}(\Omega X;\mathbb{Z}_{(2)})}$. With (4) or (10) we get :

$$\Psi^2(\omega^{(i)}) \equiv 2^{28}\, \omega^{(i)} \pmod{\mathcal{F}_{60}\widetilde{K}(\Omega X;\mathbb{Z}_{(2)})}.$$

As the A-Hss for ΩX collapses, $H^*(\Omega X;;\mathbb{Z}_{(2)}) \cong E^0(K^*(\Omega X;\mathbb{Z}_{(2)}))$ and thus $\mu_{14}^{(i)}$ is represented by $s_{14}^{(i)}$ $(i = 1,\dots, m)$, it follows from lemma 2 that $\omega^{(i)}$ is represented by $s_{56}^{(i)} \in H^{56}(\Omega X;\mathbb{Z}_{(2)})$ $(i = 1,\dots, m)$. Using one more time Atiyah's result (for the element $\omega^{(i)}$) we obtain that $Sq^2 u_{56}^{(i)} = 0$ for each $i = 1,\dots, m$ which contradicts lemma 1 and thus proves the main theorem.

Final remark. As mentioned in the introduction, the hypothesis on the $\mathcal{A}(2)$ - action on K is not essential. In fact every $\mathcal{A}(2)$ - action on K can always be described in the following way :

The relations $Sq^{15}x_{15}^{(i)} = (x_{15}^{(i)})^2$ $(i = 1,\dots, m)$, $Sq^{15} = Sq^1Sq^2Sq^4Sq^8$ and the linear independence of $\{(x_{15}^{(i)})^2 : i = 1,\dots, m\}$ allow us to change (if necessary) the generators $x_{23}^{(i)}, x_{27}^{(i)}, x_{29}^{(i)}$ such that the following Steenrod connections hold :

$$Sq^8\, x_{15}^{(i)} = x_{23}^{(i)},\ Sq^4 x_{23}^{(i)} = x_{27}^{(i)},\ Sq^2 x_{27}^{(i)} = x_{29}^{(i)},\ Sq^1 x_{29}^{(i)} = (x_{15}^{(i)})^2 \quad (i = 1,\dots, m).$$

Next let $h : X \longrightarrow \prod_{i=1}^{m} K(\mathbb{Z}/2;15)$ be the map such that in mod 2 cohomology

$h^*(\iota_{15}^{(i)}) = x_{15}^{(i)}$ $(i = 1,\dots, m)$ ($\iota_{15}^{(i)}$ beeing the canonical generator of the ith factor of $\bigotimes_{i=1}^{m} H^*(K(\mathbb{Z}/2;15);\mathbb{Z}/2)$). It is not difficult to prove that h^* factors through $\bigotimes_{i=1}^{m} A_i$:

$$\begin{array}{ccc} \bigotimes_{i=1}^{m} H^*(K(\mathbb{Z}/2;15);\mathbb{Z}/2) & \longrightarrow & H^*(X;\mathbb{Z}/2) \\ & \searrow \quad \nearrow & \\ & \bigotimes_{i=1}^{m} A_i & \end{array}$$

Therefore with the above choice of generators, A_i $(i = 1,\dots, m)$ becomes an $\mathcal{A}(2)$ - subalgebra of K. Moreover it is easy to show that $Sq^1 w_j = Sq^2 w_j = 0$ for each j such that $n_j = 4, 5$. As the argument to prove the theorem does not imply filtrations greater than 58 in $K^*(\Omega X;\mathbb{Z}_{(2)})$, it is possible to prove it without any hypothesis on the $\mathcal{A}(2)$ - action on K. This can be achieved in the following way : Let us perform the changes mentioned above. Lemma 1 and 2 can then be obtained similary. Next, we consider a subcomplex Y of X which carries the $\mathbb{Z}_{(2)}$ - homology of X up to dimension 61 (hmology approximation). The inclusion $Y \longrightarrow X$ induces a mod 2 cohomology isomorphism up to dimension 62 (note that $H^{62}(X;\mathbb{Z}/2) = 0$). One checks then that the results of the final part of section 1, concerning the $\mathbb{Z}_{(2)}$ - cohomology of X, persist up to dimension 62. The argument in K - theory is valid for Y in place of X.

REFERENCES

[Ad] J.F. ADAMS. Vector fields on spheres, *Ann. of Math.* 75 (1962), 603 - 632.

[A-W] J.F. ADAMS and C.W. WILKERSON. Finite H - spaces and algebras over the Steenrod algebra, *Ann. of Math.* 111 (1980), 95 - 143.

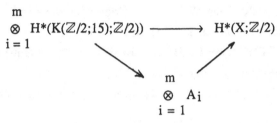

[A-T] S. ARAKI and H. TODA. Multiplicative structures in mod q cohomology theories I, *Osaka Math. J.* 2 (1965), 71 - 115.

[At] M.F. ATIYAH. Power operations in K-theory, *Quart. J. Math. Oxford* (2) 17 (1966), 165 - 193.

[A-H] M.F. ATIYAH and F. HIRZEBRUCH. Vectior bundles and homogenous spaces, *Proceedings of Symposia in Pure Mathematics* 3 A.M.S. (1961), 24 - 51.

[H] J.R. HUBBUCK. Generalized cohomology operations and H - spaces of low rank, *Trans. Amer. Math. Soc.* 141 (1969), 335 - 360.

[J-S] A. JEANNERET and U. SUTER. Réalisation topologique de certaines algèbres associées aux algèbres de Dickson, *Preprint.*

[K] R.M. KANE. The homology of Hopf spaces, *North Holland Math. Study* # 40 (1988).

[L] J.P. LIN. Steenrod connections and connectivity in H - spaces, *Memoirs Amer. Math. Soc.* # 369, 1987.

[L-W] J.P. LIN and F. WILLIAMS. On 14-connected finite H - spaces, *Israel J. Math.* 66 (1989) 274 - 288.

[S-S] L. SMITH and R.M. SWITZER. Realizability and nonrealizability of Dickson Algebras as cohomology ring, *Proc. Amer. Math. Soc.* 89 (1983), 303 - 313.

Institut de Mathématiques et d'Informatique
Université de Neuchatel
Chantemerle 20
Ch-2000 NEUCHATEL
Switzerland

The boundedness conjecture for the action of the Steenrod algebra on polynomials

D. P. Carlisle R. M. W. Wood

Dedicated to the memory of J. F. Adams.

1 Introduction

This note is concerned with the action of the Steenrod algebra $\mathcal{A}$, at the prime 2, on the polynomial ring

$$S = \mathsf{F}_2[x_1, \ldots, x_n]$$

in n variables x_i over the field F_2 of two elements. We discuss a few things which have been done on the Peterson conjecture and in particular prove conjecture 2.4 in [9]. To explain what these conjectures are all about we recall that S is identified with the cohomology ring of the product of n copies of $\mathsf{R}\mathsf{P}^\infty$ and thereby receives its module structure over $\mathcal{A}$. We write S_d for the set of homogeneous polynomials of degree d. Then $S = \bigoplus S_d$ is graded by degree as a module over $\mathcal{A}$. A purely algebraic description of the module action is given briefly at the beginning of [8].

A major problem is to find a minimal generating set of homogeneous polynomials for S as an $\mathcal{A}$ module. Alternatively expressed we would like a basis for the quotient vector space

$$C = S/\mathcal{A}^+S$$

of the polynomial algebra by the image of the positive part $\mathcal{A}^+$ of the Steenrod algebra. The vector space C inherits a grading by degree from S. In [6] Bill Singer has introduced the terminology *spike* to describe a monomial

$$x_1^{e_1} x_2^{e_2} \cdots x_n^{e_n}$$

in which all exponents e_i are powers of two minus one. It is easy to see that all spikes must be included in any generating set for S because a spike can never appear as a term of any element in the image of the Steenrod algebra. Although for $n > 1$ the spikes alone do not suffice as a generating set of S, the essence of the Peterson conjecture is that a minimal generating set can be chosen from the same gradings as the spikes. Some of the motivation for studying this problem is mentioned in [9]. It stems from various sources including Frank Peterson's work on characteristic classes of manifolds [4], Bill Singer's work on the Adams spectral sequence [5] and the Peterson conjecture [6], and also from the interrelation between modular representation theory of general linear groups and splitting theory of classifying spaces in topology, a subject which has received a lot of attention by many authors in recent years [1]. Indeed the general linear group $GL(n)$ over F_2 acts on $S = \mathsf{F}_2[x_1, \ldots, x_n]$ by matrix substitution and the action commutes with that of the Steenrod squares. Hence S_d and C_d are finite dimensional representations of $GL(n)$ occurring naturally and therefore worthy of study. Of course S_d has always been of great importance in representation theory [2]. We shall give some reasons why the modules C_d are also interesting. For one thing the dimension of S_d increases with d. In contrast we now state as a theorem the previously conjectured result [9].

Theorem 1.1 *There is a function $\delta(n)$ of n only, such that* $\dim(C_d) \leqslant \delta(n)$ *for all d.*

This is the boundedness conjecture referred to in the title. Up to $n = 3$ the best values of $\delta(n)$ are $\delta(1) = 1$, $\delta(2) = 3$, $\delta(3) = 21$. Previously one of the authors had announced 27 as an upper bound for $\delta(3)$ [9] but subsequently it has been found that 21 is the best possible. This has also been proved by Masaki Kameko in his thesis [3] where more information can be found on the dimensions of vector spaces C_d and bases for them. During the Adams Symposium further results on this problem were announced by Mike Boardman. Kameko conjectures that $\dim C_d \leqslant \prod_{i=1}^{n}(2^i - 1)$. A stronger conjecture is that the C_d are quotients of the permutation representation of $GL(n)$ on the cosets of the general linear group by its 2-Sylow subgroup. The work of Boardman in the case $n = 3$ does suggest that the C_d are related to permutation representations. We explained in [9] why every simple $GL(n)$-module occurs as a composition factor in C_d for some d.

Throughout this paper we reserve the positive integer n for the number of variables in the polynomial algebra $S = \mathsf{F}_2[x_1, \ldots, x_n]$ and use other letters like e and d for exponents of variables or degrees of monomials. As usual the function $\alpha(k)$ denotes the number of digits 1 in the dyadic expansion of k.

The proof of theorem 1.1 depends on an elementary fact about representing numbers as sums of powers of two and the following result which narrows down a generating system for the module S to a set of bounded cardinality in each degree.

Theorem 1.2 *For a fixed n, the $\mathcal{A}$-module S is generated by monomials $x_1^{e_1} x_2^{e_2} \cdots x_n^{e_n}$ where, up to permutation of the variables, $\alpha(e_i + 1) \leqslant i$.*

This result is in the same mould but tantalisingly different from the the Peterson conjecture which can be phrased as follows [9].

Theorem 1.3 *A generating set for S as an $\mathcal{A}$-module can be chosen from monomials of degree d satisfying $\alpha(d+n) \leqslant n$.*

Neither result seems to imply the other. Taken together, 1.2 and 1.3 can be used in ad hoc ways to get various formats for generating monomials of the module S but we feel that there should be a better theorem yet to be discovered which subsumes both of them and reveals the best possible value of $\delta(n)$ for all n. The proofs of theorems 1.1 and 1.2 appear in section 3.

2 The block notation

It is a formidable problem to decide whether a particular monomial is or is not in the image $\mathcal{A}^+S$ of the Steenrod algebra. The strong form of the Peterson conjecture [9] can sometimes help.

Theorem 2.1 *Let $f = x_1^{e_1} x_2^{e_2} \cdots x_n^{e_n}$ be a monomial with exactly r of the exponents e_i odd numbers. Let $d = e_1 + \cdots + e_n$ be the total degree. If $\alpha(r+d) > r$ then f is in $\mathcal{A}^+S$.*

For example the monomial $x_2^2 x_3^3 x_4^4 x_5^5 x_6^6$ is in $\mathcal{A}^+S$. On the other hand the monomial $g = x_1 x_2^2 x_3^3 x_4^4 x_5^5 x_6^6$ fails to be, as one can check by specialising the variables $x_1 = x_2$, $x_4 = x_5 = x_6$ to get the spike $x_1^3 x_3^3 x_4^{15}$. Now exploit the fact that the Steenrod squares commute with matrix substitution to see that g is not in $\mathcal{A}^+S$.

In trying to cope with the combinatorial complexity of this subject we have found it helpful to introduce a pictogram system for visualising monomials and gaining some intuitive feel for the problems. A left justified block of digits 0 or 1 with n rows represents a monomial $x_1^{e_1} \cdots x_n^{e_n}$ in which the i-th row of the block is the *reverse* binary expansion of e_i.

Note that we index the rows of the block starting at 1, because the i-th row corresponds to x_i. However we index the columns of a block starting from 0, because column j corresponds to $x_i^{2^j}$.

Two examples in the case $n = 3$ are

$$B = \begin{array}{llll} 1 & 0 & 1 & \\ 0 & 1 & 1 & 1 \\ 1 & 0 & 1 & \end{array} \qquad\qquad B' = \begin{array}{lll} 1 & 0 & 1 \\ 1 & 1 & \\ 1 & & \end{array}$$

standing respectively for the monomials $x_1^5 x_2^{14} x_3^5$ and $x_1^5 x_2^3 x_3$.

A spike corresponds to a block where, in each row, no digit 0 appears to the left of a digit 1. For example:

$$\begin{matrix} 1 & 1 & 1 & 1 \\ 1 & 1 & 1 & \\ 1 & 1 & 1 & \end{matrix}$$

It will be found convenient to arrange the rows of a spike from top to bottom in descending order of magnitude of the exponents of the corresponding variables.

The *weight vector* of a block B is the vector of integers

$$w(B) = (w_0, w_1, \ldots)$$

where w_j is the number of digits 1 in column j of the block. The blocks B, B' exhibited above have weight vectors

$$w(B) = (2, 1, 3, 1), \quad w(B') = (3, 1, 1).$$

The leading component w_0 of the weight vector is of course the number of odd exponents in the monomial.

Blocks are partially ordered by lexicographically ordering their weight vectors. For example $B < B'$. This ordering is well suited to the action of the Steenrod algebra.

Lemma 2.2 *The action of any positive Steenrod operator on a non-trivial block produces blocks of strictly lower order.*

For general information on the Steenrod algebra we refer to [7]. The Cartan formula reduces the proof of lemma 2.2 to the case $n = 1$. Since the Steenrod algebra is generated by the particular operators Sq^{2^i} as i ranges over the non-negative integers it is enough to verify the statement for such a squaring operation on a power x^d of a single variable. If the reverse binary expansion of d has digit 0 in column j then x^d is annihilated by Sq^{2^j}. Otherwise the 1 in column j is added (arithmetically) to the next column. In any case the number of digits 1 in column j is diminished. There is no effect on previous columns although subsequent columns may increase or diminish through the knock on effect of adding the 1. To put it briefly the effect of a Steenrod operation on a block tends to move digits 1 from left to right. This completes the proof.

It is also evident from the proof that the total number of digits 1 in any particular row cannot be increased by the action of a Steenrod operation. Another fact worth mentioning is that any matrix substitution in a block produces blocks of no greater order. We shall not need this fact in this paper but it does show that the order relation is well behaved with respect to the action of the general linear groups.

The juxtaposition BB' of two blocks B and B' indicates a block in which the columns of B' follow the columns of B. For instance in the above example

$$BB' = \begin{array}{ccccccc} 1 & 0 & 1 & 0 & 1 & 0 & 1 \\ 0 & 1 & 1 & 1 & 1 & 1 & \\ 1 & 0 & 1 & 0 & 1 & & \end{array}$$

If BB' is a spike then clearly B and B' are spikes. Of course there are no spikes in degree d unless $\alpha(n+d) \leqslant n$ as one can see by adding a 1 to the beginning of each row of a spike. On the other hand there may be more than one spike in some degree, as the following pictures show.

$$B_1 = \begin{array}{cccc} 1 & 1 & 1 & 1 \\ 1 & 1 & 1 & \\ 1 & 1 & 1 & \\ 1 & 1 & & \\ 1 & & & \end{array} \qquad B_2 = \begin{array}{cccc} 1 & 1 & 1 & 1 \\ 1 & 1 & 1 & 1 \\ 1 & & & \\ 1 & & & \\ 1 & & & \end{array} \qquad B_3 = \begin{array}{ccccc} 1 & 1 & 1 & 1 & 1 \\ 1 & & & & \\ 1 & & & & \end{array}$$

We call a spike *minimal* if it is minimal in the partial order with respect to other spikes of the same degree.

Note that any right vertical section of a minimal spike is again minimal. Another interesting fact about a minimal spike is that no two rows can be equal unless they are the bottom two rows in our standard arrangement. For example in the above diagram only B_3 is minimal. Indeed B_1 is transformed into the lower spike B_2 by arithmetically adding the last digit 1 of the fourth row to the 1 above it in the third row and allowing the resulting digit 1 to float up to fill the vacant slot at the end of the second row. This trick works in general except when the pair of equal rows of the spike are at the bottom. Now suppose we have a minimal spike B of degree d with n non-empty rows. For a number $r < n$ add a 1 to each of the top r rows of B. This produces a number m such that $\alpha(m) = r$. It follows that $\alpha(r+d) > r$ because the sum of the digits in the remaining $n-r$ rows of B cannot exceed the smallest power of 2 in m. This leads to the following lemma.

Lemma 2.3 *Let B be a minimal spike of degree d and B' a block of the same degree. If the leading components of the weight vectors satisfy $w_0(B') < w_0(B)$ then B' is in the image of the Steenrod algebra.*

To prove this lemma we simply take $r = w_0(B')$ in the above discussion and observe that $\alpha(r+d) > r$. Interpreting r as the number of odd exponents in the monomial B' we see that B' satisfies the conditions of the strong Peterson conjecture 2.1 and is therefore in the image of the Steenrod algebra.

The proof of the strong Peterson conjecture [9] relies on the following lemma which can be used in a variety of contexts. Here χ is the anti-automorphism of the Steenrod algebra.

Lemma 2.4 *Let u, v denote polynomials in S and let Θ be an element of $\mathcal{A}$. Then*

$$u\Theta(v) - \chi(\Theta)(u)v$$

is in the image of the Steenrod algebra.

We like to paraphrase this result by saying that the action of the Steenrod squares is *transferred* from the polynomial v to the polynomial u modulo the image of the Steenrod algebra. It could be used for example on a block, split horizontally, to bring the top part into some desired format at the cost of transferring Steenrod operations onto the bottom part. Actually our first application is in a vertical block splitting to prove the following theorem which is one of the most useful techniques yet found for handling blocks.

Theorem 2.5 *Let $B = B_1B_2$ and suppose B_2 is in the image of the Steenrod algebra. Then, modulo the image of the Steenrod algebra,*

$$B = \sum_i B_i',$$

where, for all i, $B_i' < B$ in the partial order.

To prove the result we first recall a fundamental fact concerning the fractal nature of the action of Steenrod operations on polynomials.

Lemma 2.6 *Let Θ be an element in $\mathcal{A}^+$ and let Φ be obtained from Θ by replacing each occurrence of Sq^k by Sq^{2^jk} for a prescribed j. Then for any polynomial u we have*

$$\Phi(u^{2^j}) = (\Theta u)^{2^j}.$$

To continue the proof of 2.5 let u_1 and u_2 be the monomials corresponding to B_1 and B_2. Then B corresponds to $u_1u_2^{2^j}$ if B_1 has j columns. By the hypotheses of the theorem we can write

$$u_2 = \sum_i \Theta_i v_i$$

for certain elements Θ_i in $\mathcal{A}^+$. Then with the reference to lemma 2.6 we have

$$u_2^{2^j} = \sum_i \Phi_i v_i^{2^j}.$$

Working modulo the image of the Steenrod algebra lemma 2.4 now gives

$$u_1 u_2^{2^j} = \sum_i (\chi(\Phi_i)u_1)v_i^{2^j},$$

which translates into block notation as

$$B_1 B_2 = \sum_i (\chi(\Phi_i)B_1)B_i'',$$

where the monomial v_i corresponds to the block B_i''. Now by lemma 2.2 we see that $\chi(\Phi_i)B_1$ has order strictly less order than B_1. It follows that the right hand side of the equation has order strictly less than B_1B_2 and this completes the proof.

An easy consequence of this theorem is the theorem of Bill Singer [6] which is the best rule of thumb so far devised for testing if monomials lie in $\mathcal{A}^+S$.

Theorem 2.7 *If a block B is lower in order than the lowest spike of that degree then B is in $\mathcal{A}^+S$.*

The proof goes as follows. Let $w(T) = (t_0, t_1, \ldots, t_k)$ be the weight vector of the lowest spike T. Let B have weight vector $w(B) = (b_0, b_1, \ldots, b_k)$ lower than $w(T)$. Then there is a first i such that $b_i < t_i$. Now split B into sub-blocks $B = B_1B_2$, where B_2 has the same columns as B from position i onwards (so B_1 is empty if $i = 0$). Let $T = T_1T_2$ be the corresponding splitting of T. Then T_2 is a minimal spike in its own degree and $B_2 < T_2$. In fact $w_0(B_2) < w_0(T_2)$ and lemma 2.3 shows that B_2 is in the image of the Steenrod algebra. Appealing now to theorem 2.5 we see that B can be replaced by strictly lower blocks modulo the image of the Steenrod algebra. Such blocks still satisfy the starting condition that they are lower than the least spike. Iteration of the argument therefore proves the result since the process must come to a stop after finitely many steps.

We mention in passing another result, which we shall not prove here, but which acts as a companion to Singer's theorem.

Theorem 2.8 *A generating set for S can be chosen from blocks which have the degree of a spike and which are not larger than the greatest spike of that degree.*

What this means is that generators can be chosen between the orders of the lowest and highest spikes. In particular, in a degree where there is only one spike up to row permutations, generators can be taken from just one weight class. Surprisingly this doesn't seem to help in the proof of the boundedness conjecture.

Our next task is to explain certain manipulations of blocks which have proved useful in attempting to find canonical generators for the module S.

The first of these we refer to as *splicing in a section of digits 1.* Suppose a digit 1 in the i-th row and column j of a block B_i is immediately preceded by a section of m digits 0 in the same row. Let block B be formed by removing the given digit 1 from position (i, j) and inserting a digit 1 in each position $(i, j-r)$, $1 \leqslant r \leqslant m$. Similarly let blocks B_k, $k \neq i$ be formed, where in addition to the preceding manoeuvre, we also remove a digit 1 in position $(k, j-m)$, if it exists, and add it (arithmetically) to the next position in the same row. If there is no such digit 1 then define $B_k = 0$. The following pictures illustrate a case where $i = 2$, $j = 3$, $m = 2$.

$$B_2 = \begin{matrix} 1 & 1 & 1 & 1 \\ 1 & 0 & 0 & 1 \\ 1 & 1 & & \end{matrix} \quad B = \begin{matrix} 1 & 1 & 1 & 1 \\ 1 & 1 & 1 & \\ 1 & 1 & & \end{matrix} \quad B_1 = \begin{matrix} 1 & 0 & 0 & 0 & 1 \\ 1 & 1 & 1 & & \\ 1 & 1 & & & \end{matrix} \quad B_3 = \begin{matrix} 1 & 1 & 1 & 1 \\ 1 & 1 & 1 & \\ 1 & 0 & 1 & \end{matrix}$$

Lemma 2.9 *Modulo the image of the positive Steenrod algebra and modulo blocks lower in the order than* B_i *we have*

$$B_1 + B_2 + \cdots + B_n = 0.$$

The proof follows immediately by applying Sq^{2^j-m} to the block B described above and from the observations made in the proof of lemma 2.2.

Note that we may strengthen 2.9 by removing the proviso 'modulo blocks lower in the order than B_i' in the case that $m = j$, as then $\sum_i B_i = Sq^1 B$.

Of course some of the blocks B_k may have order greater than that of B_i. In the case $m = 1$, however, where splicing effectively just moves the original digit 1 in B_i one place to the left in each of the blocks B_k, there is no weight increase in any preceding column and all the blocks B_k have order no greater than that of the initial B_i. So, by iterating this particular manoeuvre, any block can be replaced, modulo the image of the Steenrod algebra and modulo blocks of lower order, by new blocks in which a specified row has all its digits 1 adjacent, in other words the exponent of the corresponding variable is a power of two minus one. In the case $n = 1$ for example this simply proves the well known fact that S is generated by powers of the form $x^{2^\lambda-1}$. Let us continue to illustrate the ad hoc use of lemma 2.9 in seeking generators for the case $n = 2$.

Theorem 2.10 *When* $n = 2$, *monomials corresponding to blocks of the forms*

$$\begin{matrix} 1 & \cdots & 1 & 1 & \cdots & 1 \\ 1 & \cdots & 1 & & & \end{matrix} \qquad \begin{matrix} 1 & \cdots & 1 & & & \\ 1 & \cdots & 1 & 1 & \cdots & 1 \end{matrix} \qquad \begin{matrix} 1 & \cdots & 1 & 1 & & & & \\ 1 & \cdots & 1 & 0 & 1 & \cdots & 1 \end{matrix}$$

form a generating set for the $\mathcal{A}$*-module* S.

By lemma 2.4, we may start with blocks of the form

$$\begin{matrix} 1 & 1 & \cdots & 1 & & & \\ k_0 & k_1 & \cdots & k_{r-1} & k_r & \cdots & k_{r+s} \end{matrix}$$

where the first row consists of r adjacent digits 1 and each k_j is 0 or 1. If any $k_j = 1$, $j \neq r$ is preceded by a digit 0 then, modulo the image of $\mathcal{A}$, the splicing process (with $m = 1$) produces blocks of strictly lower order. The same applies to the case $k_r = 1$ if it is preceded by a section of at least two digits 0, when we take $m = 2$. The result follows by induction on the order of the monomials, the base case of the induction being a trivial case of Singer's theorem 2.7.

The first two blocks shown in 2.10 are of course spikes. An alternative format for the third block is as follows.

$$\begin{matrix} 1 & \cdots & 1 & 1 & \cdots & 1 & \\ 1 & \cdots & 1 & 0 & \cdots & 0 & 1 \end{matrix}$$

For certain degrees d these three types can exist simultaneously and represent linearly independent generators of C_d. More information on the $GL(2)$-module structure of C_d is given in [9, Proposition 2.5]. In particular $\delta(2) = 3$ is best possible.

3 The boundedness conjecture

In this section we embark on the proof of theorem 1.1. The following picture illustrates the format of a typical generator described in theorem 1.2.

$$\begin{matrix} 1 & 1 & 1 & 1 & 1 & \\ 1 & 1 & 1 & 0 & 0 & 1 \\ 1 & 0 & 1 & 0 & 1 & \\ 0 & 1 & 0 & 0 & 1 & \\ 0 & 0 & 1 & 1 & 1 & \end{matrix}$$

The exponent e_i of the i-th variable satisfies the condition $\alpha(e_i + 1) \leqslant i$. This amounts to saying that the general block in our desired format has a spike at the top left, together with some 'floating' digits 1 such that the i-th row has no more than $i - 1$ of these floating digits. For present purposes we shall call a block *regular* if it satisfies this condition after possible re-ordering of the rows. Theorem 1.2 now states that a block is a sum of regular blocks modulo the image of the Steenrod algebra.

The proof of 1.2 is by double induction on degree and number of variables but not this time on the partial ordering of blocks, which may in fact rise at certain points in the procedure to be described.

The statement of 1.2 is clearly true in the one variable case and for a block T consisting of a single column of digits 1 in the n variable case. The inductive argument proceeds in three stages. We start with a general block B.

In the first step of the proof we replace, modulo the image of $\mathcal{A}$, the given block by a sum of blocks each of which is either regular, or has at least one digit 0 in column 0.

So consider the opposite situation of a block of the form $B = TB'$, where as above T consists of a single column of digits 1. Note that the action of a positive Steenrod square on T produces blocks in which the leading column has at least one digit 0. We may assume by induction on degree that the result is proved for B'. Then, transferring the action of Steenrod squares onto T in the manner described in the the proof of theorem 2.5 B is replaced either by blocks of the form TB_1, where B_1 is regular, or by $T'B_2$, where T' is a single column block with at least one digit 0. The former case is already regular and we have achieved the first step.

The aim of the second step is to replace the original block by a sum of blocks of type B in which, after possible re-ordering of the rows, the block consisting of the first $n-1$ rows of B is regular and the last row has a leading zero.

By step one we may start with a block whose leading column has at least one digit 0. Re-order the rows of the block so that the last row has a leading 0. Let G denote the block formed by the first $n - 1$ rows. By the inductive hypothesis on the number of variables and application of lemma 2.4 we can write G as a sum of regular blocks plus elements of the Steenrod algebra acting on blocks which are also regular. Transferring the Steenrod operations onto the last row preserves the leading 0 of the last row and does not increase the number of digits 1 in that row. This concludes the second step.

The aim of the third step is to replace the original block by a sum of blocks of type B in which the block consisting of the first $n - 1$ rows of B is regular, just as in step two, but this time the last row of B has a leading 0 and at most $n - 1$ digits 1.

To achieve this we start with a block B as produced by step two, and splice into the last row a leading section of digits 1 as explained in lemma 2.9. This results in Sq^1 being transferred to the block G. As G is in regular form, each of the blocks in Sq^1G will have at least one row with leading digit 0 and at most $n - 1$ digits 1.

We now repeat step 2, choosing this row to be the last. Observing that Steenrod operations, applied to the last row, do not increase the number of digits 1, we see that the resulting blocks have their first $n-1$ rows in regular form and a last row with leading 0 and not more than $n - 1$ digits 1. The whole block is therefore regular. This completes the proof of 1.2.

We like to think of theorem 1.2 as a 'horizontal' theorem because it uses a criterion on the rows of the block whereas the strong Peterson conjecture uses 'vertical' information on the columns.

As a matter of interest the above proof can be pushed a bit further to reveal a stronger statement. Let r be the number of digits 1 in column 0 of a block. Call a block *special* if, after possible permutation of its rows, it satisfies the condition $\alpha(e_i+1) \leqslant \min(i, r+1)$. This implies that the block is regular but, more than this, the number of 'floating' digits 1 does not exceed r in any row. In particular any row starting with a digit 0 has at most r digits 1 in it. The picture at the beginning of this section has this property with $r = 3$.

We strengthen theorem 1.2 as follows:

Theorem 3.1 *A block may be written, modulo the image of $\mathcal{A}$, as a sum of special blocks.*

The proof is similar to the one above but with an additional induction on the number of digits 1 in column 0. By theorem 1.2 we may start with a regular block. If there are no zeros in column 0 then the block is already special. Assume therefore, as in step two above, that the last row has a leading zero. Now proceed as in step two with special replacing regular. This produces blocks with the first $n-1$ rows in special form and a leading 0 in the last row. For such a block let r be the number of digits 1 in column 0.

We now follow step three noting this time that the number of digits 1 in the last row of each block produced is at most r. Now let r' be the number of digits 1 in column 0 of such a block. If $r' \geqslant r$ then the block is special. If $r' < r$ then we iterate step three. Hence the result follows by induction on the number of digits 1 in column 0, noting that if all rows begin with a zero then the block is a square and therefore in the image of the Steenrod algebra.

It remains to show how theorem 1.2 implies theorem 1.1. This is a question of elementary number theory. We start with the following fact about dyadic expansions.

Lemma 3.2 *Let $\epsilon'(m,k)$ denote the number of ways of representing a positive integer k as a sum of not more than m powers of 2 allowing repetitions. Then there is a function $\epsilon(m)$ such that $\epsilon'(m,k) < \epsilon(m)$ for all k.*

The proof is by induction on m. The statement is clearly true for $m = 1$. In this case $\epsilon(m) = 1$ since every number is or is not a power of two. Now in any representation of k as a sum

$$k = k_1 + k_2 + \cdots + k_m,$$

where each k_i is zero or a power of two, at least one of the k_i must be of the form 2^l, where l lies in the range

$$\log_2(k/m) \leqslant l \leqslant \log_2(k).$$

But there are at most $\log_2(m) + 1$ integers in this range and therefore

$$\epsilon'(m,k) \leqslant [\log_2(m) + 1]\epsilon(m-1).$$

This establishes the inductive step and shows that $\epsilon'(m,k)$ is bounded as function of m independent of k.

To complete the proof of theorem 1.1, let $k = d + n$ and $f_i = e_i + 1$ in the usual notation for d, n, e_i. We want to show that the number of ordered ways of representing k in the form

$$k = f_1 + f_2 + \cdots + f_n,$$

where $\alpha(f_i) \leqslant i$, is bounded as a function of n independent of d. Taking the dyadic expansion of each k_i we see that such a representation gives rise to an expansion of k into not more than

$$1 + 2 + \cdots + n = n(n+1)/2 = m$$

powers of two as in lemma 3.2. Since the number of orderings and the number of partitions of n are functions of n only, the proof of theorem 1.1 now follows from lemma 3.2.

Of course the upper estimate for $\delta(n)$ that comes out of the above proof is hopelessly large and what is needed is more insight into the nature of the $GL(n)$ representations C_d.

We conclude this article by exhibiting a collection of blocks whose equivalence classes form set of generators for C_d in the case $n = 3$. They are obtained by the ad hoc use of the processes explained above. The numbers in brackets show how many row permutations are needed.

The blanks between digits 1 in any row denote digits 0 and the ellipsis denotes digits 1. The blocks are formed essentially by starting with a spike and moving the last two digits 1 of the first row and the last digit 1 of the second row downwards in various combinations. What we are describing here is the 'generic' case. In some degrees patterns will coalesce or not be possible. In general however all six row permutations of the last block are needed because these are spikes. On the other hand there are linear relations between permutations of the other blocks. To get an independent set we can make do with just one block of the first type, the three cyclic permutations of the second, all but one of the six permutations of the third and the three cyclic permutations of the next two types. In all this gives 21 as the generic degree for C_d.

$$\begin{array}{ccccccccccc} 1 & \cdots & 1 & 1 & \cdots & 1 & 1 & \cdots & 1 & & \\ 1 & \cdots & 1 & 1 & \cdots & 1 & & & & 1 & \\ 1 & \cdots & 1 & & & & 1 & & & & 1 \end{array} \qquad (1)$$

$$\begin{array}{ccccccccccc} 1 & \cdots & 1 & 1 & \cdots & 1 & 1 & \cdots & 1 & 1 & \\ 1 & \cdots & 1 & 1 & \cdots & 1 & & & & & 1 \\ 1 & \cdots & 1 & & & & 1 & & & & \end{array} \qquad (3)$$

$$\begin{array}{ccccccccccc} 1 & \cdots & 1 & 1 & \cdots & 1 & 1 & \cdots & 1 & 1 & \\ 1 & \cdots & 1 & 1 & \cdots & 1 & 1 & & & & \\ 1 & \cdots & 1 & & & & & & & & 1 \end{array} \qquad (5)$$

$$\begin{array}{ccccccccccc} 1 & \cdots & 1 & 1 & \cdots & 1 & 1 & \cdots & 1 & 1 & 1 \\ 1 & \cdots & 1 & 1 & \cdots & 1 & & & & & \\ 1 & \cdots & 1 & & & & 1 & & & & \end{array} \qquad (3)$$

$$\begin{array}{ccccccccccc} 1 & \cdots & 1 & 1 & \cdots & 1 & 1 & \cdots & 1 & 1 & \\ 1 & \cdots & 1 & 1 & \cdots & 1 & 1 & & & & 1 \\ 1 & \cdots & 1 & & & & & & & & \end{array} \qquad (3)$$

$$\begin{array}{ccccccccccc} 1 & \cdots & 1 & 1 & \cdots & 1 & 1 & \cdots & 1 & 1 & 1 \\ 1 & \cdots & 1 & 1 & \cdots & 1 & 1 & & & & \\ 1 & \cdots & 1 & & & & & & & & \end{array} \qquad (6)$$

References

[1] D. P. Carlisle and N. J. Kuhn. *Subalgebras of the Steenrod algebra and the action of matrices on truncated polynomial algebras.* J. Algebra **121** (1989), 370–387.

[2] G. D. James and A. Kerber *The representation theory of the symmetric group.* Encyclopaedia of mathematics, Vol **16** Addison-Wesley, London (1981).

[3] M. Kameko. *Products of projective spaces as Steenrod modules.* Thesis, John Hopkins University.

[4] F. P. Peterson *A-generators for certain polynomial algebras* Math. Proc. Camb. Phil. Soc. **105** (1989) 311–312.

[5] W. M. Singer. *The transfer in homological algebra.* Math. Z. **202** (1989), 493–525.

[6] W. M. Singer. *On the action of Steenrod squares on polynomials.* Proc. A.M.S. **111** (1991), 577–583.

[7] N. E. Steenrod and D. B. A. Epstein. *Cohomology Operations.* Annals of Math. Studies **50** Princeton University Press, 1962.

[8] R. M. W. Wood. *Steenrod squares of polynomials and the Peterson conjecture.* Math. Proc. Camb. Phil. Soc. **105** (1989) 307–309.

[9] R. M. W. Wood. *Steenrod squares of polynomials.* Advances in Homotopy Theory, LMS Lecture notes series **139** CUP (1989) 173–177.

D. P. Carlisle
Computer Science Department
Manchester University
Oxford Road
Manchester
M13 9PL

R. M. W. Wood
Mathematics Department
Manchester University
Oxford Road
Manchester
M13 9PL

Representations of the Homology of BV and the Steenrod Algebra I

Mohamed Ali Alghamdi, M. C. Crabb and J. R. Hubbuck

1. Introduction

Let $A_k^* = \mathsf{F}_2[t_1, t_2, \ldots, t_k]$ be the graded unstable algebra over the mod 2 Steenrod algebra where each t_i has grading one. Interesting properties of this ring have been obtained by Lannes, Adams-Gunawardena-Miller, Carlisle-Kuhn, Wood and others. Among them is the result [1]

$$\mathrm{End}_{\mathcal{A}(2)} A_k^* = \mathsf{F}_2[M_k(\mathsf{F}_2)],$$

that the ring of endomorphisms of A_k^* over the Steenrod algebra is isomorphic to the semi-group ring of $k \times k$-matrices over F_2. Here V_k denotes the elementary Abelian 2-group F_2^k of rank k and so A_k^* can be identified with $H^*(BV_k, \mathsf{F}_2)$. This is the first note in which we explore related ideas: the general linear group $GL_k(\mathsf{F}_2) = GL(V_k)$ acts on $H^*(BV_k, \mathsf{F}_2)$ and we reveal explicit relationships between its representations and the action of the Steenrod algebra.

It is convenient to consider the dual situation of the Pontrjagin ring $H_*(BV_k, \mathsf{F}_2)$ over $\mathcal{A}(2)$ where we define the action of the Steenrod squares Sq_i by $(Sq^i a, x) = (a, Sq_i x)$, $a \in H^*(BV_k, \mathsf{F}_2)$, $x \in H_*(BV_k, \mathsf{F}_2)$. We consider two subrings of $H_*(BV_k, \mathsf{F}_2)$. The first is that annihilated by elements of positive degree in the Steenrod algebra and the second is a subring of this. Both are stable under the action of $GL(V_k)$ but the representations of the latter are described more simply. We investigate these in detail when $k = 3$.

One of the unsolved problems in this area is to find minimal sets of generators for A_k^* as a module over $\mathcal{A}(2)$ [4]. For k equal to one or two this is a routine computation but was a well known unresolved problem for k equal to three until Masaki Kameko announced the solution at the Adams Memorial Conference in June 1990; in particular he obtained explicit formulae for the dimensions of $F_2 \otimes_{\mathcal{A}(2)} A_3^*$ [3]. The first-named author was at that time working on the latter problem in the dual formulation for his PhD thesis at Aberdeen University using different techniques and this was essentially completed in November 1990 (with the benefit of Kameko's results as a check on accuracy in the final stages). This result is needed to relate the two rings mentioned above and a version of the proof is included below. The final section of the paper describes the representation theory.

The referee has informed us that the third example we give in section 4 was conjectured some years ago by David Carlisle and verified in the dual formulation by Mike Boardman. Also Mike Boardman displayed a poster at the Adams Memorial Conference which "exhibited the module structures in the cokernel version"; unfortunately neither of the authors present read it.

We are grateful to Reg Wood for stimulating conversations on these matters.

2. The rings $M(k)$ and $L(k)$

We write $M(k)$ for the subring of $H_*(BV_k, F_2)$ annihilated by elements of positive grading in $\mathcal{A}(2)$.

As $BV_1 = RP^\infty$, it is well known that $H^*(BV_1, F_2) \simeq F_2[a_1]$ and the Pontrjagin ring $H_*(BV_1, F_2) \simeq F_2[u_{2^i}]/(u_{2^i}^2)$, $i \geq 0$. The latter is a biassociative, bicommutative Hopf algebra whose comultiplication is determined by the formula $\Delta_*(u_t) = \sum_{i=0}^{t} u_i \otimes u_{t-i}$, where $u_0 = 1$ and $u_i = u_{2^\alpha} u_{2^\beta} \dots u_{2^\omega}$ if the non zero part of the binary expansion of i is $2^\alpha + 2^\beta + \dots + 2^\omega$. The action of the Steenrod algebra is determined by the usual Cartan formula and for $i > 0$,

$$Sq_i u_{2^r} = \begin{cases} u_{2^j} u_{2^{j+1}} \dots u_{2^{r-1}} & \text{if } i = 2^j < 2^r \\ 0 & \text{otherwise} \end{cases}.$$

A routine calculation establishes that $\{u(s) = u_1 u_2 \dots u_{2^{s-1}} \ : s \geq 1\}$ is a basis for $M(1)$ (in positive dimensions) over F_2.

Each $v \in V_k - \{0\}$ defines a homomorphism $V_1 = F_2 \to V_k$ by inclusion onto $\{0, v\}$ and therefore a homomorphism $v_* : H_*(BV_1, F_2) \to H_*(BV_k, F_2)$. We set

$$a_s(v) = v_*(u(s)).$$

The ring $L(k)$ is defined to be the subring of $H_*(BV_k, F_2)$ generated by $\{a_s(v) : s \geq 1, \ v \in V - \{0\}\}$.

Then $L(1) = M(1)$ and it is not difficult to verify that $L(2) = M(2)$. Clearly $L(k) \subseteq M(k)$, but in general they are not equal although it seems probable that, in the appropriate sense, they are equal in almost all degrees. We will restrict attention for the remainder of this paper to the case $k = 3$

and establish that $L(3) = M(3)$ except in degrees $2^{s+3} + 2^{s+1} + 2^s - 3$ for $s \geq 0$ where $L(3)$ has dimension 14 as a vector space over $\mathbf{F}_2$ and $M(3)$ has dimension 15. We simplify notations by writing L, M, V for $L(3)$, $M(3)$ and V_3.

Identifying V with $\mathbf{F}_2 \oplus \mathbf{F}_2 \oplus \mathbf{F}_2$, there are seven embeddings $\mathbf{F}_2 \to V$ denoted by $(1,0,0), (0,1,0), (0,0,1), (0,1,1), (1,0,1), (1,1,0), (1,1,1)$. With this identification, we have $BV = \mathbf{R}P^\infty \times \mathbf{R}P^\infty \times \mathbf{R}P^\infty$ and $H_*(BV, \mathbf{F}_2) = \mathbf{F}_2[x_{2^i}, y_{2^i}, z_{2^i}]/(x_{2^i}^2, y_{2^i}^2, z_{2^i}^2)$, $i \geq 0$. The following computation is the key to subsequent calculations.

Proposition 2.1. *The elements $a_s(v)$ for $v \in V - \{0\}$ are*

$$x(s) = x_1 x_2 \ldots x_{2^s-1}, \quad y(s) = y_1 y_2 \ldots y_{2^s-1}, \quad z(s) = z_1 z_2 \ldots z_{2^s-1},$$

$$yz(s) = (y_1 + z_1)(y_2 + z_2) \ldots (y_{2^s-1} + z_{2^s-1}),$$

$$xz(s) = (x_1 + z_1)(x_2 + z_2) \ldots (x_{2^s-1} + z_{2^s-1}),$$

$$xy(s) = (x_1 + y_1)(x_2 + y_2) \ldots (x_{2^s-1} + y_{2^s-1}),$$

$$xyz(s) = (x_1 + y_1 + z_1)(x_2 + y_2 + z_2 + y_1 z_1 + x_1 z_1 + x_1 y_1) \ldots$$

$$\ldots (x_{2^s-1} + y_{2^s-1} + z_{2^s-1} + y_{2^s-2} z_{2^s-2} + x_{2^s-2} z_{2^s-2} + x_{2^s-2} y_{2^s-2}).$$

Proof. The first three elements correspond to $(1,0,0)$, $(0,1,0)$ and $(0,0,1)$.

We define x_i, y_i, z_i when i is not a power of 2 as was done for u_i. The key observation in the calculations below is that if $i \equiv 2^t \bmod 2^{t+1}$, then $x_i = x_{2^t} x_{2^{t_2}} \ldots x_{2^{t_s}}$ with $t < t_2 < \ldots < t_s$, as the square of each x_{2^i} is zero.

Then arguing by induction on s for $v = (0,1,1)$, we have $a_s(v) = ((c \times \Delta)\Delta)_* u(s) = yz(s-1)(\sum y_{2^{s-1}-i} z_i)$, with the summation running over $0 \leq i \leq 2^{s-1}$ and c the constant map. If $0 < i < 2^{s-1}$ and $i \equiv 2^t \bmod 2^{t+1}$, then $2^{s-1} - i \equiv 2^t \bmod 2^{t+1}$ and so $yz(s-1) y_{2^{s-1}-i} z_i = 0$. Thus $a_s(v) = yz(s)$.

To complete the proof we must show that $(1 \otimes \Delta_*)\Delta_* u(s) = xyz(s)$. Again arguing by induction on s, we can assume that

$$(1 \otimes \Delta_*)\Delta_* u(s) = xyz(s-1)(\sum x_i y_j z_k)$$

where the summation is over all non negative integer triples (i,j,k) with sum 2^{s-1}. We rewrite the terms under the summation sign as

$$w_1 + w_2 + w_4 + \ldots$$

$$\ldots + w_{2^s-3} + (y_{2^s-2} z_{2^s-2} + x_{2^s-2} z_{2^s-2} + x_{2^s-2} y_{2^s-2}) + (x_{2^s-1} + y_{2^s-1} + z_{2^s-1})$$

where w_{2^t}, $0 \le t \le s-3$ is the sum of all terms with two of $\{i,j,k\}$ congruent to 2^t mod 2^{t+1}, and therefore with precisely two such terms. Then by the symmetry present $w_{2^t} = (y_{2^t-1} z_{2^t-1} + x_{2^t-1} z_{2^t-1} + x_{2^t-1} y_{2^t-1})w'_{2^t}$. So considering $xyz(s-1)w_{2^t}$, we can replace the factor $(x_{2^t} + y_{2^t} + z_{2^t} + y_{2^t-1} z_{2^t-1} + x_{2^t-1} z_{2^t-1} + x_{2^t-1} y_{2^t-1})$ in $xyz(s-1)$ by $(x_{2^t} + y_{2^t} + z_{2^t})$. But again by symmetry and dimensional considerations $w'_{2^t} = (x_{2^t} + y_{2^t} + z_{2^t})w''_{2^t}$. So $xyz(s-1)w_{2^t} = 0$ and $xyz(s-1)(\sum x_i y_j z_k) = xyz(s)$ as required.

It is clear from Proposition 2.1 that the product of any four classes $a_s(v)$ is zero and so we deduce

Theorem 2.2. *$L_n = 0$ unless $n = 2^\alpha + 2^\beta + 2^\gamma - 3$ with $\alpha \ge \beta \ge \gamma \ge 0$.*

We now list products of generators.

Table 1

For $s > 0$,

$$y(s)z(s) = y(s)yz(s) = z(s)yz(s)$$
$$x(s)z(s) = x(s)xz(s) = z(s)xz(s)$$
$$x(s)y(s) = x(s)xy(s) = y(s)xy(s)$$

$$x(s)yz(s) = yz(s)xyz(s) = x(s)xyz(s)$$
$$y(s)xz(s) = xz(s)xyz(s) = y(s)xyz(s)$$
$$z(s)xy(s) = xy(s)xyz(s) = z(s)xyz(s)$$

$$xy(s)xz(s) = xy(s)yz(s) = xz(s)yz(s) = t(s),$$

where $t(s) = (y_1 z_1 + x_1 z_1 + x_1 y_1)(y_2 z_2 + x_2 z_2 + x_2 y_2)\dots$

$$\dots(y_{2^s-1} z_{2^s-1} + x_{2^s-1} z_{2^s-1} + x_{2^s-1} y_{2^s-1}).$$

Definition. The operator e acts on $H_*(BV, \mathsf{F}_2)$ by doubling subscripts and therefore dimensions, for example $ex_1 y_2 z_4 = x_2 y_4 z_8$.

Table 2

For $t > s > 0$ with $u = t - s$ and E the iterated operator e^s,

$$y(s)z(t) = yz(s)z(t) = y(s)z(s)Ez(u)$$
$$z(s)y(t) = yz(s)y(t) = y(s)z(s)Ey(u)$$
$$y(s)yz(t) = z(s)yz(t) = y(s)z(s)Eyz(u)$$

$$x(s)z(t) = xy(s)z(t) = x(s)z(s)Ez(u)$$
$$z(s)x(t) = xz(s)x(t) = x(s)z(s)Ex(u)$$
$$x(s)xz(t) = z(s)xz(t) = x(s)z(s)Exz(u)$$

$$x(s)y(t) = xy(s)y(t) = x(s)y(s)Ey(u)$$
$$y(s)x(t) = xy(s)x(t) = x(s)y(s)Ex(u)$$
$$x(s)xy(t) = y(s)xy(t) = x(s)y(s)Exy(u)$$

$$x(s)yz(t) = xyz(s)yz(t) = x(s)yz(s)Eyz(u)$$
$$yz(s)x(t) = xyz(s)x(t) = x(s)yz(s)Ex(u)$$
$$x(s)xyz(t) = yz(s)xyz(t) = x(s)yz(s)Exyz(u)$$

$$y(s)xz(t) = xyz(s)xz(t) = y(s)xz(s)Exz(u)$$
$$xz(s)y(t) = xyz(s)y(t) = y(s)xz(s)Ey(u)$$
$$y(s)xyz(t) = xz(s)xyz(t) = y(s)xz(s)Exyz(u)$$

$$z(s)xy(t) = xyz(s)xy(t) = z(s)xy(s)Exy(u)$$
$$xy(s)z(t) = xyz(s)z(t) = z(s)xy(s)Ez(u)$$
$$z(s)xyz(t) = xy(s)xyz(t) = z(s)xy(s)Exyz(u)$$

$$xy(s)yz(t) = xz(s)yz(t) = t(s)Eyz(u)$$
$$xy(s)xz(t) = yz(s)xz(t) = t(s)Exz(u)$$
$$xz(s)xy(t) = yz(s)xy(t) = t(s)Exy(u)$$

Definition. We define the operator

$$f : H_*(BV_k, \mathsf{F}_2) \to H_*(BV_k, \mathsf{F}_2)$$

by $f(w) = x_1 y_1 z_1 e(w)$.

Explicit formulae for triple products of generators are easily deduced from Tables 1 and 2 above. If $u(\alpha)$, $v(\beta)$, $w(\gamma)$ are three generators with $\alpha \geq \beta \geq \gamma \geq 0$, then either $u(\alpha)v(\beta)w(\gamma) = 0$ or $u(\alpha)v(\beta)w(\gamma) = f^{\gamma-1} x_1 y_1 z_1$ if $\alpha = \beta = \gamma > 0$ or $f^{\gamma} u(\alpha - \gamma)v(\beta - \gamma)$ otherwise.

Conversely a routine verification establishes that $f^{\gamma} u(\alpha)v(\beta) = u(\alpha + \gamma)v(\beta + \gamma)w(\gamma)$ for some $w(\gamma)$.

The verification of Tables 1 and 2 follows from Proposition 2.1.

We wish to interpret these results more geometrically. From Tables 1 and 2 it follows that

Proposition 2.3. *If $u, v \in V - \{0\}$, $u \neq v$, $t \geq s$, then*

$$a_s(u)a_t(v) = a_s(u+v)a_t(v).$$

Let E be a two dimensional subspace of V. By Proposition 2.3 we may write unambiguously

$$a_s(E) = a_s(u)a_s(v)$$

where $\{u, v\}$ is a basis for E, as can be checked explicitly from Table 1. For uniformity of notation, we will write $a_s(l)$ for $a_s(v)$ when l is the line containing v. Also by Proposition 2.3 if $t > s$ and $l \subset E$, $a_{s,t}(E, l) = a_s(u)a_t(v)$ is well defined when $l = \langle v \rangle$ and $E = \langle v, u \rangle$. We will require to use certain relationships satisfied by the $a_s(l)$, $a_s(E)$ and $a_{s,t}(E, l)$.

(2.4) $$\sum_{l \subset V} a_1(l) = 0$$

(2.5) $$\sum_{E \subset V} a_2(E) = 0$$

(2.6) $$\text{For fixed } E, \quad \sum_{l \subset E} a_1(l) = 0$$

(2.7) $$\text{For fixed } E, \quad \sum_{l \subset E} a_{s,s+1}(E, l) = 0 \quad (s \geq 1)$$

(2.8) For fixed l, $\sum_{E \supset l} a_1(E) = 0$

(2.9) For fixed l, $\sum_{E \supset l} a_{1,t}(E, l) = 0 \quad (t > 1)$

Equations (2.4),(2.5),(2.6) and (2.8) can be checked directly from the definitions; (2.7) and (2.9) can then be read off from Table 2.

We now write out bases for L_n when (a) $n = 2^s - 2$, (b) $n = 2^s - 1$, (c) $n = 2^t + 2^s - 2$, $t > s > 0$ and finally (d) $2^\alpha + 2^\beta + 2^\gamma - 3 \neq 2^s - 1$ with $\alpha \geq \beta \geq \gamma > 0$.

Proposition 2.10(a) L_{2^s-2}, $s \geq 2$.
(i) *If $s > 3$, L_{2^s-2} has dimension 7 and a basis of elements as described in Table 1 or equivalently as $\{a_{s-1}(E) : E \subset V\}$.*
(ii) *If $s = 3$, L_6 has dimension 6 and a basis $\{y(2)z(2), x(2)z(2), x(2)y(2), x(2)yz(2), y(2)xz(2), z(2)xy(2)\}$.*

$$L_6 = \langle a_2(E) : E \subset V \rangle / \{\sum_{E \subset V} a_2(E)\},$$

the quotient of the free vector space on $\{a_2(E)\}$ by the subspace generated by $\{\sum_{E \subset V} a_2(E)\}$.
(iii) *If $s = 2$, L_2 has dimension 3 and a basis $\{y(1)z(1), x(1)z(1), x(1)y(1)\}$.*

$$L_2 = \langle a_1(E) : E \subset V \rangle / \{\sum_{E \supset l} a_1(E) : l \subset V\}$$

Proof. (a) The only solution of $2^\alpha + 2^\beta + 2^\gamma - 3 = 2^s - 2$ with $\alpha \geq \beta \geq \gamma \geq 0$ is $\alpha = \beta = s - 1$, $\gamma = 0$. So $L_{2^s-2} = L_{2^{s-1}-1} . L_{2^{s-1}-1}$. Also the solutions of $2^\alpha + 2^\beta + 2^\gamma - 3 = 2^{s-1} - 1$ with $\alpha \geq \beta \geq \gamma \geq 0$ are $\alpha = \beta = s - 1$, $\gamma = 0$ or if $s > 2$, $\alpha = \beta = s - 2$, $\gamma = 1$. Let $\tilde{L}_{2^{s-1}-1} =$

$$\langle x(s-1), y(s-1), z(s-1), yz(s-1), xz(s-1), xy(s-1), xyz(s-1) \rangle.$$

Using Theorem 2.2, $L_{2^s-2} = \tilde{L}_{2^{s-1}-1} . \tilde{L}_{2^{s-1}-1}$. Thus the elements listed in Table 1 span L_{2^s-2}. Routine computations establish the appropriate linear independence conditions.

We define $\tilde{L}_{2^s-1}$ as in the proof above.

Proposition 2.10(b) L_{2^s-1}, $s \geq 1$.
(i) *For* $s > 2$, $L_{2^s-1} = \tilde{L}_{2^s-1} \oplus f L_{2^{s-1}-2}$ *and* $L_3 = \tilde{L}_3$. *For* $s > 1$, $\tilde{L}_{2^s-1}$ *has dimension* 7 *and a basis consisting of the elements described in Proposition 2.1 or equivalently* $\{a_s(l) : l \subset V\}$. L_{2^s-1} *has dimension* 7 *for* $s = 2$, 10 *if* $s = 3$, 13 *if* $s = 4$ *and* 14 *if* $s > 4$.
(ii) $L_1 = V$ *has dimension* 3. *Alternatively*

$$L_1 = \langle a_1(l) : l \subset V \rangle / \{\sum_{l \subset E} a_1(l) : E \subset V\}.$$

Proof. (b) As mentioned in the proof of (a), for $s > 2$ the two solutions of $2^\alpha + 2^\beta + 2^\gamma - 3 = 2^s - 1$ with $\alpha \geq \beta \geq \gamma \geq 0$ are $\alpha = s$, $\beta = \gamma = 0$ and $\alpha = \beta = s-1$, $\gamma = 1$. So $L_{2^s-1} = \tilde{L}_{2^s-1} \oplus f L_{2^{s-1}-2}$. Checking linear independence is again routine.

Proposition 2.10(c) $L_{2^t+2^s-2}$, $t > s \geq 1$.
(i) *For* $t > s+1 > 2$, $L_{2^t+2^s-2}$ *has dimension* 21 *with basis as described in Table 2 or as* $\{a_{s,t}(E,l) : l \subset E \subset V\}$.
(ii) *For* $t = s+1 > 2$, $L_{2^{s+1}+2^s-2}$ *has dimension* 14 *and a basis is* $\{y(s)z(t), z(s)y(t), x(s)z(t), z(s)x(t), x(s)y(t), y(s)x(t), x(s)yz(t), y(s)xz(t), z(s)xy(t), x(t)yz(s), y(t)xz(s), z(t)xy(s), t(s)e^s xz(1), t(s)e^s yz(1)\}$. *Alternatively*

$$L_{2^{s+1}+2^s-2} = \langle a_{s,s+1}(E,l) : l \subset E \subset V \rangle / \{\sum_{l \subset E} a_{s,s+1}(E,l) : E \subset V\}.$$

(iii) *For* $t > 2$, $s = 1$, L_{2^t} *has dimension* 14 *with basis* $\{y(1)z(t), z(1)y(t), x(1)z(t), z(1)x(t), x(1)y(t), y(1)x(t), x(1)yz(t), y(1)xz(t), z(1)xy(t), x(1)xz(t), y(1)yz(t), z(1)xz(t), y(1)xyz(t), z(1)xyz(t)\}$.

$$L_{2^t} = \langle a_{1,t}(E,l) : l \subset E \subset V \rangle / \{\sum_{E \supset l} a_{1,t}(E,l) : l \subset V\}.$$

(iv) *For* $t = 2$, $s = 1$, L_4 *has dimension* 8 *and basis* $\{y(1)z(2), z(1)y(2), x(1)z(2), z(1)x(2), x(1)y(2), y(1)x(2), x(1)yz(2), y(1)xz(2)\}$.

$$L_4 =$$

$$\langle a_{1,2}(E,l) : l \subset E \subset V \rangle / \{\sum_{E \supset l} a_{1,2}(E,l) : l \subset V;\ \sum_{l \subset E} a_{1,2}(E,l) : E \subset V\}.$$

Proof. (c) If $2^\alpha + 2^\beta + 2^\gamma - 3 = 2^t + 2^s - 2$ with $\alpha \geq \beta \geq \gamma \geq 0$, $t > s \geq 1$, then $\gamma = 0$, $\beta = s$ and $\alpha = t$. Part (i) follows easily by inspecting Table 2. If $t = s+1 > 3$ in (ii) again the result is clear and the case $t = 3$, $s = 2$ can be verified by direct computation.

If $t > s = 1$, then $L_{2^s} = L_{2^s-1} . L_1 = \tilde{L}_{2^s-1} . L_1$ and again (iii) and (iv) are routine computations.

Proposition 2.10(d) *L_n with $n = 2^\alpha + 2^\beta + 2^\gamma - 3$, $\alpha \geq \beta \geq \gamma > 0$, $n \neq 2^s - 1$.*
(i) *If $\alpha = \beta = \gamma$, $L_n = f^{\gamma-1} L_3$,*
(ii) *Otherwise, $L_n = f^\gamma L_{2^{\alpha-\gamma}+2^{\beta-\gamma}-2}$.*

Bases in all cases can be read off from parts (a), (b) and (c) using the monomorphism f.

Proof. The result follows from $L_n = L_{2^\alpha-1} . L_{2^\beta-1} . L_{2^\gamma-1} = \tilde{L}_{2^\alpha-1} . \tilde{L}_{2^\beta-1} . \tilde{L}_{2^\gamma-1}$ and the comments following Table 2.

A scrutiny of the table in section 8 of [**3**] which gives the dimension of M_n (with corrections provided by Kameko) and of Theorem 2.3 and Proposition 2.10(a),(b),(c),(d) above which give the dimension of L_n yields the following.

Theorem 2.11. *$L_n = M_n$ unless $n = 2^{s+3} + 2^{s+1} + 2^s - 3$, $s \geq 0$ when L_n has dimension* 14 *and M_n has dimension* 15.

We will give in the next section a proof of this theorem. It calculates the dimension of M_n in a way which at least superficially is very different from the methods of [**3**]. We provide some details as although cumbersome calculations are not avoided in the second half of the proof, the first part is sufficiently conceptual to suggest how one might generalize for $k > 3$.

3. The proof of Theorem 2.11

The general homogeneous $w \in H_*(BV, \mathbb{F}_2)$ will be written in the form

$$w = \sum_{\delta \in \Delta(w)} x_1^{\alpha_0} y_1^{\beta_0} z_1^{\gamma_0} x_2^{\alpha_1} y_2^{\beta_1} z_2^{\gamma_1} \ldots x_{2^s}^{\alpha_s} y_{2^s}^{\beta_s} z_{2^s}^{\gamma_s}$$

where $\Delta(w)$ is a finite set of sequences $\delta = (\alpha_0, \beta_0, \gamma_0, \alpha_1, \beta_1, \gamma_1, \ldots)$ with each α_i, β_i, γ_i equal to 0 or 1. So the dimension $|w|$ of w is $\sum (\alpha_i + \beta_i + \gamma_i) 2^i$ for any $\delta \in \Delta(w)$. The *type* and *weight* of the monomial associated with δ are $(\alpha_0 + \beta_0 + \gamma_0, \alpha_1 + \beta_1 + \gamma_1, \ldots)$ and $\sum (\alpha_i + \beta_i + \gamma_i)$ respectively. On occasions we will write u_{2^i} in place of any of x_{2^i}, y_{2^i} or z_{2^i}.

Proposition 3.1. *$w \in M$ if and only if $fw \in M$.*

Proof. The formula for $Sq_i u_{2^r}$ implies that for $i > 0$, $j > 0$, $Sq_{2j} u_{2^i} = e(Sq_j u_{2^{i-1}})$. It follows by linearity and the Cartan formula that for $i > 0$

$Sq_{2i}\, ew = eSq_i w$ modulo the ideal generated by x_1, y_1 and z_1. So $Sq_{2i}\, fw = x_1 y_1 z_1 Sq_{2i}\, ew = x_1 y_1 z_1\, e(Sq_i w)$.

Therefore if $w \in M$, $Sq_{2i}\, fw = 0$ and clearly $Sq_{2i+1}\, fw = Sq_{2i} Sq_1\, fw = 0$ for $i \geq 0$. Conversely if $fw \in M$, $fSq_i w = Sq_{2i}\, fw = 0$ for $i > 0$ and so $Sq_i w = 0$.

For notational convenience, if $\alpha_0 + \beta_0 + \gamma_0 = 0$ for each $\delta \in \Delta(w)$, we will write v, v' , v_1, etc, in place of w and more generally $v(s)$, $v'(s)$, $v_1(s)$, etc will denote a w with $\alpha_i + \beta_i + \gamma_i = 0$, $0 \leq i < s$ for each $\delta \in \Delta(w)$.

Definition. The operator r is defined by

$$r(x_1^{\alpha_0}\, y_1^{\beta_0}\, z_1^{\gamma_0} \ldots x_{2^s}^{\alpha_s}\, y_{2^s}^{\beta_s}\, z_{2^s}^{\gamma_s}) = x_0^{\alpha_0}\, y_0^{\beta_0}\, z_0^{\gamma_0}\, x_1^{\alpha_1}\, y_1^{\beta_1}\, z_1^{\gamma_1} \ldots x_{2^s-1}^{\alpha_s}\, y_{2^s-1}^{\beta_s}\, z_{2^s-1}^{\gamma_s})$$

and extending linearly where $x_0 = y_0 = z_0 = 0$.

So $re(w) = w$ and $er(v) = v$. Proposition 3.1 can now be restated as

(3.2) $$r(v) \in M \quad \text{if and only if} \quad x_1 y_1 z_1 v \in M.$$

The next lemma is useful and illustrates the types of arguments to be used.

Lemma 3.3. *Let $w \in M$. Then for each $\delta \in \Delta(w)$, $\alpha_0 + \beta_0 + \gamma_0 \neq 0$.*

Proof. We will argue by contradiction and suppose that for some $w \in M$ there exists $\delta \in \Delta(w)$ with $\alpha_0 + \beta_0 + \gamma_0 = 0$. We choose a $\delta \in \Delta(w)$ with $\alpha_i + \beta_i + \gamma_i = 0$ for $0 \leq i < s$ where s is chosen to be as large as possible; so the monomial associated with δ has type $(0, 0, \ldots, 0, \alpha_s + \beta_s + \gamma_s, *, * \ldots)$. We show first that without loss of generality we can assume that $s = 1$.

Clearly $x_1 y_1 z_1 w \in M$ and $x_1 y_1 z_1 w = x_1 y_1 z_1 v$ for some $v \neq 0$. Then $r(v) \in M$ by (3.2) and contains a monomial of type $(0, 0, \ldots, 0, \alpha_s + \beta_s + \gamma_s, *, * \ldots)$ where now $\alpha_s + \beta_s + \gamma_s$ is in the s-th position not the $(s+1)$-st. Repeating the operation, we can assume that $s = 1$ and this will be the largest possible s which can occur satisfying the condition in the paragraph above.

So without loss of generality

$$w = y_1 z_1 v_1 + x_1 z_1 v_2 + y_1 z_1 v_3 + x_2 v(2)_1 + y_2 v(2)_2 + z_2 v(2)_3 + x_2 y_2 z_2 v(2)_4,$$

if $|w| \equiv 2 \bmod 4$ or if $|w| \equiv 0 \bmod 4$

$$w = y_1 z_1 v_1 + x_1 z_1 v_2 + x_1 y_1 v_3 + y_2 z_2 v(2)_1 + x_2 z_2 v(2)_2 + x_2 y_2 v(2)_3.$$

We consider $Sq_1 w = 0$, using the Cartan formula. In the first alternative if any of $v(2)_1$, $v(2)_2$, $v(3)_3$ is non-zero, we obtain a non vanishing monomial in $Sq_1 w$ of type $(1, 0, *, *, \ldots)$, which is impossible. So $v(2)_4 \neq 0$ and in $Sq_1 w$ we have a non-vanishing monomial of type $(1, 2, *, *, \ldots)$ which again is impossible. In the second alternative if any of $v(2)_1$, $v(2)_2$ or $v(3)_3$ is non zero, we obtain a non-zero monomial in $Sq_1 w$ of type $(1, 1, *, *, \ldots)$ which contradicts w being a member of M. Thus the lemma is established.

Definition. A monomial w is *regular* if $\alpha_i + \beta_i + \gamma_i \geq \alpha_{i+1} + \beta_{i+1} + \gamma_{i+1}$ for all $i \geq 0$. Otherwise it is *irregular.* A general non-zero element w is *totally irregular* if for each $\delta \in \Delta(w)$ the associated monomial is irregular. A general element w is *weakly irregular* if there exists $\delta \in \Delta(w)$ with $\alpha_0 + \beta_0 + \gamma_0 = 1$ and each δ with $\alpha_0 + \beta_0 + \gamma_0 = 1$ has an irregular associated monomial.

Theorem 3.4. *Let w be weakly or totally irregular. Then $w \notin M$.*

An immediate corollary obtained by calculating the dimension of a regular monomial is

Corollary 3.5. *$M_n = 0$ unless $n = 2^\alpha + 2^\beta + 2^\gamma - 3$ with $\alpha \geq \beta \geq \gamma \geq 0$.*

As $L_n \subseteq M_n$, it follows from the results of section 2 that L_n and M_n are zero for the same values of n.

We prove Theorem 3.4 by induction on degree. This begins with the lemma

Lemma 3.6. *Let $w \in M$ be a sum of monomials in x_i, y_i, z_i, $1 \leq i \leq 2$. Then each monomial is regular.*

The proof is a simple computation. (In dimensions 5, 7 and 8 use Lemma 3.3.)

Suppose first that $|w| \equiv 1 \bmod 2$ with w weakly irregular. Then $w = x_1 v_1 + y_1 v_2 + z_1 v_3 + x_1 y_1 z_1 v_4$.

If $|v_i| \equiv 2 \bmod 4$, then $v_i = x_2 v(2)_i + y_2 v(2)'_i + z_2 v(2)''_i + x_2 y_2 z_2 v(2)'''_i$. As $y_1 z_1 w \in M$, $x_1 r v(2)_1 + y_1 r v(2)'_1 + z_1 r v(2)''_1 + x_1 y_1 z_1 r v(2)'''_1 \in M$ by (3.3), and is either weakly irregular, which is impossible by the induction hypothesis, or equals $x_1 y_1 z_1 r v(2)'''_1$. In a similar manner considering $x_1 z_1 w$ and $x_1 y_1 w$, we deduce that $w = x_1 x_2 y_2 z_2 v(2)'''_1 + y_1 x_2 y_2 z_2 v(2)'''_2 + z_1 x_2 y_2 z_2 v(2)'''_3 + x_1 y_1 z_1 v_4$. Considering $Sq_1 z_1 w = 0$, we deduce that

$$x_1 y_1 z_1 (x_2 z_2 v(2)'''_1 + y_2 z_2 v(2)'''_2) = 0.$$

Thus $v(2)'''_1 = v(2)'''_2 = 0$ and similarly $v(2)'''_3 = 0$. Hence $w = x_1 y_1 z_1 v_4$ is not weakly irregular.

If $|v_i| \equiv 0 \bmod 4$, then using (3.2) and Lemma 3.3 we see that

$$v_i = y_2 z_2 v(2)_i + x_2 z_2 v(2)'_i + x_2 y_2 v(2)''_i.$$

Then by considering monomials of type $(3,1,*,*,\ldots)$ in $Sq_1 z_1 w = 0$, we deduce that $x_1 y_1 z_1 (z_2 v(2)_1 + x_2 v(2)''_1 + z_2 v(2)'_2 + y_2 v(2)''_2) = 0$. So $v(2)''_1 = v(2)''_2 = 0$ and $v(2)_1 = v(2)'_2$. Repeating the argument for $y_1 w$ and $x_1 w$ yields that $w = (x_1 y_2 z_2 + y_1 x_2 z_2 + z_1 x_2 y_2) v(2)_1$. Considering only terms of type $(3,0,*,*,\ldots)$ in $Sq_2 w = 0$ we see that $x_1 y_1 z_1 v(2)_1 = 0$. Hence $w = 0$ which is impossible as w is weakly irregular.

If $|w| \equiv 1 \bmod 2$ and w is totally irregular but not weakly irregular, $w = x_1 y_1 z_1 v$ and $r(v) \in M$ is totally irregular which is not possible by the inductive hypothesis.

We now assume that $|w| \equiv 0 \bmod 2$ and that w is totally irregular. Then $w = y_1 z_1 v_1 + x_1 z_1 v_2 + x_1 y_1 v_3$.

If $|v_i| \equiv 2 \bmod 4$, then arguing as above $r(v_i) \in M$ is either weakly irregular, which is not possible, or $v_i = x_2 y_2 z_2 v(2)_i$. So $w = y_1 z_1 x_2 y_2 z_2 v(2)_1 + x_1 z_1 x_2 y_2 z_2 v(2)_2 + x_1 y_1 x_2 y_2 z_2 v(2)_3$. As $Sq_1 w = 0$ it follows that $v(2)_1 = v(2)_2 = v(2)_3 = 0$. Thus $w = 0$ which is not possible.

If $|v_i| \equiv 0 \bmod 4$, then $r(v_i) = y_1 z_1 v'_i + x_1 z_1 v''_i + x_1 y_1 v'''_i \in M$ and is totally irregular or zero. So $r(v_i) = 0$ and $w = 0$, which again is not possible. This completes the proof of Theorem 3.4.

We now begin the proof of Theorem 2.11. The approach is naïve but has been set up in such a way that it must succeed. Let $w \in M_{2t}$. Then $w = y_1 z_1 v_1 + x_1 z_1 v_2 + x_1 y_1 v_3$. Then as in the last proof it follows that $r(v_i) \in M_{t-1}$ and we assume inductively that we have constructed a basis of M_{t-1} of dimension q. Thus we can express w in terms of $3q$ coordinates. We now equate $Sq_1 w$ to zero. In fact $w \in M$ if and only if $Sq_1 w = 0$ and this leads to linear relations among the $3q$ coordinates. (This is not obviously true for $w \in M_{2t-1}$ but we are then only concerned with $2t-1 = 2^s - 1$.) If the rank of the system of equations is r, dim $M_{2t} = 3q - r$. With the one exception, when $2t = 8$, this equals the dimension of L_{2t} and so $L_{2t} = M_{2t}$ and we already have a basis for L_{2t}. Most of the computations are straightforward, though there are many of them, for one can neglect elements of higher weight. A few calculations in dimensions $2^t + 2^s - 2$ when s or $t - s$ is small are more involved. We consider the modules M_n in approximately the same order as in Proposition 2.10.

Proposition 3.7(a) *If* $n = 2^s - 2$, $s > 1$, *then* $M_n = L_n$.

Proof. It is clear that $M_2 = L_2$.

Let $w \in M_6$ and set $A_i = a_i y_1 z_1 + b_i x_1 z_1 + c_i x_1 y_1$. Then, arguing as above, $w = y_1 z_1 e A_1 + x_1 z_1 e A_2 + x_1 y_1 e A_3$. As $Sq_1 w = 0$, we obtain

$x_1y_1z_1e\{b_1z_1 + c_1y_1 + a_2z_1 + c_2x_1 + a_3y_1 + b_3x_1\} = 0$. Thus $c_2 + b_3 = 0$, $c_1 + a_3 = 0$, $b_1 + a_2 = 0$, a system of rank 3 in 9 coordinates. So dim $M_6 = 6$ and therefore, by Proposition 2.10(a), $M_6 = L_6$.

In dimension 14, we need to use the formulae $Sq_1e(y(2)z(2)) = (y_1z_2 + z_1y_2)y_4z_4$ and $Sq_1e(x(2)yz(2)) = (x_1y_2 + y_1x_2 + x_1z_2 + z_1x_2)(x_4y_4 + x_4z_4) + (y_1 + z_1)x_2y_2z_2x_4$, and the symmetric relations obtained by permuting x, y and z. Let $A_i = a_iy(2)z(2) + b_ix(2)z(2) + c_ix(2)y(2) + d_ix(2)yz(2) + e_iy(2)xz(2) + f_iz(2)xy(2)$. Then for $w \in M_{14}$, $w = y_1z_1eA_1 + x_1z_1eA_2 + x_1y_1eA_3$. So $Sq_1w = 0$ implies that $x_1y_1z_1e\{b_1z_1x_2z_2 + c_1y_1x_2y_2 + d_1x_2(y_1 + z_1)(y_2 + z_2) + e_1y_1y_2(x_2 + z_2 + x_1z_1) + f_1z_1z_2(x_2 + y_2 + x_1y_1) + a_2z_1y_2z_2 + c_2x_1x_2y_2 + d_2x_1x_2(y_2 + z_2 + y_1z_1) + e_2y_2(x_1 + z_1)(x_2 + z_2) + f_2z_1z_2(x_2 + y_2 + x_1y_1) + a_3y_1y_2z_2 + b_3x_1x_2z_2 + d_3x_1x_2(y_2 + z_2 + y_1z_1) + e_3y_1y_2(x_2 + z_2 + x_1z_1) + f_3z_2(x_1 + y_1)(x_2 + y_2)\} = 0$. Equating coefficients, we obtain a system of rank 11

$$\begin{aligned} b_1 + d_1 + f_1 + f_2 &= 0 & \qquad d_1 + e_2 &= 0 \\ a_2 + e_2 + f_1 + f_2 &= 0 & e_2 + f_3 &= 0 \\ c_1 + d_1 + e_1 + e_3 &= 0 & d_2 + d_3 &= 0 \\ a_3 + f_3 + e_1 + e_3 &= 0 & e_1 + e_3 &= 0 \\ c_2 + e_2 + d_2 + d_3 &= 0 & f_1 + f_2 &= 0. \\ b_3 + f_3 + d_2 + d_3 &= 0 \end{aligned}$$

So dim $M_{14} = 18 - 11 = 7$ and $L_{14} = M_{14}$.

We can now proceed to the general case, although there is a small additional point to note when $s = 5$. Let $A_i(s) = a_iy(s)z(s) + b_ix(s)z(s) + c_ix(s)y(s) + d_ix(s)yz(s) + e_iy(s)xz(s) + f_iz(s)xy(s) + g_it(s)$. Then if $w \in M_{2^s-2}$, then $w = y_1z_1eA_1(s-2) + x_1z_1eA_2(s-2) + x_1y_1eA_3(s-2)$ and $Sq_1w = 0$. It is not necessary to write this out. We neglect all monomials of weight greater than $2s-2$ and consider $Sq_1w = x_1y_1z_1(x_2v(2)_1 + y_2v(2)_2 + z_2v(2)_3)$. So, for example, $Sq_1(x_1y_1ex(s-2)yz(s-2)) = x_1y_1z_1ex(s-2)e^2yz(s-3)$ and the contribution to $v(2)_1$ is $e^2x(s-3)yz(s-3)$. So $v(2)_1 = e^2\{c_2x(s-3)y(s-3) + d_2x(s-3)yz(s-3) + e_2y(s-3)xz(s-3) + g_2t(s-3) + b_3x(s-3)z(s-3) + d_3x(s-3)yz(s-3) + f_3z(s-3)xy(s-3) + g_3(t-3)\} = 0$. If $s > 5$, the description of the basis for $L_{2^{s-3}-2}$ implies that $c_2 = 0$, $d_2 = d_3$, $e_2 = 0$, $g_2 = g_3$, $b_3 = 0$, $f_3 = 0$. If $s = 5$ one has the same conclusion as $t(2)$ is not linearly dependent on the other terms as $y(2)z(2)$ does not appear. Now appealing to symmetry one has

$$\begin{array}{lllll} d_2 = d_3, & c_2 = 0, & b_3 = 0, & f_3 = 0, & g_1 = g_2, \\ e_1 = e_3, & a_3 = 0, & c_1 = 0, & e_2 = 0, & g_2 = g_3, \\ f_1 = f_2, & a_2 = 0, & b_2 = 0, & d_1 = 0. & \end{array}$$

There are 21 coordinates and the system of equations has rank 14. So M_{2^s-2}, $s > 4$, has dimension 7 and by Proposition 2.10(a), $M_{2^s-2} = L_{2^s-2}$.

The proofs of the other parts of Proposition 3.7 below use arguments similar to those in (a) above. We will provide fewer details for the corresponding steps (complete details can be found in [**2**]) and concentrate on the parts of the proofs which are different.

Proposition 3.7(b) *If* $n = 2^s - 1$, $s > 0$, *then* $M_n = L_n$.

Proof. Clearly $M_1 = L_1$ and a short computation shows that $M_3 = L_3$.

We set $A_i(s) = a_i x(s) + b_i y(s) + c_i z(s) + d_i yz(s) + e_i xz(s) + f_i xy(s) + g_i xyz(s)$. We assume inductively that $M_{2^r-1} = L_{2^r-1}$ for $r < s$. If $w \in M_{2^s-1}$, $w = x_1 v_1 + y_1 v_2 + z_1 v_3 + x_1 y_1 z_1 v_4$ where $v_i = eA_i(s-1) + efu_i$, $1 \leq i \leq 3$, for some $u_i \in M_{2^{s-2}-2}$ if $s \geq 4$ and with $u_i = 0$ for $s = 3$. One considers $Sq_1 z_1 w = 0$ where $z_1 w = x_1 z_1 v_1 + y_1 z_1 v_2$. Arguing as in part (a), one obtains the equations $b_1 = d_1 = a_2 = e_2 = 0$, $f_1 = f_2$, $g_1 = g_2$. For $s = 3$ it is necessary to consider terms of all weights but for $s > 3$ only terms of minimum weight. Appealing to symmetry it follows that $b_1 = c_1 = d_1 = a_2 = c_2 = e_2 = a_3 = b_3 = f_3 = 0$, $f_1 = f_2$, $e_1 = e_3$, $d_2 = d_3$ and $g_1 = g_2 = g_3$. Thus $w = a_1 x(s) + b_2 y(s) + c_3 z(s) + d_2 yz(s) + e_1 xz(s) + f_1 xy(s) + g_1 xyz(s) + x_1 y_1 z_1 v_4' + x_1 efu_1 + y_1 efu_2 + z_1 efu_3$. (We mention that unless $g_1 = 0$, $v_4 \neq v_4'$.) Hence $x_1 y_1 z_1 v_4' + x_1 efu_1 + y_1 efu_2 + z_1 efu_3 \in M$. But this is weakly irregular unless $u_1 = u_2 = u_3 = 0$. Therefore by Theorem 3.4 for $s > 3$, $u_1 = u_2 = u_3 = 0$, $x_1 y_1 z_1 v_4' \in M$ and $r(v_4') \in M_{2^{s-1}-2}$. This establishes that $M_{2^s-1} = L_{2^s-1}$ using Proposition 2.10(b).

Proposition 3.7(c,1) *If* $n = 2^s$, $s > 1$, $s \neq 3$, $M_n = L_n$. *If* $n = 8$,

$$M_8 = \langle L_8, q\rangle \text{ with } q = x_1 z_1 (y_2 z_4 + z_2 y_4) + x_1 y_1 z_2 z_4 + y_1 z_1 x_2 y_2 z_2 \notin L_8.$$

Proof. A short computation verifies that $M_4 = L_4$.

Let $A_i(s)$ be as in (b) above. Because of the exceptional result we consider M_8 in detail. If $w \in M_8$, $w = y_1 z_1 eA_1(2) + x_1 z_1 eA_2(2) + x_1 y_1 eA_3(2)$. Then $Sq_1 w = 0$ yields

$$\begin{aligned} x_1 y_1 z_1 e\{ & a_1 x_2 + e_1(x_2 + z_2 + x_1 z_1) + f_1(x_2 + y_2 + x_1 y_1) + g_1 y_1 z_1 \\ & + b_2 y_2 + d_2(y_2 + z_2 + y_1 z_1) + f_2(x_2 + y_2 + x_1 y_1) + g_2 x_1 z_1 \\ & + c_3 z_2 + d_3(y_2 + z_2 + y_1 z_1) + e_3(x_2 + z_2 + x_1 z_1) + g_3 x_1 y_1\} = 0. \end{aligned}$$

So

$$\begin{aligned} a_1 + (e_1 + e_3) + (f_1 + f_2) &= 0 \qquad & g_1 + (d_2 + d_3) &= 0 \\ b_2 + (d_2 + d_3) + (f_1 + f_2) &= 0 \qquad & g_2 + (e_1 + e_3) &= 0 \\ c_3 + (d_2 + d_3) + (e_1 + e_3) &= 0 \qquad & g_3 + (f_1 + f_2) &= 0. \end{aligned}$$

As $w \in M_8$ if and only if $Sq_1 w = 0$ and the system of equations has rank 6 in the 21 coordinates, dim $M_8 = 15 >$ dim $L_8 = 14$. A choice for an element in M_8 but not L_8 is $x_1 z_1 (y_2 z_4 + z_2 y_4) + x_1 y_1 z_2 z_4 + y_1 z_1 x_2 y_2 z_2$.

The proof that $M_{2^s} = L_{2^s}$ for $s > 3$ is similar to those used earlier. If $w \in M_{2^s}$, $w = y_1 z_1 e(A_1(s-1) + fu_1) + x_1 z_1 e(A_2(s-1) + fu_2) + x_1 y_1 e(A_3(s-1) + fu_3)$ using Proposition 2.10(b). $Sq_1 w = 0$ yields the relations $a_1 = b_2 = c_3 = 0$, $d_2 = d_3$, $e_1 = e_3$, $f_1 = f_3$ and $g_1 + g_2 + g_3 = 0$. These are sufficient to deduce that $w' = y_1 z_1 efu_1 + x_1 z_1 efu_2 + x_1 y_1 efu_3 \in M$ which must then be zero as it is totally irregular otherwise.

So there are 7 independent relations in 21 variables and hence $M_{2^s} = L_{2^s}$. Again when $s > 4$ one need only consider terms of minimum non-vanishing weight.

Proposition 3.7(c,2) *If $n = 2^{s+1} + 2^s - 2$, $s > 1$, then $M_n = L_n$.*

Proof. Using the fact that $M_4 = L_4$, it is easily established that $M_{10} = L_{10}$.

Set $B_i(s) = a_i y(s-1)z(s) + b_i y(s)z(s-1) + c_i x(s-1)z(s) + d_i x(s)z(s-1) + e_i x(s-1)y(s) + f_i y(s-1)x(s) + g_i x(s-1)yz(s) + h_i y(s-1)xz(s) + k_i z(s-1)xy(s) + l_i x(s)yz(s-1) + m_i y(s)xz(s-1) + n_i z(s)xy(s-1) + p_i t(s-1)(y_{2^s} + z_{2^s}) + q_i t(s-1)(x_{2^s} + z_{2^s})$.

If $w \in M_{2^{s+1}+2^s-2}$ with $s > 2$, $w = y_1 z_1 eB_1(s) + x_1 z_1 eB_2(s) + x_1 y_1 eB_3(s)$. We consider $Sq_1 w = 0$, beginning with $s = 3$ where it is necessary to consider terms of weight greater than the minimum. An efficient way of performing the computation is to first consider monomials of the form $x_1 y_1 z_1 x_2 y_2 z_2 v(2)$ in $Sq_1 w$. These alone yield the equations $g_2 = g_3$, $l_2 = l_3$, $h_1 = h_3$, $m_1 = m_3$, $k_1 = k_2$, $n_1 = n_2$, $p_1 = p_2 = p_3$, $q_1 = q_2 = q_3$. Using these relations all the named coefficients can be removed from the expression for w. It is then easier to find the additional equations $a_2 = a_3 = b_2 = b_3 = c_1 = c_3 = d_1 = d_3 = e_1 = e_2 = f_1 = f_2 = 0$, $g_1 = l_1 = h_2 = m_2 = k_3 = n_3 = 0$. The 28 independent relations in 42 coordinates imply that M_{22} has dimension 14 and therefore $M_{22} = L_{22}$. When $s > 3$ one obtains the same relations working with monomials of minimum weight only by arguing in a similar way as was done in the final part of the proof of Proposition 3.7(a), taking some care with symmetry arguments. Thus $M_n = L_n$ in all cases.

Proposition 3.7(c,3) *If $n = 2^t + 2^s - 2$ with $s = 2$, $t \geq 4$, then $M_n = L_n$.*

Proof. The case $s = 2$, $t = 4$ gives unique problems as $M_8 \neq L_8$.

Set $C_i(s) = a_i y(1)z(s) + b_i z(1)y(s) + c_i x(1)z(s) + d_i z(1)x(s) + e_i x(1)y(s) + f_i y(1)x(s) + g_i x(1)yz(s) + h_i y(1)xz(s) + k_i z(1)xy(s) + l_i x(1)xy(s) + m_i y(1)yz(s) + n_i z(1)xy(s) + r_i y(1)xyz(s) + p_i z(1)xyz(s)$ and $u = x_1 z_1 (y_2 z_4 +$

$z_2y_4) + x_1y_1z_2z_4 + y_1z_1x_2y_2z_2$. Let $w \in M_{18}$. So $w = y_1z_1e(C_1(3) + q_1u) + x_1z_1(C_2(3) + q_2u) + x_1y_1s(C_3(3) + q_3u)$. We consider $Sq_1w = 0$ systematically equating all coefficients of monomials of the form $x_1y_1z_1w'$, where each monomial in w' has factors x_i, y_j and z_k for some i, j and k. It then follows that $q_1 = q_2 = q_3 = 0$ and $g_2 + g_3 = 0$, $h_1 + h_3 = 0$, $k_1 + k_2 = 0$, $r_1 + r_3 = 0$, $p_1 + p_2 = 0$, $r_2 + p_3 = 0$. These equations can be used to simplify the form of w in a manner similar to that used above. The complete solution where t_i, $1 \le i \le 21$, are arbitrary is

$a_1 = t_1$	$b_1 = t_3$	$c_1 = t_2$	$d_1 = 0$	$e_1 = t_4$	$f_1 = 0$	$g_1 = t_{10}$	$h_1 = t_{12}$
$a_2 = t_2$	$b_2 = 0$	$c_2 = t_5$	$d_2 = t_6$	$e_2 = 0$	$f_2 = t_7$	$g_2 = t_{11}$	$h_2 = t_{13}$
$a_3 = 0$	$b_3 = t_4$	$c_3 = 0$	$d_3 = t_7$	$e_3 = t_8$	$f_3 = t_9$	$g_3 = t_{11}$	$h_3 = t_{12}$

$k_1 = t_{14}$	$l_1 = t_{15}$	$m_1 = t_{17}$	$n_1 = t_{13}$	$r_1 = t_{19}$	$p_1 = t_{21}$	$q_1 = 0$
$k_2 = t_{14}$	$l_2 = t_{15}$	$m_2 = t_{10}$	$n_2 = t_{18}$	$r_2 = t_{20}$	$p_2 = t_{21}$	$q_2 = 0$
$k_3 = t_{15}$	$l_3 = t_{16}$	$m_3 = t_{10}$	$n_3 = t_{13}$	$r_3 = t_{19}$	$p_3 = t_{20}$	$q_3 = 0.$

If $t > 4$ one obtains the same solution (without q_1, q_2 and q_3) but the argument is much shorter using the linear independence of $\{x(t-2), y(t-2), z(t-2), yz(t-2), xz(t-2), xy(t-2), xyz(t-2)\}$ and working only with elements of lowest non vanishing weight.

Thus $M_n = L_n$ in all cases.

Proposition 3.7(c,4) *If* $n = 2^t + 2^s - 2$, $u = t - s \ge 2$, $s \ge 3$, *then* $M_n = L_n$.

Proof. We refer to Table 2 giving seven blocks of three elements which we label $A, B, \dots, G$ and let $A_i(s), B_i(s), \dots, G_i(s)$ denote arbitrary linear combinations of elements in the block, for each u. Thus $A_i(s) = y(s)z(s)\{a_iEz(u) + a_i'Ey(u) + a_i''Eyz(u)\}$, $B_i(s) = x(s)z(s)\{b_iEz(u) + b_i'Ex(u) + b_i''Exz(u)\}$, etc. We write $Q_i = A_i(s) + B_i(s) + \cdots + G_i(s)$.

We argue by induction on s keeping u fixed, using Proposition 3.7(c,3). If $w \in M_{2^t+2^s-2}$, $w = y_1z_1eQ_1(s-1) + x_1z_1eQ_2(s-1) + x_1y_1eQ_3(s-1)$. We again consider $Sq_1w = 0$ but as e^{s-1} $\{x(u), y(u), z(u), yz(u), xz(u), xy(u), xyz(u)\}$ are linearly independent, the computation breaks up into seven separate cases, or using symmetry, just three calculations.

Let

$$\begin{aligned}
\alpha_i &= b_i'x(s-1)y(s-1) + c_i'x(s-1)y(s-1) + d_i'x(s-1)yz(s-1),\\
\beta_i &= a_i''y(s-1)z(s-1) + d_ix(s-1)yz(s-1) + g_it(s-1),\\
\gamma_i &= d_i''x(s-1)yz(s-1) + e_i''y(s-1)xz(s-1) + f_i''z(s-1)xy(s-1),
\end{aligned}$$

arising from $Ex(u)$, $Eyz(u)$ and $Exyz(u)$ respectively. Then $Sq_1w = 0 \bmod (x_1y_1z_1x_2y_2z_2x_4y_4z_4)$ implies that $Sq_1\{y_1z_1e\theta_1 + x_1z_1e\theta_2 + x_1y_1e\theta_3\}$

$= 0$ where $\theta_i = \alpha_i$, $1 \le i \le 3$ or β_i or γ_i. The " mod $(x_1y_1z_1x_2y_2z_2x_4y_4z_4)$" condition for $s > 3$ is equivalent to working modulo elements of higher weight and is the appropriate condition for $s = 3$. We complete the calculation in this latter case as again the argument is routine for $s > 3$.

When $s = 3$ and $\theta = \alpha$ we obtain $x_1y_1z_1\{b'_1z_2x_4z_4 + c'_1y_2x_4y_4 + d'_1x_4(y_2+z_2)(y_4+z_4) + b'_2x_2x_4z_4 + c'_2x_2x_4y_4 + d'_2x_2x_4(y_4+z_4+x_2y_2) + d'_3x_2x_4(y_4+z_4+y_2z_2)\} = 0$. So $b'_1 = c'_1 = b'_2 = c'_2 = 0$, $d'_1 = 0$, $d'_2 + d'_3 = 0$. Similarly when $\theta = \beta$ or ,γ the corresponding equations yield $g_1 = g_2 = g_3$, $a''_2 = a''_3 = 0$, $d_1 = 0$, $d_2 + d_3 = 0$ and $d''_1 = e''_2 = f''_3 = 0$, $e''_1 + e''_3 = 0$, $f''_1 + f''_2 = 0$, $d''_2 + d''_3 = 0$. Permuting the variables and using symmetry, we see that all 63 coordinates are determined by arbitrary choices of $a_1, a'_1, a''_1, b_2, b'_2, b''_2, c_3, c'_3, c''_3, d_2, d'_2, d''_2, e_1, e'_1, e''_1, f_1, f'_1, f''_1, g_1, g'_1, g''_1$. Thus $M_n = L_n$ as required.

The proof of Theorem 2.11. By Theorem 2.3 and Corollary 3.5, both L_n and M_n are trivial unless $n = 2^\alpha + 2^\beta + 2^\gamma - 3$ where $\alpha \ge \beta \ge \gamma \ge 0$. If $\gamma = 0$, then the conclusions of Proposition 3.7 imply that $L_n = M_n$. So let $\gamma > 0$. If $w \in M_n$, $w = x_1v_1 + y_1v_2 + z_1v_3 + x_1y_1z_1v_4$. So by Theorem 3.4 either $x_1v_1 + y_1v_2 + z_1v_3$ contains a regular monomial and so has dimension of the form $2^s - 1$ or $w = x_1y_1z_1v_4$. The former case is considered in Proposition 3.7(b). Otherwise $f : M_{2^{\alpha-1}+2^{\beta-1}+2^{\gamma-1}-3} \to M_{2^\alpha+2^\beta+2^\gamma-3}$ is an isomorphism and either $f^\gamma : M_{2^{\alpha-\gamma}+2^{\beta-\gamma}-2} \to M_{2^\alpha+2^\beta+2^\gamma-3}$ is an isomorphism if α, β and γ are not all equal or $f^{\gamma-1} : M_3 \to M_{3.2^\gamma-3}$ is an isomorphism if $\alpha = \beta = \gamma > 0$. The theorem follows by using Proposition 2.10(d).

4. The action of $GL(V)$ on L_n and M_n

The action of $GL(V) = GL_3(\mathsf{F}_2)$ on V leads to compatible actions on $L_n \subseteq M_n$ and $f : L_n \to L_{2n+3}$ is a monomorphism of $GL(V)$-spaces. So we need just describe the $GL(V)$-spaces L_n when $n = 2^\alpha + 2^\beta - 2$ with $\alpha > \beta \ge 0$ or $\alpha \ge \beta > 0$. Also $GL(V)$ acts transitively on $G_1(V)$ the set of one-dimensional subspaces, on $G_2(V)$ the set of two-dimensional subspaces and on $G_{1,2}(V)$, the set of flags $\{l \subset E \subset V\}$.

We can read off a description of L_n as a $GL(V)$-space from Proposition 2.10(a),(b) and (c). We do this explicitly in the algebraically stable situation, for it is this which we seek to generalize to higher dimensions. The remaining cases for L_n are quotients of these as the relations (2.4)-(2.9) are clearly preserved by the $GL(V)$-action. We recall that the order of $GL(V)$ is $(2^3-1)(2^3-2)(2^3-4) = 168$.

As $GL(V)$-modules,

(i) L_n, $n = 2^s - 2$, $s > 3$

$$L_n = M_n = \mathsf{F}_2[G_2(V)] \simeq \mathsf{F}_2[GL(V)/H_2]$$

where H_2 is the subgroup of order 24 consisting of all invertible matrices of the form

$$\begin{bmatrix} a_{11} & a_{12} & a_{13} \\ a_{21} & a_{22} & a_{23} \\ 0 & 0 & 1 \end{bmatrix}$$

and $GL(V)/H_2$ is the coset $GL(V)$-space;

(ii) L_n, $n = 2^s - 1$, $s > 4$

$$L_n = M_n = \mathsf{F}_2[G_1(V)] \oplus \mathsf{F}_2[G_2(V)] \simeq \mathsf{F}_2[GL(V)/H_1 \sqcup GL(V)/H_2]$$

where H_1 is the subgroup of order 24 consisting of all invertible matrices of the form

$$\begin{bmatrix} 1 & a_{12} & a_{13} \\ 0 & a_{22} & a_{23} \\ 0 & a_{32} & a_{33} \end{bmatrix};$$

(iii) L_n, $n = 2^t + 2^s - 2$, $t > s + 1 > 2$

$$L_n = M_n = \mathsf{F}_2[G_{1,2}(V)] \simeq \mathsf{F}_2[GL(V)/H_3]$$

where H_3 is the Sylow 2-subgroup of upper triangular matrices.

References

[1] J. F. ADAMS, J. H. GUNAWARDENA and H. R. MILLER, *The Segal conjecture for elementary abelian p-groups*, Topology **24** (1985), 435–460.

[2] MOHAMED ALI ALGHAMDI, Ph.D. Thesis, University of Aberdeen, March 1991.

[3] M. KAMEKO, Products of projective spaces as Steenrod modules, Preprint (version 1.2), June 1990.

[4] R. M. W. WOOD, *Steenrod squares of polynomials*, Advances in Homotopy Theory, LMS lecture note series **139** (1989), 173–177.

Department of Mathematical Sciences
University of Aberdeen
Aberdeen AB9 2TY
Scotland

Generic Representation Theory and Lannes' T-functor

Nicholas J. Kuhn[1]
Department of Mathematics
University of Virginia
Charlottesville, VA 22903

§1. Introduction

I remember Frank Adams describing certain recent developments in topology as follows: It is the business of homotopy theorists to compute [X,Y], and, while traditionally X has been a finite complex, now we can let X = BG. This was, of course, initiated by H. Miller in [M], and is ongoing.

Underlying the topological theorems were some wonderful new results about $H^*(V)$, the mod p cohomology of an elementary abelian p-group V, viewed as an object in $\mathcal{U}_p$, the category of unstable modules over the Steenrod algebra A_p. These results were ultimately given an elegant treatment by J. Lannes, who considers the functor $T_V: \mathcal{U}_p \to \mathcal{U}_p$ left adjoint to $H^*(V) \otimes$ ____. Many of us have learned the mantra "T_V is exact and takes tensor products to tensor products." (Most people have been content to believe

---

[1]Research partially funded by the N.S.F.

the survey [L1], but the nontrivial details have finally appeared in [L2].)

At the Manchester conference, I lectured on a representation theoretic framework for understanding Steenrod algebra "technology," as in [HLS] and [K1]. Here I wish to give an exposition of T_V and its properties from this point of view.

The key observation is as follows. Let $\mathcal{F}(q)$ be the category with objects functors

$$F: \text{finite } \mathbb{F}_q\text{-vector spaces} \to \mathbb{F}_q\text{-vector spaces},$$

where $\mathbb{F}_q$ is the field with q elements. Morphisms in $\mathcal{F}(q)$ are the natural transformations. Following [HLS], to an unstable A_p-module M, one defines a functor $\ell(M)$ in $\mathcal{F}(p)$ by

$$\ell(M)(W) = \mathrm{Hom}_{A_p}(M, H^*(W))',$$

where "′" denotes the continuous dual of a profinite vector space. A straightforward calculation then shows that

$$\ell(T_V M)(W) = \ell(M)(V \oplus W),$$

i.e., the following diagram commutes:

$$\begin{array}{ccc} \mathcal{U}_p & \xrightarrow{T_V} & \mathcal{U}_p \\ \ell \downarrow & & \ell \downarrow \\ \mathcal{F}(p) & \xrightarrow{\overline{T}_V} & \mathcal{F}(p) \end{array},$$

where $\overline{T}_V$ is the shift operator: $(\overline{T}_V F)(W) = F(V \oplus W)$.

It is obvious that $\overline{T}_V$ is exact and takes tensor products to tensor products. A main point of [K1] is that considerations of the category $\mathcal{F}(p)$ lead naturally to the category $\mathcal{U}_p$ (or more accurately, the "Bockstein free" version $\mathcal{U}(p)$), making it tempting to try to "lift" these obvious properties of $\overline{T}_V$ to T_V. As we will see, to some extent this can be done.

Sections 2 and 3 are an introduction to $\mathcal{F}(q)$ and our approach to $\mathcal{U}(q)$, surveying results in [K1,K2]. In §4 we state the four (!) properties of T_V that I feel are fundamental, and relate "weak" versions to results in §3. Section 5 contains the main original results of the paper: some new insights into the interdependence of the properties. Section 6 has some observations that may help T_V-connoiseurs-to-be confront the details of the "doubling" construction (and its q-analogue). A short appendix on "q-Boolean algebras" is provided as a service to readers of [LZ2], [L2], [HLS], etc.

§2. Generic representation theory

The category $\mathcal{F}(q)$ is an abelian category in the

obvious way, e.g., $F \to G \to H$ is exact means that $F(V) \to G(V) \to H(V)$ is exact for all V. As such, we can define F in $\mathcal{F}(q)$ to be *simple* if it has only 0 and F as subobjects, *finite* if it has a finite composition series with simple subquotients, and *locally finite* if it is the union of its finite subobjects. We let $\mathcal{F}_{\omega}(q)$ denote the full subcategory of locally finite objects.

Example 2.1. Λ^d, the dth exterior power functor, is a simple object. This follows from the well-known fact $\Lambda^d(\mathbb{F}_q^n)$ is a simple $GL_n(\mathbb{F}_q)$-module for all $n \geq d$. More generally, it turns out that if F is *any* simple functor then $F(\mathbb{F}_q^n)$ is *always* a simple $GL_n(\mathbb{F}_q)$-module (or 0 if n is small). Conversely, any simple $GL_n(\mathbb{F}_q)$-module M extends uniquely to a simple functor F such that $F(\mathbb{F}_q^n) \simeq M$ and $F(\mathbb{F}_q^m) = 0$ for $m < n$. See [K2].

Example 2.2. Let S_d and S^d be defined by $S_d(V) = (V^{\otimes d})^{\Sigma_d}$ and $S^d(V) = V^{\otimes d}/\Sigma_d$. These are finite. Very generally, there is a very practical criterion: F is finite if and only if the assignment

$$n \longmapsto \dim_{\mathbb{F}_q} F(\mathbb{F}_q^n)$$

is a polynomial function of n. See [K1].

Example 2.3. Given a finite-dimensional $\mathbb{F}_q$-vector space W, let P_W be defined by $P_W(V) = \mathbb{F}_q[\mathrm{Hom}(W,V)]$, the vector space with basis Hom(W,V). By Yoneda's lemma,

$$\mathrm{Hom}_{\mathcal{F}}(P_W, F) = F(W),$$

and thus the P_W are projective generators for $\mathcal{F}(q)$. If $W \neq 0$, P_W is not locally finite [K1, Appendix B].

Example 2.4. Dually, let I_W be defined by

$$I_W(V) = \mathbb{F}_q[\mathrm{Hom}(W,V)]^* = \mathbb{F}_q^{\mathrm{Hom}(W,V)}.$$

Then we have

$$\mathrm{Hom}_{\mathcal{F}}(F, I_W) = F(W)^*,$$

and thus the I_W are injective cogenerators for $\mathcal{F}(q)$. (I_W is called $I_W{*}$ in [K1].) The I_W are locally finite, the key point being the observation that there is a natural isomorphism

$$S^*(V)/(v^q - v) \xrightarrow{\sim} I_{\mathbb{F}_q}(V),$$

visibly exhibiting $I_{\mathbb{F}_q}$ as a direct limit of finite objects. It follows formally that, in $\mathcal{F}_\omega(q)$, any object F has an injective resolution

$$0 \to F \to I_0 \to I_1 \to I_2 \to \cdots,$$

in which each I_S is a (possibly infinite) sum of I_W's. See [K1].

Two basic operations in $\mathcal{F}(q)$ are the tensor product

$$\otimes : \mathcal{F}(q)\times\mathcal{F}(q) \to \mathcal{F}(q),$$

defined by $(F\otimes G)(V) = F(V)\otimes G(V)$, and duality

$$D: \mathcal{F}(q)^{op} \to \mathcal{F}(q),$$

defined by $(DF)(V) = F(V^*)^*$. Note that $I_V\otimes I_W \simeq I_{V\otimes W}$, $DS^d = S_d$, and $DP_W = I_{W^*}$.

Finally, we explain why one should regard objects in $\mathcal{F}(q)$ as "generic representations." Note that there are exact evaluation functors

$$e_n: \mathcal{F}(q) \to GL_n(\mathbb{F}_q)\text{-modules},$$

defined by $e_nF = F(\mathbb{F}_q^n)$. Via these evaluations, an object in $\mathcal{F}(q)$ can be regarded as a compatible family of $GL_n(\mathbb{F}_q)$-modules, one for each n. The philosophy is that to study $GL_n(\mathbb{F}_q)$-modules, one should (and must) consider $\mathcal{F}(q)$. Example 2.1 illustrates this. This is the theme of [K2].

§3. Generic representation theory and $\mathcal{U}(q)$

Let $\mathcal{S}^*$ (respectively $\mathcal{S}_*$) denote the full subcategory of $\mathcal{F}(q)$ with objects S^d (S_d). Duality induces an isomorphism $\mathcal{S}_*^{op} \simeq \mathcal{S}^*$. The category $\mathcal{S}^*$ is additive over $\mathbb{F}_q$, and we let $\mathcal{U}(q)$ denote the category of "representations" of $\mathcal{S}^*$, i.e., additive functors

$$M: \mathcal{S}^* \to \mathbb{F}_q\text{-vector spaces.}$$

With $M_n = M(S^n)$, such an M should be regarded as a graded vector space with extra structure. Inspection of $\mathrm{Hom}_{\mathcal{F}}(S^n, S^m)$ then reveals

__Proposition 3.1__. [K1] $\mathcal{U}(q)$ is the category of unstable A(q) modules.

Here A(p) is the algebra of Steenrod reduced p^{th} powers (sorry, no Bocksteins), with grading divided by 2 if $p > 2$, and, if $q = p^s$, $A(q) \subseteq A(p)\otimes\mathbb{F}_q$ is the sub-Hopf algebra dual to $\mathbb{F}_q[\xi_1, \xi_2, \cdots]/(\xi_i \mid i \not\equiv 0 \bmod s)$.

__Examples 3.2__.

(1) Let F(n) be defined by $F(n)_m = \mathrm{Hom}_{\mathcal{F}}(S^n, S^m)$. By Yoneda's lemma

$$\mathrm{Hom}_{\mathcal{U}}(F(n), M) = M_n,$$

and thus the F(n) are projective generators for $\mathcal{U}(q)$.

(2) Let J(n) be defined by $J(n)_m = \mathrm{Hom}_{\mathcal{F}}(S^m, S^n)^*$. Then

$$\mathrm{Hom}_{\mathcal{U}}(M, J(n)) = M_n^*,$$

and thus the $J(n)$ are injective cogenerators for $\mathcal{U}(q)$.

(3) $S^*(V)$ is the symmetric algebra on V. Note that, if $q = 2$, $S^*(V^*) = H^*(V)$.

Now let $\mathcal{U}(q) \underset{r}{\overset{\ell}{\rightleftarrows}} \mathcal{F}_\omega(q)$ be defined by letting ℓ be left adjoint to r where

$$r(F)_n = \mathrm{Hom}_{\mathcal{F}}(S_n, F).$$

Suitably interpreted, this left adjoint ℓ is given by the formula

$$\ell(M)(V) = M \otimes_{\mathcal{S}_*} S_*(V),$$

which can be rearranged to read

$$\ell(M)(V) = \mathrm{Hom}_{\mathcal{U}}(M, S^*(V^*))'.$$

We encourage the reader to verify that $r(S_n) = F(n)$, $\ell(F(n)) = S_n$ and $r(I_V) = S^*(V^*)$. The finiteness of S_n insures that r takes direct sums to direct sums (as does ℓ, being a left adjoint).

The following is a fundamental theorem from [K1].

<u>Theorem 3.3</u>. Any finite F embeds in a sum of the form $\bigoplus_{i=1}^{k} S^{d_i}$. Equivalently, the family S_d , $d \geq 0$, generates $\mathcal{F}_\omega(q)$.

The equivalence is via duality.

This is a "generic" version of the following theorem in group representation theory (see, e.g., [A, p. 45]): Any finitely generated GL(V)-module M embeds in a sum of the form $\bigoplus_{i=1}^{k} S^{d_i}(V)$. Its importance in the study of $\mathcal{U}(q)$ lies in the fact that, via an appropriate "one-sided" Morita theorem [K1], we have

<u>Theorem 3.4</u>. [K1] The following are equivalent.

(1) Theorem 3.3.

(2) (a) $S^*(V)$ is injective in $\mathcal{U}(q)$, for all V.

(b) The natural map $\mathbb{F}_q[\mathrm{Hom}(V,W)] \to \mathrm{Hom}_{\mathcal{U}}(S^*(V),S^*(W))$ is an isomorphism, for all V and W.

(3) ℓ is exact and r is fully faithful.

Of course, (2)(a) is the q-analogue of results by G. Carlsson and H. Miller in [C,M], and (2)(b) is the analogue of a result in [AGM]. A reformulation of (3) is that ε_F: $\ell r(F) \to F$ is an isomorphism, and ℓ and r induce an equivalence of abelian categories

$$\mathcal{U}(q)/\mathcal{N}(q) \simeq \mathcal{F}_\omega(q),$$

where $\mathcal{N}(q)$ is the full subcategory of $\mathcal{U}(q)$ with objects M such that $\mathrm{Hom}_{\mathcal{U}}(M,S^*(V)) = 0$ for all V. Stated this way, (3) is the q-analogue of the main result of [HLS, Part I]. We remark that M is said to be *nilclosed* if $\eta_M: M \to r\ell(M)$ is an isomorphism. This terminology is justified by a theorem of Lannes and Schwartz identifying $\mathcal{N}(p)$ with the "nilpotent" modules [LS].

There are further consequences to be reaped. There is a tensor product in $\mathcal{U}(q)$, and one has

<u>Theorem 3.5</u>. [K1] Both r and ℓ take tensor products to tensor products. More precisely, the natural maps

$$\mu(F,G): r(F)\otimes r(G) \to r(F\otimes G), \text{ and}$$
$$\mu(M,N): \ell(M\otimes N) \to \ell(M)\otimes \ell(N)$$

are isomorphisms for all F, G in $\mathcal{F}(q)$, and M, N in $\mathcal{U}(q)$.

Here the proof goes as follows. Because $I_V\otimes I_W \simeq I_{V\oplus W}$, the statement that $\mu(I_V, I_W)$ is an isomorphism becomes the identity

$$S^*(V)\otimes S^*(W) \simeq S^*(V\oplus W).$$

Resolving both F and G by injectives then leads to $\mu(F,G)$ being an isomorphism in general. Carefully using statement (3) of Theorem 3.4, the statement that $\mu(M,N)$ is an isomorphism then formally follows.

That $\mu(M,N)$ is an isomorphism can be restated as

(3.6) $\quad \mathrm{Hom}_{\mathcal{U}}(M,S^*(V))\hat{\otimes}\mathrm{Hom}_{\mathcal{U}}(N,S^*(V)) \simeq \mathrm{Hom}_{\mathcal{U}}(M\otimes N,S^*(V)).$

This is the q-analogue of Lannes' theorem [LZ2, Appendix].

§4. The four properties of T_V

Definition 4.1. Let T_V: $\mathcal{U}(q) \to \mathcal{U}(q)$ be left adjoint to $S^*(V^*) \otimes$ ____, so that $\mathrm{Hom}_{\mathcal{U}}(T_V M, N) \simeq \mathrm{Hom}_{\mathcal{U}}(M, S^*(V^*) \otimes N)$.

From the "mantra" of the introduction, one would assume that there are *two* properties of T_V to be highlighted. The main algebraic theorem of Lannes' detailed preprint [L2, Theorem 0.1] has *three* parts, and I assert that there are *four* basic properties of T_V.

Property A. T_V is exact.

Property B. $T_V S^*(W^*) = \mathbb{F}_q^{\mathrm{Hom}(V,W)} \otimes S^*(W^*)$.

Property C. The natural map $\mu(M,N)$: $T_V(M \otimes N) \to T_V M \otimes T_V N$ is an isomorphism for all M, N.

Property D. If K in $\mathcal{U}(q)$ is an unstable algebra, so is $T_V K$.

This last property needs some explanation. The qth power map ξ: $S^n(V) \to S^{nq}(V)$ induces ξ: $M_n \to M_{qn}$ for any M in $\mathcal{U}(q)$. (If $q = 2$, $\xi = Sq_0$.) Then a commutative algebra in the category $\mathcal{U}(q)$ is an *unstable algebra* if, in addition, $\xi(x) = x^q$ for all x in K. We say K is in $\mathcal{K}(q)$.

When q = p, Properties A and B were proved by Lannes and Zarati [LZ1, Theorem 0 and Corollary 9.1.3], Property B being an unstable version of the main result of [AGM]. Properties C and D, of course, appear in [L2]. The Lannes-Zarati proof of Property A generalizes easily to all q, while a key detail of their proof of Property B doesn't seem to (the proof of Lemma 9.1.2 of [LZ1]). As will be seen in the next two sections, if Property B could be proved for all q, then Properties C and D would follow.

When restricted to dimension 0, all four properties imply interesting "weak" versions:

(a) Property A yields statement (2)(a) of Theorem 3.4;

(b) Property B yields statement (2)(b) of Theorem 3.4;

(c) Property C yields (3.6) (or, equivalently, Theorem 3.5);

(d) Property D yields Theorem 4.2.

<u>**Theorem 4.2**</u>. If K is in $\mathcal{K}(q)$, then $\ell(K)$ takes values in q-Boolean algebras.

Here an $\mathbb{F}_q$-algebra B is <u>q-Boolean</u> if $x^q = x$ for all x in B. In the appendix we offer readers a short proof of a general structure theorem for such objects: such a B is always isomorphic to $\mathbb{F}_q^{\mathrm{Spec}(B)}$ where Spec(B) is the profinite set of algebra maps from B to $\mathbb{F}_q$. As in [LZ1], one then deduces

<u>Corollary 4.3</u>. If K is in $\mathcal{K}(q)$, the natural map

$$\mathbb{F}_q[\mathrm{Hom}_{\mathcal{K}}(K, S^*(V))] \to \mathrm{Hom}_{\mathcal{U}}(K, S^*(V))$$

is an isomorphism.

Here the left side is the profinite vector space with "basis" the profinite Hom-set.

<u>Remark 4.4</u>. We note that the proofs of Properties A and B in [LZ1] both rely on inductions starting from the corresponding weak versions.

§5. <u>New relationships between Properties A, B, and C</u>

Now let us try to understand the properties of T_V via $\mathcal{F}_\omega(q)$. As in the introduction, let $\overline{T}_V$: $\mathcal{F}(q) \to \mathcal{F}(q)$ be defined by $(\overline{T}_V F)(W) = F(V \oplus W)$. It is elementary to check that the diagram

(5.1)
$$\begin{array}{ccc} \mathcal{U}(q) & \xrightarrow{T_V} & \mathcal{U}(q) \\ \ell\downarrow & & \ell\downarrow \\ \mathcal{F}_\omega(q) & \xrightarrow{\overline{T}_V} & \mathcal{F}_\omega(q) \end{array}$$

commutes, i.e., there is a natural isomorphism $\alpha(M)\colon \ell T_V(M) \xrightarrow{\sim} \overline{T}_V \ell(M)$.

Note that all of the following are obvious (although the reader is encouraged to verify (b)):

(a) $\overline{T}_V$ is exact;

(b) $\overline{T}_V I_W = \mathbb{F}_q^{\mathrm{Hom}(V,W)} \otimes I_W$;

(c) $\overline{T}_V$ takes tensor products to tensor products;

(d) if F takes values in q-Boolean algebras, so does $\overline{T}_V F$.

This is clearly promising. The main results of this section are the following observations.

<u>Theorem 5.2</u>. Assuming any of the equivalent statements in Theorem 3.4 (e.g., the weak versions of Properties A and B), the following are equivalent.

(1) If M is nilclosed, so is $T_V M$.

(2) The diagram

$$\begin{array}{ccc} \mathcal{U}(q) & \xrightarrow{T_V} & \mathcal{U}(q) \\ r\uparrow & & r\uparrow \\ \mathcal{F}_\omega(q) & \xrightarrow{\overline{T}_V} & \mathcal{F}_\omega(q) \end{array}$$

commutes, i.e., the natural map $\beta(F)\colon T_V r(F) \to r\overline{T}_V(F)$ is an isomorphism for all F.

(3) (a) Property A holds.

(b) Property B holds.

<u>Theorem 5.3</u>. Properties A and B imply Property C.

<u>Proof of Theorem 5.2</u>. <u>(1) ⇒ (2)</u> Assuming (1), both the domain and range of $\beta(F)$ will be nilclosed. Thus to show $\beta(F)$ is an isomorphism, it suffices to show that $\ell\beta(F)$ is an isomorphism. Now use the diagram

$$\begin{array}{ccc} \ell T_V r(F) & \xrightarrow{\ell\beta(F)} & \ell r \overline{T}_V(F) \\ {\scriptstyle \alpha r(F)}\downarrow & & \downarrow{\scriptstyle \varepsilon} \\ \overline{T}_V \ell r(F) & \underset{\sim}{\xrightarrow{\varepsilon}} & \overline{T}_V(F) \end{array}.$$

<u>(2) ⇒ (1)</u>. To say that M is nilclosed is to say that it is isomorphic to a module in the image of r. Assuming (2), if $M \simeq r(F)$, $T_V M \simeq r(\overline{T}_V F)$.

<u>(2) ⇒ (3)(a)</u>. Observe that T_V is right exact, since it is a left adjoint. Thus T_V will be exact if its first left derived functor is trivial, or equivalently, if T_V preserves the exactness at P_1 of an exact sequence of the form

$$0 \to N \to P_1 \to P_0 \to M \to 0,$$

where P_0 and P_1 are "free" unstable A(q)-modules (i.e., direct sums of F(n)'s).

To show this, we note that P_0 and P_1 are nilclosed, thus so is N. It follows then that

$$0 \to N \to P_1 \to P_1$$

is obtained by applying r to an exact sequence

$$0 \to F_2 \to F_1 \to F_0.$$

Now assuming (2), we conclude that

$$0 \to T_V N \to T_V P_1 \to T_V P_0$$

can be identified with

$$0 \to r\overline{T}_V F_2 \to r\overline{T}_V F_1 \to r\overline{T}_V F_0.$$

This last sequence is exact because $\overline{T}_V$ is exact and r is left exact.

<u>(2) ⇒ (3)(b)</u>. This is clear, recalling that $S^*(W^*) = r(I_W)$.

<u>(3) ⇒ (2)</u>. Statement (3)(b) says that $\beta(I_W)$ is an isomorphism for all W, and thus $\beta(I)$ is an isomorphism for all injectives I in $\mathcal{F}_\omega(q)$. (Both $T_V r$ and $r\overline{T}_V$ commutes with direct sums.) The composite $r\overline{T}_V$ is clearly left exact. Assuming (3)(a), so is $T_V r$. Then a 5-lemma argument using an injective resolution $0 \to F \to I_0 \to I_1$ shows that if $\beta(I_0)$ and $\beta(I_1)$ are isomorphisms, so is $\beta(F)$.

<u>Proof of Theorem 5.3</u>. Assuming Properties A and B, we know that statement (2) of Theorem 5.2 holds. Since both $\overline{T}_V$ and r preserve tensor products, so does

$T_V r = r\overline{T}_V$. Thus $\mu(M,N)\colon T_V(M\otimes N) \to T_V M \otimes T_V N$ is an isomorphism for all nilclosed M and N, in particular for M = F(m) and N = F(n). For a general M and N, $\mu(M,N)$ is seen to be an isomorphism by resolving both M and N by sums of F(n)'s, then using the right exactness of T_V and a 5-lemma argument (twice).

§6. Doubling and Property D

In this section we try to put the q-analogue of the doubling construction into the framework of generic representation theory. Then, following roughly the same path as in [L2], we deduce that Property D holds if Properties A and C do.

We have the elementary but fundamental observation.

Lemma 6.1. The following diagram commutes:

$$\begin{array}{ccc} S_q(V) & \hookrightarrow & V^{\otimes q} \\ {\scriptstyle D\xi}\downarrow & & \downarrow \\ V & \overset{\xi}{\hookrightarrow} & S^q(V) \end{array} .$$

Here ξ is the qth power map and $D\xi$ its dual.

Corollary 6.2. For any F in $\mathcal{F}(q)$, the image of the composite $S_q F \hookrightarrow F^{\otimes q} \twoheadrightarrow S^q F$ is naturally isomorphic to F.

The importance of Lemma 6.1 in the general theory of $\mathcal{F}(q)$ will be made clear in [K3]. For example, we have

<u>Theorem 6.3</u>. [K3] Every natural map $h: S_m(V) \to S^n(V)$ is a sum $\sum_{i=1}^{k} g_i f_i$, where g_i and f_i are maps

$$S_m(V) \xrightarrow{f_i} V^{\otimes d_i} \xrightarrow{g_i} S^n(V).$$

Furthermore, the consequences of Lemma 6.1 precisely determine the lack of uniqueness in the representation $h = \sum_{i=1}^{k} g_i f_i$.

We now attempt to "lift" Lemma 6.1 and Corollary 6.2 to $\mathcal{U}(q)$. We need two lemmas.

<u>Lemma 6.4</u>.

(1) Any map $a: S^{mq} \to S^{nq}$ defines a (necessarily unique) $\varphi(a)$ making the diagram

$$\begin{array}{ccc} S^{mq} & \xrightarrow{a} & S^{nq} \\ \xi\uparrow & & \xi\uparrow \\ S^m & \xrightarrow[\varphi(a)]{} & S^n \end{array}$$

commute.

(2) If $d \neq nq$ for any n, then every composite of the form $S^m \xrightarrow{\xi} S^{mq} \xrightarrow{a} S^d$ is zero. (We say $\varphi(a) = 0$ in this case.)

Lemma 6.5.

(1) Given M in $\mathcal{U}(q)$, and a: $S^{mq} \to S^{nq}$, the diagram

$$\begin{array}{ccc} S^q(M)_{mq} & \xrightarrow{a} & S^q(M)_{nq} \\ \uparrow & & \uparrow \\ M_m & \xrightarrow[\varphi(a)]{} & M_n \end{array}$$

commutes.

(2) If $d \neq nq$ for any n, then the composite $M_m \to S^q(M)_{mq} \xrightarrow{a} S^q(M)_d$ is zero, for all M in $\mathcal{U}(q)$ and a: $S^{mq} \to S^d$.

Both lemmas can easily be proved by computation, letting a: $S^{mq} \to S^d$ run through the Milnor basis for $A(q)_{d-mq}$. One learns, e.g., that $\varphi(Sq^{2i}) = Sq^i$ when $q = 2$. Presumably, the second lemma is a formal consequence of the first, but I have had trouble verifying this.

Recall that $\mathcal{S}^*$ was defined as the full subcategory of $\mathcal{F}(q)$ with objects S^d. It is convenient to make the zero object of $\mathcal{F}(q)$ an object in $\mathcal{S}^*$. Armed with the last two lemmas, we can make the following definitions.

Definitions 6.6.

(1) Let $\varphi: \mathcal{S}^* \to \mathcal{S}^*$ be the functor defined on objects

by $\varphi(S^{mq}) = S^m$ and $\varphi(S^d) = 0$ if $d \neq mq$ for any m. On morphisms, let φ be defined as in Lemma 6.4.

(2) Let $\Phi: \mathcal{U}(q) \to \mathcal{U}(q)$ be defined by $\Phi M = M \circ \varphi$.

(3) $\xi: M_m \to M_{mq}$ induces a natural transformation

$$\lambda: \Phi M \to M.$$

(4) $M_m \hookrightarrow S^q(M)_{mq}$ induces a natural inclusion

$$i: \Phi M \to S^q M.$$

<u>Proposition 6.7</u>.

(1) Via i, ΦM is isomorphic to the image of the composite

$$S_q M \hookrightarrow M^{\otimes q} \twoheadrightarrow S^q M.$$

(2) $\ell(\lambda): \ell(\Phi M) \to \ell(M)$ is an isomorphism for all M in $\mathcal{U}(q)$.

(3) For all M in $\mathcal{U}(q)$, there is a commutative diagram

$$\begin{array}{ccc} \ell(\Phi M) & \xrightarrow{\ell(i)} & \ell(S^q M) \\ {\scriptstyle \ell(\lambda)}\downarrow & & \downarrow \\ \ell(M) & \xrightarrow{\xi} & S^q \ell(M) \end{array},$$

where the right vertical map is an isomorphism since ℓ is exact and preserves tensor products.

(4) For all F in $\mathcal{F}(q)$, there is a commutative diagram

$$\begin{array}{ccc} \Phi r(F) & \xrightarrow{i} & S^q r(F) \\ \lambda \downarrow & & \downarrow \\ r(F) & \xrightarrow{r(\xi)} & r(S^q F) \end{array},$$

where the right map can be regarded as the nilclosure map η.

(5) There is a natural isomorphism $\Phi(M\otimes N) \simeq \Phi M \otimes \Phi N$ and a commutative diagram

$$\begin{array}{ccccc} M\otimes N & \xleftarrow{\lambda} & \Phi(M\otimes N) & \xrightarrow{i} & S^q(M\otimes N) \\ \| & & \wr\| & & \downarrow \\ M\otimes N & \xleftarrow{\lambda\otimes\lambda} & \Phi M\otimes\Phi N & \xrightarrow{i\otimes i} & S^q M\otimes S^q N. \end{array}$$

<u>Proof of (1)</u>. This is obvious, given Lemma 6.1.

<u>Proof of (2)</u>. Because $\xi : S^*(V) \to S^*(V)$ is monic, it is clear that both $\ker(\lambda)$ and $\operatorname{coker}(\lambda)$ are in $\mathcal{N}(q)$, so $\ell(\lambda)$ is an isomorphism.

<u>Proof of (3)</u>. Recalling the definition of ℓ from §3, we need to check that

$$\begin{array}{ccc} S_q(\mathrm{Hom}_{\mathcal{U}}(M, S^*(V))) & \longrightarrow & \mathrm{Hom}_{\mathcal{U}}(M, S^*(V)) \\ \downarrow & & \downarrow \\ \mathrm{Hom}_{\mathcal{U}}(S^q M, S^*(V)) & \longrightarrow & \mathrm{Hom}_{\mathcal{U}}(\Phi M, S^*(V)) \end{array}$$

commutes. This is straightforward to verify, noting that $\xi(x) = x^q$ for all x in $S^*(V)$.

<u>Proof of (4)</u>. This follows by adjointing (3).

Proof of (5). This is straightforward to verify.

Note that this proposition roughly says that if $F = \ell(M)$, then applying ℓ to the diagram

$$\begin{array}{ccc} S_q M & \hookrightarrow & M^{\otimes q} \\ \downarrow & & \downarrow \\ \Phi M & \hookrightarrow & S^q M \end{array} \quad \text{yields} \quad \begin{array}{ccc} S_q F & \hookrightarrow & F^{\otimes q} \\ \downarrow & & \downarrow \\ F & \hookrightarrow & S^q F \end{array},$$

i.e., we have "lifted" the situation of Lemma 6.1 to $\mathcal{U}(q)$.

Following Lannes, we can reformulate the definition of $\mathcal{K}(q)$.

Definitions 6.8.

(1) Let $\mathcal{K}(q)$ be the category of associative, unital, commutative algebras K in the category $\mathcal{U}(q)$ such that the diagram

$$\begin{array}{ccc} \Phi(K) & \xrightarrow{i} & S^q(K) \\ & \searrow^{\lambda} & \downarrow m \\ & & K \end{array}$$

commutes. (Here m is induced by the multiplication $K \otimes K \to K$.)

(2) Let $\mathcal{B}(q)$ be the category of functors F in $\mathcal{F}(q)$

taking values in q-Boolean algebras, i.e., the diagram

$$\begin{array}{ccc} F & \xrightarrow{\xi} & S^qF \\ & \searrow\!\!\!\!= & \downarrow m \\ & & F \end{array}$$

commutes.

<u>Theorem 6.9</u>. If $K \in \mathcal{K}(q)$, then $\ell(K) \in \mathcal{B}(q)$. If $F \in \mathcal{B}(q)$, then $r(F) \in \mathcal{K}(q)$.

This follows immediately from Proposition 6.7. Note that the first statement is Theorem 4.2, the weak version of Property D. To obtain the strong version, we need to study the natural map

$$T_V\Phi(M) \to \Phi T_V(M)$$

defined to be adjoint to the composite

$$\Phi M \to \Phi(S^*(V^*)\otimes T_VM) \simeq \Phi S^*(V^*)\otimes\Phi T_VM \to S^*(V^*)\otimes\Phi T_VM.$$

<u>Proposition 6.10</u>.

(1) The following diagram commutes for all M:

$$\begin{array}{ccccc} T_VM & \xleftarrow{T_V\lambda} & T_V\Phi M & \xrightarrow{T_Vi} & T_VS^qM \\ \| & & \downarrow & & \downarrow \\ T_VM & \xleftarrow{\lambda} & \Phi T_VM & \xrightarrow{i} & S^qT_VM \end{array}.$$

(2) Assuming Properties A and C, $T_V\Phi M \to \Phi T_VM$ is an isomorphism for all M.

Proof of (1). Both squares are checked by adjointing. For example, for the right square we have

$$\begin{array}{ccccccc}
\Phi M & \to & \Phi(S^*(V^*)\otimes T_V M) & \to & \Phi S^*(V^*)\otimes \Phi T_V M & \xrightarrow{\lambda\otimes 1} & S^*(V^*)\otimes \Phi T_V M \\
i\downarrow & & i\downarrow & & i\otimes i\downarrow & & 1\otimes i\downarrow \\
S^q M & \to & S^q(S^*(V^*)\otimes T_V M) & \to & S^q S^*(V^*)\otimes S^q T_V M & \to & S^*(V^*)\otimes S^q T_V M,
\end{array}$$

where the right square commutes since $S^*(V^*)$ is in $\mathcal{K}(q)$.

Proof of (2). If T_V is exact and commutes with tensor products, the result follows from Proposition 6.7(1).

Theorem 6.11. If Properties A and C hold, so does Property D.

Proof. We wish to show:

$$\text{if } \begin{array}{ccc} \Phi K & \xrightarrow{i} & S^q K \\ & {\scriptstyle\lambda}\searrow & \downarrow m \\ & & K \end{array} \text{ commutes, so does } \begin{array}{ccc} \Phi T_V K & \xrightarrow{i} & S^q T_V K \\ & {\scriptstyle\lambda}\searrow & \downarrow m \\ & & T_V K \end{array}.$$

To see this consider the following diagram.

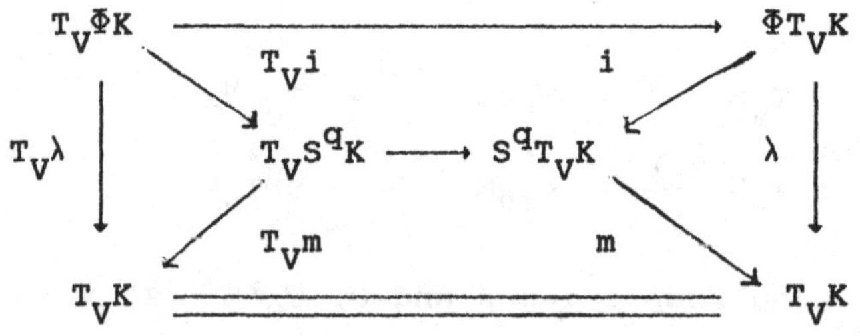

The left triangle commutes by assumption. The outside rectangle and upper quadrilateral commute by the last proposition, which also shows that the horizontal maps are isomorphisms. The lower quadrilateral commutes by definition. Thus so does the right triangle.

Appendix: The structure of q-Boolean algebras

Let B be a q-Boolean algebra. Recall that Spec(B) is the set of algebra homomorphisms from B to $\mathbb{F}_q$. Filtering B by its finitely generated subalgebras B_α induces a bijection $\mathrm{Spec}(B) = \varprojlim \mathrm{Spec}(B_\alpha)$, giving Spec(B) the structure of a profinite set. We let $\mathbb{F}_q^{\mathrm{Spec}(B)}$ denote the continuous functions from Spec(B) to $\mathbb{F}_q$. There is a natural algebra map

$$\gamma_B\colon B \to \mathbb{F}_q^{\mathrm{Spec}(B)}$$

defined by $\gamma_B(x)(\alpha) = \alpha(x)$.

The goal of this appendix is to give an elementary proof of

Theorem A.1. γ_B is an isomorphism.

Readers of the papers of Lannes, et al., might assume that this has already been proved (when q = p) in one of [LZ2], [L2], or [HLS], but "précisons un peu." (The author would like to thank the referee for insisting that this "elementary" proof be truly so!)

To prove the theorem, we first observe that if $\varinjlim B_\alpha = B$, then $\varinjlim \mathbb{F}_q^{\mathrm{Spec}(B_\alpha)} = \mathbb{F}_q^{\mathrm{Spec}(B)}$. Thus it suffices to show that γ_B is an isomorphism in the special case when B is finitely generated.

Note that γ_B is clearly an isomorphism if B is isomorphic to $\mathbb{F}_q \times \cdots \times \mathbb{F}_q$. The proof of the theorem is then completed with

<u>Proposition A.2</u>. If B is finitely generated, then B is isomorphic to $\mathbb{F}_q \times \cdots \times \mathbb{F}_q$.

<u>Proof</u>. B is certainly finite dimensional: there are clearly at most q^k distinct words that one can make with any k elements of B.

Now observe that for any x in B, $e = x^{q-1}$ is idempotent. Furthermore, e = 0 only when x = 0, and e = 1 only when x is invertible. If neither of these alternatives hold, there will be a nontrivial decomposition $B \simeq eB \times (1-e)B$. Thus B is isomorphic to a finite product of fields.

Finally we note that if B is a field, then B is isomorphic to $\mathbb{F}_q$, as B is an extension of $\mathbb{F}_q$ having the property that every nonzero element is a solution to $x^{q-1} = 1$.

References

[AGM] J. F. Adams, J. H. Gunawardena, and H. R. Miller, the Segal conjecture for elementary abelian p-groups, Topology 24 (1985), 435-460.

[A] J. L. Alperin, Local Representation Theory, Cambridge University Press, Cambridge, 1986.

[C] G. Carlsson, G. B. Segal's Burnside ring conjecture for $(\mathbb{Z}/2)^k$, Topology 22 (1983), 83-103.

[HLS] H.-W. Henn, J. Lannes, and L. Schwartz, The categories of unstable modules and unstable algebras over the Steenrod algebra modulo nilpotent objects, preprint, 1990.

[K1,2,3] N. J. Kuhn, Generic representations of the finite general linear groups and the Steenrod algebra: I, II, III. Part I is a preprint, Parts II, III are in preparation.

[L1] J. Lannes, Sur la cohomologie modulo p des p-groupes abéliens élémentaires, Proc. Durham Symposium on Homotopy Theory 1985, L.M.S., Camb. Univ. Press, 1987, 97-116.

[L2] J. Lannes, Sur les espaces fonctionnels dont la source est le classifiant d'un p-groupe abélien élémentaire, preprint, 1990.

[LS] J. Lannes and L. Schwartz, Sur la structure des A-modules instables injectifs, Topology 28 (1989), 153-169.

[LZ1] J. Lannes and S. Zarati, Sur les $\mathcal{U}$-injectifs, Ann. Scient. Ec. Norm. Sup. 19 (1986), 303-333.

[LZ2] J. Lannes and S. Zarati, Sur les foncteurs dérivés de la déstabilisation, Math. Zeit. 194 (1987), 25-59.

[M] H. R. Miller, The Sullivan conjecture on maps from classifying spaces, Ann. Math. 120 (1984), 39--87.

SOME CHROMATIC PHENOMENA IN THE HOMOTOPY OF MSp

ANDREW BAKER

Introduction.

In this paper, we derive formulæ in Brown-Peterson homology at the prime 2 related to the family of elements $\varphi_n \in \mathrm{MSp}_{8n-3}$ of N. Ray, whose central rôle in the structure of MSp has been highlighted by recent work of V. Vershinin and other Russian topologists. In effect, we give explicit "chromatic" representatives for these elements, which were known to be detected in KO and mod 2 KU-homology, and are thus "v_1-periodic" in the parlance of [4] and [5]. In future work we will investigate further the v_1 periodic part of MSp and discuss the relationship of our work with that of B. Botvinnik.

I would like to thank Nigel Ray for many helpful discussions and large amounts of advice on MSp (including severe warnings!) over many years; in particular, §5 in this paper was prompted by his suggestions about the detection of φ_n in the classical Adams spectral sequence. I would also like to thank Boris Botvinnik, Vassily Gorbunov and Vladimir Vershinin for discussions on the material of earlier versions of this paper both during and after the J. F. Adams Memorial Symposium and in particular for bringing to my attention Buhštaber's article [2] which contains related results.

§1 Some algebraic results on $E_*(\mathrm{MSp})$.

Let E be a commutative ring spectrum, and let $x^E \in E^2(\mathbb{C}\mathrm{P}^\infty)$ be a complex orientation in the sense of [1]. Then the results of the following Theorem are well known.

Theorem (1.1).

a) *The natural map* $j_1 : \mathbb{C}\mathrm{P}^\infty \longrightarrow \mathbb{H}\mathrm{P}^\infty$ *induces a split monomorphism*

$$E^*(\mathbb{H}\mathrm{P}^\infty) \xrightarrow{j_1{}^*} E^*(\mathbb{C}\mathrm{P}^\infty),$$

Key words and phrases. Symplectic bordism, Brown-Peterson theory.

The author would like to acknowledge the support of the Science and Engineering Research Council whilst this research was conducted .

and a split epimorphism

$$E_*(\mathbb{C}\mathrm{P}^\infty) \xrightarrow{j_{1*}} E_*(\mathbb{H}\mathrm{P}^\infty),$$

of modules over E_*.

b) *As an* E_* *algebra,*

$$E_*(\mathrm{MSp}) = E_*[Q_k^E : k \geqslant 1],$$

and moreover the natural morphism of ring spectra $j\colon \mathrm{MSp} \longrightarrow \mathrm{MU}$ *induces an embedding of* E_* *algebras*

$$j_*\colon E_*(\mathrm{MSp}) \xrightarrow{j_*} E_*(\mathrm{MU}).$$

We will need explicit sets of generators for the E homology and cohomology of $\mathbb{H}\mathrm{P}^\infty$ and MSp. Recall the canonical complex line bundle $\eta \longrightarrow \mathbb{C}\mathrm{P}^\infty$ and its E-theory 1st Chern class $c_1^E(\eta) = x^E$; we have $E^*(\mathbb{C}\mathrm{P}^\infty) = E_*[[x^E]]$. Then the map $j_1\colon \mathbb{C}\mathrm{P}^\infty \longrightarrow \mathbb{H}\mathrm{P}^\infty$ classifies the quaternionic line bundle $\eta \otimes_{\mathbb{C}} \mathbb{H} \longrightarrow \mathbb{C}\mathrm{P}^\infty$, which has as its underlying complex bundle $\eta + \overline{\eta}$, where $\overline{(\ )}$ denotes complex conjugation. We define *the E-theory 1st symplectic Pontrjagin class* of a symplectic bundle $\zeta \longrightarrow X$ to be the negative of the E-theory 2nd Chern class of the complex bundle ζ' underlying ζ,

$$\wp^E(\zeta) = -c_2(\zeta') \in E^4(X).$$

In particular we set $\wp^E = \wp^E(\xi) \in E^4(\mathbb{H}\mathrm{P}^\infty)$, where $\xi \longrightarrow \mathbb{H}\mathrm{P}^\infty$ is the canonical symplectic line bundle; this gives

$$j_1^*\wp^E = \wp^E(\eta + \overline{\eta}) = -c_1^E(\eta)c_1^E(\overline{\eta}) \in E^4(\mathbb{C}\mathrm{P}^\infty).$$

We also have (as graded algebras over E_*)

$$E^*(\mathbb{H}\mathrm{P}^\infty) = E_*[[\wp^E]]. \tag{1.2}$$

Now let the elements $\beta_n^E \in E_{2n}(\mathbb{C}\mathrm{P}^\infty)$, $n \geqslant 0$, form the standard E_* basis for $E_*(\mathbb{C}\mathrm{P}^\infty)$ as detailed in [1]; this basis is dual to that of the monomials $(x^E)^n$ in $E^*(\mathbb{C}\mathrm{P}^\infty)$ under the Kronecker pairing $\langle\ ,\ \rangle$:

$$\langle (x^E)^r, \beta_s^E \rangle = \delta_{r,s},$$

where δ is the Kronecker delta function. Also for any T, let $\overline{T} = [-1]_E(T)$ be the -1 series for the formal group law $F^E(X,Y) \in E_*[[X,Y]]$ associated to the orientation x^E as described in [1]; notice that $\overline{x^E} = c_1^E(\overline{\eta})$.

Now in $E_*(\mathbb{H}\mathrm{P}^\infty)$ we can define a sequence of elements $\gamma_n \in E_{4n}(\mathbb{H}\mathrm{P}^\infty)$, $n \geqslant 0$, by requiring that these are dual to the $(\wp^E)^n$:

$$\langle (\wp^E)^r, \gamma_s \rangle = \delta_{r,s}. \tag{1.3}$$

It is easily verified that a generating function for these elements is the series

$$\sum_{n\geqslant 0} \gamma_n(-T\overline{T})^n = \sum_{n\geqslant 0} j_{1*}\beta_n^E T^n. \tag{1.4}$$

Let $i_1\colon \mathbb{H}\mathrm{P}^\infty \simeq \mathrm{MSp}(1) \longrightarrow \Sigma^4\mathrm{MSp}$ be the standard map, then we have

Theorem (1.5).

a) *The elements* γ_n *for* $n \geqslant 0$ *form an* E_* *basis for* $E_*(\mathbb{H}\mathrm{P}^\infty)$.

b) *The elements* $i_{1*}\gamma_{n+1}$ *for* $n \geqslant 1$ *form a set of polynomial generators for the* E_* *algebra* $E_*(\mathrm{MSp})$.

From now on we set $Q_n^E = i_{1*}\gamma_{n+1} \in E_{4n}(\mathrm{MSp})$.

We also need some information on the image of the ring homomorphism $j_*: E_*(\mathrm{MSp}) \longrightarrow E_*(\mathrm{MU})$. We have the two generating functions

$$Q^E(T) = \sum_{n\geqslant 0} Q_n^E T^{n+1} \in E_*(\mathrm{MSp})[[T]],$$

$$j_*Q^E(T) = \sum_{n\geqslant 0} j_*Q_n^E T^{n+1} \in E_*(\mathrm{MU})[[T]].$$

Recall from [1] the standard generators $B_n^E \in E_{2n}(\mathrm{MU})$, (for $n \geqslant 0$ and $B_0 = 1$) for which

$$E_*(\mathrm{MU}) = E_*[B_n^E : n \geqslant 1],$$

and let $B^E(T) = \sum_{n\geqslant 0} B_n^E T^{n+1}$. The proof of the next result is left to the reader.

Proposition (1.6). *The series* $j_*Q^E(-T\overline{T}) \in E_*(\mathrm{MU})$ *satisfies the equation*

$$Q^E(-T\overline{T}) = -B^E(T)B^E(\overline{T}).$$

§2 Symplectic Pontrjagin classes in $E \wedge \mathrm{MSp}$ theory.

Recall that the ring spectrum MSp is universal for orientations for quaternionic bundles. The universal orientation is induced from the class

$$\wp^{\mathrm{MSp}}: \mathbb{H}\mathrm{P}^\infty \simeq \mathrm{MSp}(1) \longrightarrow \Sigma^4\mathrm{MSp} \in \mathrm{MSp}^4(\mathbb{H}\mathrm{P}^\infty).$$

We also have

$$\text{(2.1)} \qquad \mathrm{MSp}^*(\mathbb{H}\mathrm{P}^\infty) = \mathrm{MSp}_*[[\wp^{\mathrm{MSp}}]].$$

Now let E be a complex oriented ring spectrum as in §1. Then the class $\wp^E$ will serve as a universal orientation for quaternionic line bundles in E theory. We can consider the representable cohomology theory $(E\wedge\mathrm{MSp})^*(\ )$ on either of the categories of CW spectra or spaces. As $E_*(\mathrm{MSp})$ is free over E_*, we have a Boardman isomorphism

$$\text{(2.2)} \qquad (E \wedge \mathrm{MSp})^*(\ ) \cong E_*(\mathrm{MSp})\widehat{\otimes}_{E_*} E_*(\ )$$

where $\widehat{\otimes}$ denotes the completed tensor product with respect to the skeletal topology for infinite complexes. From §1 and standard arguments about the map $i_1: \mathbb{H}\mathrm{P}^\infty \longrightarrow \Sigma^4\mathrm{MSp}$, we have

Proposition (2.3). *In* $(E \wedge \mathrm{MSp})^*(\mathbb{H}\mathrm{P}^\infty) \cong E_*(\mathrm{MSp})\widehat{\otimes}_{E_*}E^*(\mathbb{H}\mathrm{P}^\infty)$ *we have the identity*

$$\wp^{\mathrm{MSp}} = Q^E(\wp^E).$$

For later convenience we also introduce the series

$$N^E(T) = \sum_{n\geqslant 0} N_n^E T^{n+1} \in E_*(\mathrm{MSp})[[T]]$$

determined by the equation

$$N^E(Q^E(T)) = T. \tag{2.4}$$

Notice that $N_n^E \equiv -Q_n^E$ modulo decomposables, and hence we can take the N_n^E to be polynomial generators for $E_*(\mathrm{MSp})$.

§3 Detecting Ray's element φ_n with a complex oriented cohomology theory.

In this section we again let E be a complex oriented commutative ring spectrum. We will also require the following assumption to hold:

TF: E_* is torsion free.

This condition is satisfied for the following spectra: $E = \mathrm{MU}, \mathrm{BP}, \mathrm{KU}, \mathrm{H}\,\mathbb{Z}$, which include all those which we will be explicitly considering in this paper.

Now consider the space $\mathbb{R}\mathrm{P}^\infty$. Let $\rho \longrightarrow \mathbb{R}\mathrm{P}^\infty$ be the canonical real line bundle and $\lambda = \rho \otimes_{\mathbb{R}} \mathbb{C} \longrightarrow \mathbb{R}\mathrm{P}^\infty$ be its complexification. Let $w^E = c_1^E(\lambda) \in E^2(\mathbb{R}\mathrm{P}^\infty)$. Then we have

$$E^*(\mathbb{R}\mathrm{P}^\infty) = E_*[[w^E]]/([2]_E(w^E)), \tag{3.1}$$

where $[2]_E T$ denotes the 2 series of the formal group law associated to x^E.

Now we will consider the space $\mathbb{R}\mathrm{P}^\infty \times \mathbb{H}\mathrm{P}^\infty$ and apply the cohomology theory $(E \wedge \mathrm{MSp})^*(\ )$ to it. Since both $E_*(\mathrm{MSp})$ and $E^*(\mathbb{H}\mathrm{P}^\infty)$ are free modules over E_*, we have the following isomorphisms

$$(E \wedge \mathrm{MSp})^*(\mathbb{R}\mathrm{P}^\infty \times \mathbb{H}\mathrm{P}^\infty) \cong E_*(\mathrm{MSp})\widehat{\otimes}_{E_*}E^*(\mathbb{R}\mathrm{P}^\infty)\widehat{\otimes}_{E_*}E^*(\mathbb{H}\mathrm{P}^\infty),$$
$$(E \wedge \mathrm{MSp})^*(\mathbb{R}\mathrm{P}^\infty \times \mathbb{C}\mathrm{P}^\infty) \cong E_*(\mathrm{MSp})\widehat{\otimes}_{E_*}E^*(\mathbb{R}\mathrm{P}^\infty)\widehat{\otimes}_{E_*}E^*(\mathbb{C}\mathrm{P}^\infty)$$

Let $\xi \longrightarrow \mathbb{H}\mathrm{P}^\infty$ be the canonical quaternionic line bundle. Then the quaternionic bundle

$$\rho \otimes_{\mathbb{R}} \xi = \lambda \otimes_{\mathbb{C}} \xi \longrightarrow \mathbb{R}\mathrm{P}^\infty \times \mathbb{H}\mathrm{P}^\infty$$

is defined and so has a 1st symplectic Pontrjagin class in each of the theories represented by the spectra MSp, E and $E \wedge \mathrm{MSp}$.

In the ring $(E \wedge \mathrm{MSp})^*(\mathbb{RP}^\infty \times \mathbb{HP}^\infty)$, we have expressions of the form

(3.2) $$\wp^E(\rho \otimes_{\mathbb{R}} \xi) = \wp^E + w^2 + \sum_{n\geqslant 1} \widehat{\theta}_n^E (\wp^E)^{n+1},$$

and

(3.3) $$\wp^{\mathrm{MSp}}(\rho \otimes_{\mathbb{R}} \xi) = \wp^{\mathrm{MSp}} + w^2 + \sum_{n\geqslant 1} \widehat{\theta}_n^{\mathrm{MSp}} (\wp^{\mathrm{MSp}})^{n+1},$$

where $\widehat{\theta}_n^E \in E^{4n}(\mathbb{RP}^\infty)$ and $\widehat{\theta}_n^{\mathrm{MSp}} \in \mathrm{MSp}^{4n}(\mathbb{RP}^\infty)$. Upon applying the split monomorphism j_1^* these yield the following equations in $(E\wedge\mathrm{MSp})^*(\mathbb{RP}^\infty \times \mathbb{CP}^\infty)$:

(3.2′)

$$\wp^E((\lambda \otimes_{\mathbb{C}} \eta) \otimes_{\mathbb{C}} \mathbb{H}) = j_1^* \left[\wp^E + w^2 + \sum_{n\geqslant 1} \widehat{\theta}_n^E (\wp^E)^{n+1} \right],$$

(3.3′)

$$\wp^{\mathrm{MSp}}((\lambda \otimes_{\mathbb{C}} \eta) \otimes_{\mathbb{C}} \mathbb{H}) = j_1^* \left[\wp^{\mathrm{MSp}} + w^2 + \sum_{n\geqslant 1} \widehat{\theta}_n^{\mathrm{MSp}} (\wp^{\mathrm{MSp}})^{n+1} \right],$$

Now recall from (2.3) that $\wp^{\mathrm{MSp}} = Q^E(\wp^E)$. Thus we obtain

$$\wp^{\mathrm{MSp}}((\lambda \otimes_{\mathbb{C}} \eta) \otimes_{\mathbb{C}} \mathbb{H}) = Q^E(j_1^*\wp^E + w^2 + \sum_{n\geqslant 1} \widehat{\theta}_n^E (j_1^*\wp^E)^{n+1}).$$

Recall from [6] the following construction. Consider the inclusion of the bottom cell $S^1 \cong \mathbb{RP}^1 \hookrightarrow \mathbb{RP}^\infty$. Then we can restrict $\rho, \lambda \longrightarrow \mathbb{RP}^\infty$ to $\rho_1, \lambda_1 \longrightarrow \mathbb{RP}^1$ and obtain the classes

$$\wp^{\mathrm{MSp}}(\rho_1 \otimes_{\mathbb{R}} \xi) \in \mathrm{MSp}^4(\mathbb{RP}^1 \times \mathbb{HP}^\infty),$$

and its image under j_1^*,

$$\wp^{\mathrm{MSp}}(\lambda_1 \otimes_{\mathbb{C}} (\eta + \overline{\eta})) \in \mathrm{MSp}^4(\mathbb{RP}^1 \times \mathbb{CP}^\infty).$$

By definition, Ray's elements $\theta_n \in \mathrm{MSp}_{4n-3}$ are given by

$$\wp^{\mathrm{MSp}}(\rho_1 \otimes_{\mathbb{R}} \xi) - \wp^{\mathrm{MSp}} = \sum_{n\geqslant 1} \theta_n (\wp^{\mathrm{MSp}})^n$$

in $\mathrm{MSp}^*(\mathbb{R}\mathrm{P}^1 \times \mathbb{H}\mathrm{P}^\infty)$, and we also have

$$\wp^{\mathrm{MSp}}((\lambda_1 \otimes_{\mathbb{C}} \eta) \otimes_{\mathbb{C}} \mathbb{H}) - \wp^{\mathrm{MSp}} = \sum_{n \geqslant 1} \theta_n (\wp^{\mathrm{MSp}})^n$$

in $(E \wedge \mathrm{MSp})^*(\mathbb{R}\mathrm{P}^1 \times \mathbb{C}\mathrm{P}^\infty)$. Stably, the smash product $\mathbb{R}\mathrm{P}^1 \wedge \mathbb{H}\mathrm{P}^\infty$ is a retract of $\mathbb{R}\mathrm{P}^1 \times \mathbb{H}\mathrm{P}^\infty$, and this allows us to identify the above expressions with elements of $\mathrm{MSp}^*(\mathbb{R}\mathrm{P}^1 \wedge \mathbb{H}\mathrm{P}^\infty)$ and $(E \wedge \mathrm{MSp})^*(\mathbb{R}\mathrm{P}^1 \wedge \mathbb{C}\mathrm{P}^\infty)$. We will also use the notation $\varphi_n = \theta_{2n}$ of [6].

We can factor the inclusion $\mathbb{R}\mathrm{P}^1 \hookrightarrow \mathbb{R}\mathrm{P}^\infty$ as $\mathbb{R}\mathrm{P}^1 \hookrightarrow \mathbb{R}\mathrm{P}^2 \hookrightarrow \mathbb{R}\mathrm{P}^\infty$, and since $\mathbb{R}\mathrm{P}^2$ is a $\mathbb{Z}/2$ Moore space, we see that $2\theta_n = 0$. Under our assumption **TF** together with (3.1), we thus have that $\theta_n \longmapsto 0$ under the E theory Hurewicz homomorphism $\mathrm{MSp}_* \longrightarrow E_*(\mathrm{MSp})$. Equivalently, the class $\wp^{\mathrm{MSp}}((\lambda_1 \otimes_{\mathbb{C}} \eta) \otimes_{\mathbb{C}} \mathbb{H}) - \wp^{\mathrm{MSp}}$ maps to 0 under the natural homomorphism $\mathrm{MSp}^*(\mathbb{R}\mathrm{P}^1 \wedge \mathbb{H}\mathrm{P}^\infty) \longrightarrow (E \wedge \mathrm{MSp})^*(\mathbb{R}\mathrm{P}^1 \wedge \mathbb{H}\mathrm{P}^\infty)$. Let

$$\wp^{\mathrm{MSp}}(\lambda_2 \otimes_{\mathbb{C}} \xi) = \wp^{\mathrm{MSp}} + \sum_{n \geqslant 1} \widetilde{\theta}_n (\wp^{\mathrm{MSp}})^n \in (E \wedge \mathrm{MSp})^*(\mathbb{R}\mathrm{P}^2 \times \mathbb{H}\mathrm{P}^\infty)$$

where

$$\widetilde{\theta}_n \in \mathrm{MSp}^*(\mathbb{R}\mathrm{P}^2 \wedge \mathbb{H}\mathrm{P}^\infty) \subset (E \wedge \mathrm{MSp})^*(\mathbb{R}\mathrm{P}^2 \wedge \mathbb{H}\mathrm{P}^\infty)$$

is the image of $\widehat{\theta}^{\mathrm{MSp}}$ under the map induced by the inclusion $\mathbb{R}\mathrm{P}^2 \hookrightarrow \mathbb{R}\mathrm{P}^\infty$. Now for any spectrum F with F_* torsion free there is an isomorphism $F^k(\mathbb{R}\mathrm{P}^2) \cong F_{2-k}/2F_{2-k}$ induced from Spanier-Whitehead duality. In particular, this applies to the cases $F = E \wedge \mathrm{MSp}, \mathrm{MSp}$ that we are dealing with, and thus we can interpret $\widetilde{\theta}_n$ as an element of $\mathrm{MSp}_{4n-2}/(2) \subset E_{4n-2}(\mathrm{MSp})/(2)$. Let $w_2 \in E^2(\mathbb{R}\mathrm{P}^2)$ be the restriction of the generator $w^E \in E^2(\mathbb{R}\mathrm{P}^\infty)$; note that $w_2 = c_1^E(\lambda_2)$.

We can now give our main calculational result.

Theorem (3.4). *Let* $u = x^E$ *and* $\overline{u} = \overline{x^E}$, *satisfying*

$$-u\overline{u} = N^E(j_1^* \wp^{\mathrm{MSp}})$$

and

$$-u\overline{u}, u + \overline{u} \in \mathrm{im}\,[E^*(\mathbb{H}\mathrm{P}^\infty) \longrightarrow (E \wedge \mathrm{MSp})^*(\mathbb{H}\mathrm{P}^\infty)].$$

Then under assumption **TF** *together with (3.1), we have the following identity in* $(E \wedge \mathrm{MSp})_*(\mathbb{R}\mathrm{P}^2 \wedge \mathbb{C}\mathrm{P}^\infty)$:

$$\wp^{\mathrm{MSp}}(((\lambda_2 \otimes_{\mathbb{C}} \eta) \otimes_{\mathbb{C}} \mathbb{H})) + u\overline{u} = \frac{w_2}{N^{E\prime}(j_1^* \wp^{\mathrm{MSp}})} \left[\frac{u}{\log^{E\prime}(\overline{u})} + \frac{\overline{u}}{\log^{E\prime}(u)} \right]$$

and moreover this element is equal to $j_1^*\wp^{\mathrm{MSp}}(\rho_2 \otimes_{\mathbb{R}} \xi) - j_1^*\wp^{\mathrm{MSp}}$ *and lies in the image of the natural map* $\mathrm{MSp}^*(\mathbb{R}\mathrm{P}^2 \wedge \mathbb{H}\mathrm{P}^\infty) \longrightarrow (E \wedge \mathrm{MSp})^*(\mathbb{R}\mathrm{P}^2 \wedge \mathbb{H}\mathrm{P}^\infty)$.

Proof. We need to recall some relevant facts. Firstly, we have

$$w_2^2 = 0 = 2w_2.$$

Hence only the first order terms in w_2 are required. Secondly, we have the following formula for formal derivatives:

$$Q^{E\prime}(N^E(T)) = \frac{1}{N^{E\prime}(T)}.$$

Finally, by definition,

$$\wp^E((\lambda_2 \otimes_{\mathbb{C}} \eta) \otimes_{\mathbb{C}} \mathbb{H}) = -c^E(\lambda_2 \otimes_{\mathbb{C}} \eta + \lambda_2 \otimes_{\mathbb{C}} \overline{\eta}).$$

But expanding this using the Cartan formula, we obtain

$$\begin{aligned}\wp^E((\lambda_2 \otimes_{\mathbb{C}} \eta) \otimes_{\mathbb{C}} \mathbb{H}) &= -c_1^E(\lambda_2 \otimes_{\mathbb{C}} \eta)c_1^E(\lambda_2 \otimes_{\mathbb{C}} \overline{\eta}) \\ &= -F^E(w_2, u)F^E(\overline{w}_2, \overline{u}) \\ &= -F^E(w_2, u)F^E(w_2, \overline{u}),\end{aligned}$$

using the fact that λ_2^2 is a trivial line bundle. In this, we are setting $u = c_1^E(\eta)$ and $\overline{u} = c_1^E(\overline{\eta})$.

We also require the following well known Lemma.

Let $\log^E(X) \in E\otimes\mathbb{Q}[[X]]$ denote the logarithm series of the formal group law F^E.

Lemma (3.5). *The formula*

$$\log^{E\prime}(X) = \frac{1}{F_2^E(X,0)}$$

holds in $E_*[[X]]$, *where* $F_2^E(X,Y) = \frac{\partial}{\partial Y}F^E(X,Y)$. *Hence the logarithm of* F^E, $\log^E(X)$, *has coefficients in* E_*.

Proposition (3.4) now follows from these observations. □

As a sample application of this, we consider the case of $E = \mathrm{KU}$, i.e., complex KU-theory. This allows us to determine the KO-theory Hurewicz images of Ray's elements $\theta_n \in \mathrm{MSp}_{8n-3}$. Recall that we have $\mathrm{KU}_* = \mathbb{Z}[t, t^{-1}]$, where $t \in \mathrm{KU}_2$ is the Bott generator; this theory clearly satisfies the condition **TF**. Corresponding to the standard complex orientation in

$\mathrm{KU}^*(\ )$, the formal group law $F^{\mathrm{KU}}(X,Y) = X + Y + tXY$; this has as its logarithm the series.

$$\log^{\mathrm{KU}}(X) = t^{-1}\ln(1+tX) = \sum_{k\geqslant 1} \frac{(-t)^{k-1}X^k}{k}$$

Thus we have

$$\log^{\mathrm{KU}\,\prime}(X) = \frac{1}{(1+tX)}.$$

Now the formula of (3.4) becomes

$$\wp^{\mathrm{MSp}}(((\lambda_2 \otimes_{\mathbb{C}} \eta) \otimes_{\mathbb{C}} \mathbb{H})) = \\ \wp^{\mathrm{MSp}} + \frac{w_2}{N^{\mathrm{KU}\,\prime}(\wp^{\mathrm{MSp}})} \left[\frac{u}{\log^{\mathrm{KU}\,\prime}(\overline{u}\,)} + \frac{\overline{u}}{\log^{\mathrm{KU}\,\prime}(X)} \right].$$

We also have

$$[-1]_{\mathrm{KU}}(X) = \frac{-X}{(1+tX)}.$$

Thus we obtain

$$\begin{aligned} \wp^{\mathrm{MSp}}(((\lambda_2 \otimes_{\mathbb{C}} \eta) \otimes_{\mathbb{C}} \mathbb{H})) = \\ & \wp^{\mathrm{MSp}} + \frac{w_2}{N^{\mathrm{KU}}(\wp^{\mathrm{MSp}})} [u(1+t\overline{u}\,) + \overline{u}\,(1+tu)] \\ &= \wp^{\mathrm{MSp}} + \frac{w_2}{N^{\mathrm{KU}\,\prime}(\wp^{\mathrm{MSp}})} \cdot \frac{-tu^2}{(1+tu)} \\ &= \wp^{\mathrm{MSp}} + \frac{tw_2\wp^{\mathrm{KU}}}{N^{\mathrm{KU}\,\prime}(\wp^{\mathrm{MSp}})} \\ &= \wp^{\mathrm{MSp}} + \frac{tw_2 N^{\mathrm{KU}}(\wp^{\mathrm{MSp}})}{N^{\mathrm{KU}\,\prime}(\wp^{\mathrm{MSp}})}. \end{aligned}$$

The coefficient of $w_2(\wp^{\mathrm{MSp}})^n$ in this series is $\widetilde{\theta}_n$. Notice that we can write

$$(3.6) \qquad \frac{N(\wp^{\mathrm{MSp}})}{N^{\mathrm{KU}\,\prime}(\wp^{\mathrm{MSp}})} \equiv \wp^{\mathrm{MSp}} + \frac{\sum_{r\geqslant 1} N_{2r-1}(\wp^{\mathrm{MSp}})^{2r}}{\sum_{s\geqslant 0} N_{2s}(\wp^{\mathrm{MSp}})^{2s}} \pmod 2.$$

From this result, we can immediately deduce that $\widetilde{\theta}_{2r-1} = 0$ if $r > 1$. This suffices to prove the following result which was conjectured in [6] and also follows from an unpublished result of F. Roush which actually shows that $\theta_{2r-1} = 0$ when $r > 1$. Let $\underline{\mathrm{ko}}\colon \mathrm{MSp} \to \mathrm{KO}_*(\mathrm{MSp})$ be the KO-theory Hurewicz homomorphism.

Proposition (3.7). *For $r > 1$, we have in* $\mathrm{KO}_{8r-7}(\mathrm{MSp})$,

$$\underline{\mathrm{ko}}(\theta_{2r-1}) = 0.$$

Proof. Let $M(2) = S^0 \cup_2 \mathrm{e}^1$ denote the mod 2 Moore spectrum and let $\eta\colon S^1 \to S^0$ denote the non-trivial element of π_1^S together with the map $S^1 \wedge \mathrm{KO} \to \mathrm{KO}$ it induces; let $M(\eta) = S^0 \cup_\eta \mathrm{e}^2$ be the mapping cone. Recall from [1] "the Theorem of Reg Wood": there is a cofibre sequence

$$S^1 \wedge \mathrm{KO} \xrightarrow{\eta} \mathrm{KO} \to \mathrm{KO} \wedge M(\eta) \simeq \mathrm{KU}.$$

Now consider the exact diagram of abelian groups in which the rows are induced from the cofibre sequence for η and the columns from the cofibre sequence for multiplication by 2:

$$\begin{array}{ccccc}
\mathrm{KO}_{8r-7}(\mathrm{MSp}) & \longrightarrow & \mathrm{KO}_{8r-7}(\mathrm{MSp} \wedge M(2)) & \longrightarrow & \mathrm{KO}_{8r-8}(\mathrm{MSp}) \\
\eta \big\downarrow \text{epi} & & \eta \big\downarrow & & \eta \big\downarrow \text{epi} \\
\mathrm{KO}_{8r-6}(\mathrm{MSp}) & \longrightarrow & \mathrm{KO}_{8r-6}(\mathrm{MSp} \wedge M(2)) & \longrightarrow & \mathrm{KO}_{8r-7}(\mathrm{MSp}) \\
 & & \big\downarrow & & \\
 & & \mathrm{KU}_{8r-6}(\mathrm{MSp} \wedge M(2)) & &
\end{array}$$

Now $\widetilde{\theta}_{2r-1} = 0$ in the group $\mathrm{KU}_{8r-6}(\mathrm{MSp} \wedge M(2))$ appearing in this diagram, and therefore we can lift $\underline{\mathrm{ko}}(\theta_{2r-1})$ in the group $\mathrm{KO}_{8r-7}(\mathrm{MSp})$ to an element of the torsion group $\mathrm{KO}_{8r-7}(\mathrm{MSp} \wedge M(2))$. But this means that $\underline{\mathrm{ko}}(\theta_{2r-1})$ is the image of an element of $\mathrm{KO}_{8r-8}(\mathrm{MSp})$, a torsion free group. The only way that both of these can be true is if $\underline{\mathrm{ko}}(\theta_{2r-1}) = 0$. □

We can also see that for each $n \geqslant 1$,

$$\widetilde{\varphi}_n = \widetilde{\theta}_{2n} \equiv tN_{2n-1} \quad \text{(mod decomposables)}.$$

This of course shows that in the ring $\mathrm{KU}_*(\mathrm{MSp})/(2)$, the elements $\widetilde{\varphi}_n$ are algebraically independent over $\mathrm{KU}_*/(2)$.

§4 Some calculations in $\mathrm{BP}_*(\mathrm{MSp})$.

We now consider the case of $E = \mathrm{BP}$, the Brown-Peterson spectrum at the prime 2; we will assume the reader to be familiar with [1] and [5] which contains detailed information on BP.

We begin with some algebraic Lemmas on the formal group law for BP. Recall that the logarithm of F^{BP} is the series

$$\text{(4.1)} \qquad \log^{\mathrm{BP}}(X) = \sum_{n \geqslant 0} \ell_n X^{2^n} \in (\mathrm{BP}_* \otimes \mathbb{Q})[[X]],$$

where $\ell_n \in \mathrm{BP}_{2(2^n-1)} \otimes \mathbb{Q}$. The Hazewinkel generators are then defined recursively by the formula

$$v_n = 2\ell_n - \sum_{1\leqslant k\leqslant n-1} \ell_k v_{n-k}^{2^k} \quad \text{for } n \geqslant 1. \tag{4.2}$$

We also have the following relation for the 2-series $[2]_{\mathrm{BP}}(X)$:

$$[2]_{\mathrm{BP}}(X) \equiv \sum_{n\geqslant 1}^{\mathrm{BP}} v_n X^{2^n} \pmod 2, \tag{4.3}$$

where as usual the symbol $\sum^{\mathrm{BP}}$ indicates formal group summation. We will need the following well known facts about the formal derivative of the logarithm. The next result may be known to others but we know of no reference.

Lemma (4.4). *We have the following congruence in* $\mathrm{BP}_*[[X]]$*:*

$$\log^{\mathrm{BP}\,\prime}(X) \equiv \sum_{n\geqslant 0} v_1^{2^n-1} X^{2^n-1} \pmod 2.$$

Proof. Let

$$\log^{\mathrm{BP}\,\prime}(X) = \sum_{n\geqslant 0} L_n X^{2^n-1}$$

where $L_n = 2^n\ell_n$ is an element of $\mathrm{BP}_{2^{n+1}-2}$. We will show by induction upon n that

$$L_n \equiv v_1^{2^n-1} \pmod 2.$$

For $n = 0$ and 1, the result is clear. Suppose that it holds for some n. Then we have

$$\begin{aligned} 2^n v_{n+1} &= L_{n+1} - \sum_{1\leqslant k\leqslant n} 2^{n-k} L_k v_{n+1-k}^{2^k} \\ &\equiv v_1^{2^{n+1}-1} \pmod 2 \quad \text{by the inductive hypothesis.} \end{aligned}$$

Hence, the desired result holds for $n+1$ and the Lemma follows. □

Our next result allows us to estimate the series $[-1]_{\mathrm{BP}}(X)$ modulo 2. Again it is quite possibly known to others.

Lemma (4.5). *The following congruences hold in* $\mathrm{BP}_*[[X]]$*:*

$$[-1]_{\mathrm{BP}}(X) \equiv \sum_{n\geqslant 0}^{\mathrm{BP}} [2^n]_{\mathrm{BP}}(X) \pmod 2$$

$$\equiv X +_{\mathrm{BP}} \sum_{\substack{n\geqslant 1\\ r_1,\dots,r_n\geqslant 1}}^{\mathrm{BP}} v_{r_1} v_{r_2}^{2^{r_1}} \cdots v_{r_2}^{2^{r_1+\cdots+r_{n-1}}} X^{2^{r_1+\cdots+r_n}} \pmod 2$$

where in each case the right hand side is an X*-adically convergent series.*

Proof. The sequence $[2^n-1]_{\mathrm{BP}}(X)$ is Cauchy with respect to the X-adic topology on $\mathrm{BP}_*/(2)[[X]]$ since the formal group differences of successive terms have the form $[2^n]_{\mathrm{BP}}(X)$ and by (4.3) the leading term of this is that in X^{2^n}. Hence, we see that the limit of this sequence is

$$\sum_{n\geqslant 0}^{\mathrm{BP}} [2^n]_{\mathrm{BP}}(X),$$

giving the desired result. □

We are now in a position to calculate $\wp^{\mathrm{MSp}}(((\lambda_2 \otimes_{\mathbb{C}} \eta) \otimes_{\mathbb{C}} \mathbb{H}))$ as an element of $(\mathrm{BP} \wedge \mathrm{MSp})^*(\mathbb{RP}^2 \times \mathbb{HP}^\infty)$. Recall that we have from §3,

$$\wp^{\mathrm{MSp}}(((\lambda_2 \otimes_{\mathbb{C}} \eta) \otimes_{\mathbb{C}} \mathbb{H})) = \wp^{\mathrm{MSp}} + \frac{w_2}{N^{\mathrm{BP}\,\prime}(\wp^{\mathrm{MSp}})}\left[\frac{u}{\log^{\mathrm{BP}\,\prime}(\overline{u})} + \frac{\overline{u}}{\log^{\mathrm{BP}\,\prime}(X)}\right].$$

The ring of power series in u and $\overline{u}$ which is symmetric with respect to the automorphism interchanging these two elements is easily seen to be the power series ring on $-u\overline{u}$, hence we can express this last quantity as an element of $(\mathrm{BP} \wedge \mathrm{MSp})^*(\mathbb{RP}^2)[[\wp^{\mathrm{BP}}]]$ by making use of the fact that $\wp^{\mathrm{BP}} = -u\overline{u}$. Of course, this will be hard to do explicitly in general, but we can extract sufficient information from our formula to enable us to make some interesting deductions. Amongst these we have the following, which we leave the reader to verify. Recall that in BP_*, the ideal $I_r = (v_0, v_1, \dots, v_r)$ is invariant and prime (here as usual we set $v_0 = 2$).

Theorem (4.6). *For* $r \geqslant 1$ *we have in* $(\mathrm{BP} \wedge \mathrm{MSp})^*(\mathbb{RP}^2 \times \mathbb{HP}^\infty)$ *the congruence:*

$$\wp^{\mathrm{MSp}}(((\lambda_2 \otimes_{\mathbb{C}} \eta) \otimes_{\mathbb{C}} \mathbb{H})) \equiv \wp^{\mathrm{MSp}} + \frac{w_2}{N^{\mathrm{BP}\,\prime}(\wp^{\mathrm{MSp}})}\left[v_r (\wp^{\mathrm{MSp}})^{2^{r-1}} + v_{r+1}(\wp^{\mathrm{MSp}})^{2^r}\right] \pmod{I_r + (\wp^{\mathrm{MSp}})^{2^r+1}}.$$

This allows us to deduce that

$$\begin{aligned}\widetilde{\varphi}_1 &= v_2 + v_1 N_1\\ \widetilde{\varphi}_2 &= v_3 + v_2 N_2 \mod I_2 + \text{decomposables}\end{aligned}$$

and in general

$$\widetilde{\varphi}_{2^r+1} = v_{r+1} + v_r N_{2^r-1} \mod I_{r-1} + \text{decomposables}$$

where the term "decomposables" refers to decomposables in the BP_* algebra $\mathrm{BP}_*[N_k : k \geqslant 1]$.

§5 Detecting Ray's elements φ_n in mod 2 homology.

In this section we will give a formula for the element in the group

$$\mathrm{Ext}^{1\,4m-2}_{\mathcal{A}_*}(\mathbb{F}_2, \mathrm{H}\,\mathbb{F}_{2*}(\mathrm{MSp}))$$

(a part of the E_2 term of the classical Adams Spectral Sequence for $\pi_*(\mathrm{MSp})$) which detects N. Ray's element $\theta_m \in \mathrm{MSp}_{4m-3}$; here, $\mathcal{A}_*$ denotes the dual Steenrod algebra at the prime 2, and $\mathrm{H}\,\mathbb{F}_2$ is mod 2 ordinary homology. For the case of $m = 2n$, we have $\varphi_n = \theta_{2n}$. Our approach is to use the formulæ we obtained in §4 to determine the image of $\widetilde{\theta}_m$ under the composition

$$\mathrm{Ext}^{0\,*}_{\mathrm{BP}_*(\mathrm{BP})}(\mathrm{BP}_*, \mathrm{BP}_*(\mathrm{MSp})/(2)) \xrightarrow{\delta} \mathrm{Ext}^{1\,*}_{\mathrm{BP}_*(\mathrm{BP})}(\mathrm{BP}_*, \mathrm{BP}_*(\mathrm{MSp})) \xrightarrow{\rho} \mathrm{Ext}^{1\,*}_{\mathcal{A}_*}(\mathbb{F}_2, \mathrm{H}\,\mathbb{F}_{2*}(\mathrm{MSp})),$$

in which δ denotes the coboundary (of bidegree $(1,0)$) induced from the exact sequence of comodules

$$0 \longrightarrow \mathrm{BP}_*(\mathrm{MSp}) \xrightarrow{\times 2} \mathrm{BP}_*(\mathrm{MSp}) \to \mathrm{BP}_*(\mathrm{MSp})/(2) \longrightarrow 0,$$

and ρ denotes the reduction map which has bidgree $(0,0)$ and is induced by the morphism of comodules

$$(\mathrm{BP}_*(\mathrm{BP}), \mathrm{BP}_*(\mathrm{MSp})) \longrightarrow (\mathrm{H}\,\mathbb{F}_{2*}(\mathrm{H}\,\mathbb{F}_2) = \mathcal{A}_*, \mathrm{H}\,\mathbb{F}_{2*}(\mathrm{MSp})).$$

From §4, we have as the generating function of the $\widetilde{\theta}_m$ the series

$$\begin{aligned}\widetilde{\theta}(\wp^{\mathrm{MSp}}) &= \sum_{m\geqslant 1} \widetilde{\theta}_m (\wp^{\mathrm{MSp}})^m\\ &= \frac{1}{N^{\mathrm{BP}\,\prime}(\wp^{\mathrm{MSp}})}\left[\frac{u}{\log^{\mathrm{BP}\,\prime}(\overline{u})} + \frac{\overline{u}}{\log^{\mathrm{BP}\,\prime}(u)}\right]\end{aligned}$$

with the notational conventions of our earlier sections. The coefficients of this series in $\wp^{\mathrm{MSp}}$ lie in $\mathrm{Ext}^{0\,*}_{\mathrm{BP}_*(\mathrm{BP})}(\mathrm{BP}_*, \mathrm{BP}_*(\mathrm{MSp})/(2))$, and our first task is to calculate the series whose coefficients are the elements

$$\psi(\widetilde{\theta}_m) \in \mathrm{BP}_*(\mathrm{BP}) \otimes_{\mathrm{BP}_*} \mathrm{BP}_*(\mathrm{MSp})/(2) \pmod{I^2},$$

where $\psi\colon \mathrm{BP}_*(\mathrm{MSp}) \longrightarrow \mathrm{BP}_*(\mathrm{BP}) \otimes_{\mathrm{BP}_*} \mathrm{BP}_*(\mathrm{MSp})$ is the (left) coaction. To do this, we view the above series as lying in the ring $(\mathrm{BP} \wedge \mathrm{MSp})^*(\mathbb{HP}^\infty)/(2)$ and the coaction as being a ring homomorphism

$$\begin{aligned}(\mathrm{BP} \wedge \mathrm{MSp})^*(\mathbb{HP}^\infty) \xrightarrow{\psi} &(\mathrm{BP} \wedge \mathrm{BP} \wedge \mathrm{MSp})^*(\mathbb{HP}^\infty) \\ &\cong \mathrm{BP}_*(\mathrm{BP}) \otimes_{\mathrm{BP}_*} (\mathrm{BP} \wedge \mathrm{MSp})^*(\mathbb{HP}^\infty).\end{aligned}$$

We view the two factors of BP in the latter object as left (L) and right (R) indexed by these letters. Then we obtain

$$\begin{aligned}\psi\widetilde{\theta}(\wp^{\mathrm{MSp}}) &= \sum_{m\geqslant 1} \widetilde{\theta}_m(\wp^{\mathrm{MSp}})^m \\ &= \frac{1}{N^{\mathrm{BP}_L\,\prime}(\wp^{\mathrm{MSp}})} \left[\frac{u^L}{\log^{\mathrm{BP}_L\,\prime}(\overline{u}^L)} + \frac{\overline{u^L}}{\log^{\mathrm{BP}_L\,\prime}(u^L)}\right].\end{aligned}$$

But now in $(\mathrm{BP} \wedge \mathrm{BP})^*(\mathbb{CP}^\infty)$, the left x^{BP_L} and right orientations x^{BP_R} are related by the following formula to be found in [5]:

$$\begin{aligned} x^{\mathrm{BP}_L} &= \sideset{}{^{\mathrm{BP}_L}}\sum_{k\geqslant 0} t_k (x^{\mathrm{BP}_R})^{2^k} \\ &\equiv \sum_{k\geqslant 0} t_k (x^{\mathrm{BP}_R})^{2^k} \pmod{I}. \end{aligned} \tag{5.1}$$

Hence we have

$$u^L = \sideset{}{^{\mathrm{BP}_L}}\sum_{k\geqslant 0} t_k (u^R)^{2^k}.$$

Now we also need to estimate $\log^{\mathrm{BP}\,\prime}(X)$ modulo I^2, where

$$I = (v_k : k \geqslant 0) \lhd \mathrm{BP}_*$$

is the ideal generated by all the v_k .

Proposition (5.2). *In the ring* $\mathrm{BP}_*[[X]]$ *we have the congruence*

$$\log^{\mathrm{BP}\,\prime}(X) \equiv 1 + v_1 X \pmod{I^2}.$$

Proof. We proceed as in the proof of Lemma (4.4) and use the same notation. We must show that for $n \geqslant 2$, we have

$$L_n \equiv 0 \pmod{I^2}.$$

For $n = 2$, this is an easy consequence of the formula

$$2v_2 = L_2 - v_1^3$$

defining the Hazewinkel generator v_2. An induction on n now gives general result. □

At this stage we have established that

$$(5.3)\quad \psi\widetilde{\theta}(\wp^{\mathrm{MSp}}) \equiv \frac{1}{N^{\mathrm{BP}_L\,\prime}(\wp^{\mathrm{MSp}})}\left[\frac{u^L}{(1+v_1\overline{u^L})} + \frac{\overline{u^L}}{(1+v_1 u^L)}\right] \equiv$$
$$\frac{1}{N^{\mathrm{BP}_L\,\prime}(\wp^{\mathrm{MSp}})}\left[\frac{u^L}{(1+v_1\sum_{k\geqslant 0} t_k(\overline{u^R})^{2^k})} + \frac{\overline{u^L}}{(1+v_1\sum_{k\geqslant 0} t_k(u^R)^{2^k})}\right] \pmod{I^2},$$

where the second congruence is a consequence of (5.1). Notice that we have

$$(5.4)\quad \psi\widetilde{\theta}(\wp^{\mathrm{MSp}}) \equiv$$
$$\frac{1}{N^{\mathrm{BP}_L\,\prime}(\wp^{\mathrm{MSp}})}\left[\frac{\sum_{k\geqslant 0} t_k(u^R)^{2^k}}{(1+v_1\sum_{k\geqslant 0} t_k(\overline{u^R})^{2^k})} + \frac{\sum_{k\geqslant 0} t_k(\overline{u^R})^{2^k}}{(1+v_1\sum_{k\geqslant 0} t_k(u^R)^{2^k})}\right] \pmod{I},$$

since

$$\overline{X} \equiv X + \sum_{k\geqslant 0} v_k X^{2^k} \pmod{I^2}$$

by (4.5).

We still need to deal with the term $N^{\mathrm{BP}_L\,\prime}(\wp^{\mathrm{MSp}})$, which has to be expressed in terms of $N^{\mathrm{BP}_R\,\prime}(\wp^{\mathrm{MSp}})$. By the formula of (1.6), we have

$$\begin{aligned} Q^{\mathrm{BP}}(T^2) &\equiv Q^{\mathrm{BP}}(-T\overline{T}) \\ &\equiv B^{\mathrm{BP}}(T)^2 \pmod{I} \\ &\equiv \sum_{k\geqslant 0} B_k^{\mathrm{BP}\,2} T^{2k+2} \pmod{I} \end{aligned}$$

in the ring $\mathrm{BP}_*(\mathrm{MU})[[T]]$. The coaction on this series is given by

$$\psi(Q^{\mathrm{BP}}(T^2)) = \sum_{k\geqslant 0} B^{\mathrm{BP},\mathrm{BP}}(T)^{2k+2} \otimes_{\mathrm{BP}_*} B_k^{\mathrm{BP}\,2}$$

where the series $B^{\mathrm{BP},\mathrm{BP}}(T) \in \mathrm{BP}_*(\mathrm{BP})[[T]]$ is composition inverse of

$$\sum_{k\geqslant 0}^{\mathrm{BP}} t_k T^{2^k} \equiv \sum_{k\geqslant 0} t_k T^{2^k} \pmod{I}.$$

As $N^{\mathrm{BP}}(Z)$ is the composition inverse of $Q^{\mathrm{BP}}(Z)$, we have

$$\psi(N^{\mathrm{BP}}(Z)) = \sum_{k\geqslant 0} t_k^2 \otimes (N^{\mathrm{BP}}(Z))^{2^k} \pmod{I},$$

and hence

$$\psi(N^{\mathrm{BP}\,\prime}(Z)) = N^{\mathrm{BP}\,\prime}(Z) \pmod{I}.$$

Using this result together with (5.4), we have

$$\psi\widetilde{\theta}(\wp^{\mathrm{MSp}}) \equiv \frac{1}{N^{\mathrm{BP}_R\,\prime}(\wp^{\mathrm{MSp}})}\left[\frac{u^L}{(1+v_1\sum_{k\geqslant 0} t_k(u^R)^{2^k})} + \frac{\overline{u^L}}{(1+v_1\sum_{k\geqslant 0} t_k(u^R)^{2^k})}\right] \pmod{I^2}.$$

Expanding carefully the term inside the square brackets yields

$$u^L + \overline{u^L} \equiv \sum_{k\geqslant 0} v_k(u^R)^{2^k} \pmod{I^2}$$

and therefore

$$\psi\widetilde{\theta}(\wp^{\mathrm{MSp}}) \equiv \frac{\sum_{k\geqslant 0} v_k(N^{\mathrm{BP}}(u^R))^{2^k}}{N^{\mathrm{BP}_R\,\prime}(\wp^{\mathrm{MSp}})} \pmod{I^2}$$

since $N^{\mathrm{BP}_R}(\wp^{\mathrm{MSp}}) = \wp^{\mathrm{BP}_R} \equiv (u^R)^2 \mod I^2$.

Having calculated this, we are at last in a position to determine

$$\delta(\widetilde{\theta}_m) = \left[\frac{1}{2}(\psi - 1\otimes \mathrm{Id})(\widetilde{\theta}_m)\right] \pmod{I}.$$

We need one further fact, namely that the right unit on the v_n has the form

$$\eta_R(v_n) \equiv \sum_{0\leqslant k\leqslant n} v_k t_{n-k}^{p^k} \pmod{I^2}.$$

We therefore obtain the following generating function:

$$\sum_{m\geqslant 1} \delta(\widetilde{\theta}_m)(\wp^{\mathrm{MSp}})^m = \frac{\sum_{k\geqslant 0} t_k(u^R)^{2^k}}{1\otimes N^{\mathrm{BP}\,\prime}(\wp^{\mathrm{MSp}})} \pmod{I}$$

which when pushed into $\mathcal{A}_* \otimes \mathrm{H}\,\mathbb{F}_{2*}(\mathrm{MSp})$ yields

$$\frac{\sum_{k\geqslant 0} \zeta_k^2 \otimes N^{\mathrm{H}\,\mathbb{F}_2}(\wp^{\mathrm{MSp}})^{2^k}}{1\otimes N^{\mathrm{H}\,\mathbb{F}_2\,\prime}(\wp^{\mathrm{MSp}})},$$

where $\mathcal{A}_* = \mathbb{F}_2[\zeta_k : k \geqslant 1]$ with ζ_k being the conjugate of Milnor's generator $\xi_k \in \mathcal{A}_{2^k-1}$.

The element θ_m is detected in the classical mod 2 Adams spectral sequence for $\pi_*(\mathrm{MSp})$ by an element which originates in the E_2 term in the group $\mathrm{Ext}_{\mathcal{A}_*}^{1\,4m-2}(\mathbb{F}_2, \mathrm{H}\,\mathbb{F}_{2*}(\mathrm{MSp}))$. For low even values of m, these detecting elements are given in the following **Table**, obtained using the symbolic algebra package MAPLE. Note that $\theta_1 = [\zeta_1^2 \otimes 1]$ and $\theta_{2n} = \varphi_n$. Of course, for $n \geqslant 1$, θ_{2n+1} is zero and hence detected by 0! The notation is that implied by the cobar construction.

TABLE

$\varphi_1 : [\zeta_1^2 \otimes N_1 + \zeta_2^2 \otimes 1]$

$\varphi_2 : [\zeta_1^2 \otimes N_1N_2 + \zeta_1^2 \otimes N_3 + \zeta_2^2 \otimes N_2 + \zeta_2^2 \otimes N_1^2 + \zeta_3^2 \otimes 1]$

$\varphi_3 : [\zeta_1^2 \otimes N_1N_4 + \zeta_1^2 \otimes N_1N_2^2 + \zeta_1^2 \otimes N_5 + \zeta_1^2 \otimes N_3N_2$
$\quad + \zeta_2^2 \otimes N_1^2N_2 + \zeta_2^2 \otimes N_4 + \zeta_3^2 \otimes N_2]$

$\varphi_4 : [\zeta_1^2 \otimes N_1N_6 + \zeta_1^2 \otimes N_1N_2^3 + \zeta_1^2 \otimes N_5N_2 + \zeta_1^2 \otimes N_3N_4 + \zeta_1^2 \otimes N_3N_2^2$
$\quad + \zeta_1^2 \otimes N_7$
$\quad + \zeta_2^2 \otimes N_6 + \zeta_2^2 \otimes N_1^2N_4 + \zeta_2^2 \otimes N_1^2N_2^2 + \zeta_2^2 \otimes N_3^2$
$\quad + \zeta_3^2 \otimes N_4 + \zeta_3^2 \otimes N_2^2 + \zeta_3^2 \otimes N_1^4 + \zeta_4^2 \otimes 1]$

TABLE *(continued)*

$$
\begin{aligned}
\varphi_5 : [&\zeta_1^2 \otimes N_5N_4 + \zeta_1^2 \otimes N_5N_2^2 + \zeta_1^2 \otimes N_9 + \zeta_1^2 \otimes N_1N_8 + \zeta_1^2 \otimes N_1N_4^2 \\
&+ \zeta_1^2 \otimes N_1N_2^2N_4 + \zeta_1^2 \otimes N_1N_2^4 + \zeta_1^2 \otimes N_3N_6 + \zeta_1^2 \otimes N_3N_2^3 \\
&+ \zeta_1^2 \otimes N_7N_2 \\
&+ \zeta_2^2 \otimes N_8 + \zeta_2^2 \otimes N_1^2N_6 + \zeta_2^2 \otimes N_1^2N_2^3 + \zeta_2^2 \otimes N_3^2N_2 \\
&+ \zeta_3^2 \otimes N_6 + \zeta_3^2 \otimes N_2^3 + \zeta_3^2 \otimes N_1^4N_2 + \zeta_4^2 \otimes N_2]
\end{aligned}
$$

$$
\begin{aligned}
\varphi_6 : [&\zeta_1^2 \otimes N_7N_2^2 + \zeta_1^2 \otimes N_1N_2^2N_6 + \zeta_1^2 \otimes N_1N_2N_4^2 + \zeta_1^2 \otimes N_3N_2^2N_4 \\
&+ \zeta_1^2 \otimes N_1N_{10} + \zeta_1^2 \otimes N_3N_8 + \zeta_1^2 \otimes N_{11} + \zeta_1^2 \otimes N_7N_4 + \zeta_1^2 \otimes N_9N_2 \\
&+ \zeta_1^2 \otimes N_5N_6 + \zeta_1^2 \otimes N_3N_2^4 + \zeta_1^2 \otimes N_3N_4^2 + \zeta_1^2 \otimes N_5N_2^3 + \zeta_1^2 \otimes N_1N_2^5 \\
&+ \zeta_2^2 \otimes N_{10} + \zeta_2^2 \otimes N_1^2N_2^2N_4 + \zeta_2^2 \otimes N_5^2 + \zeta_2^2 \otimes N_1^2N_2^4 \\
&+ \zeta_2^2 \otimes N_3^2N_4 + \zeta_2^2 \otimes N_3^2N_2^2 + \zeta_2^2 \otimes N_1^2N_4^2 + \zeta_2^2 \otimes N_1^2N_8 \\
&+ \zeta_3^2 \otimes N_2^2N_4 + \zeta_3^2 \otimes N_4^2 + \zeta_3^2 \otimes N_1^4N_4 + \zeta_3^2 \otimes N_1^4N_2^2 + \zeta_3^2 \otimes N_8 \\
&+ \zeta_4^2 \otimes N_2^2 + \zeta_4^2 \otimes N_4]
\end{aligned}
$$

$$
\begin{aligned}
\varphi_7 : [&\zeta_1^2 \otimes N_{11}N_2 + \zeta_1^2 \otimes N_9N_4 + \zeta_1^2 \otimes N_9N_2^2 + \zeta_1^2 \otimes N_1N_2^6 \\
&+ \zeta_1^2 \otimes N_1N_2^4N_4 + \zeta_1^2 \otimes N_1N_6^2 + \zeta_1^2 \otimes N_3N_2^5 + \zeta_1^2 \otimes N_3N_{10} \\
&+ \zeta_1^2 \otimes N_7N_6 + \zeta_1^2 \otimes N_5N_8 + \zeta_1^2 \otimes N_1N_{12} + \zeta_1^2 \otimes N_1N_2^2N_8 \\
&+ \zeta_1^2 \otimes N_3N_2^2N_6 + \zeta_1^2 \otimes N_3N_2N_4^2 + \zeta_1^2 \otimes N_5N_2^2N_4 + \zeta_1^2 \otimes N_5N_2^4 \\
&+ \zeta_1^2 \otimes N_5N_4^2 + \zeta_1^2 \otimes N_{13} + \zeta_1^2 \otimes N_1N_4^3 + \zeta_1^2 \otimes N_7N_2^3 \\
&+ \zeta_2^2 \otimes N_5^2N_2 + \zeta_2^2 \otimes N_1^2N_2N_4^2 + \zeta_2^2 \otimes N_1^2N_2^5 + \zeta_2^2 \otimes N_1^2N_{10} \\
&+ \zeta_2^2 \otimes N_3^2N_6 + \zeta_2^2 \otimes N_{12} + \zeta_2^2 \otimes N_3^2N_2^3 + \zeta_2^2 \otimes N_1^2N_2^2N_6 \\
&+ \zeta_3^2 \otimes N_2N_4^2 + \zeta_3^2 \otimes N_{10} + \zeta_3^2 \otimes N_2^2N_6 + \zeta_3^2 \otimes N_1^4N_6 \\
&+ \zeta_3^2 \otimes N_1^4N_2^3 + \zeta_4^2 \otimes N_2^3 + \zeta_4^2 \otimes N_6]
\end{aligned}
$$

TABLE *(continued)*

$$\begin{aligned}
\varphi_8 : [&\zeta_1^2 \otimes N_3N_2^6 + \zeta_1^2 \otimes N_{13}N_2 + \zeta_1^2 \otimes N_9N_2^3 + \zeta_1^2 \otimes N_9N_6 \\
&+ \zeta_1^2 \otimes N_7N_2^2N_4 + \zeta_1^2 \otimes N_7N_8 + \zeta_1^2 \otimes N_5N_{10} + \zeta_1^2 \otimes N_{15} \\
&+ \zeta_1^2 \otimes N_7N_4^2 + \zeta_1^2 \otimes N_7N_2^4 + \zeta_1^2 \otimes N_5N_2N_4^2 + \zeta_1^2 \otimes N_5N_2^5 \\
&+ \zeta_1^2 \otimes N_{11}N_2^2 + \zeta_1^2 \otimes N_{11}N_4 + \zeta_1^2 \otimes N_3N_2^2N_8 + \zeta_1^2 \otimes N_5N_2^2N_6 \\
&+ \zeta_1^2 \otimes N_3N_{12} + \zeta_1^2 \otimes N_1N_2^4N_6 + \zeta_1^2 \otimes N_1N_4^2N_6 + \zeta_1^2 \otimes N_1N_2^2N_{10} \\
&+ \zeta_1^2 \otimes N_1N_2N_6^2 + \zeta_1^2 \otimes N_3N_4^3 + \zeta_1^2 \otimes N_3N_6^2 + \zeta_1^2 \otimes N_1N_2^7 \\
&+ \zeta_1^2 \otimes N_1N_{14} + \zeta_1^2 \otimes N_3N_2^4N_4 \\
&+ \zeta_2^2 \otimes N_1^2N_{12} + \zeta_2^2 \otimes N_3^2N_4^2 + \zeta_2^2 \otimes N_3^2N_2^2N_4 + \zeta_2^2 \otimes N_3^2N_8 \\
&+ \zeta_2^2 \otimes N_5^2N_4 + \zeta_2^2 \otimes N_3^2N_2^4 + \zeta_2^2 \otimes N_5^2N_2^2 + \zeta_2^2 \otimes N_1^2N_2^4N_4 \\
&+ \zeta_2^2 \otimes N_1^2N_2^2N_8 + \zeta_2^2 \otimes N_1^2N_4^3 + \zeta_2^2 \otimes N_1^2N_6^2 + \zeta_2^2 \otimes N_1^2N_2^6 \\
&+ \zeta_2^2 \otimes N_{14} + \zeta_2^2 \otimes N_7^2 \\
&+ \zeta_3^2 \otimes N_3^4 + \zeta_3^2 \otimes N_1^4N_4^2 + \zeta_3^2 \otimes N_1^4N_8 + \zeta_3^2 \otimes N_1^4N_2^2N_4 \\
&+ \zeta_3^2 \otimes N_1^4N_2^4 + \zeta_3^2 \otimes N_4^3 + \zeta_3^2 \otimes N_2^2N_8 + \zeta_3^2 \otimes N_6^2 + \zeta_3^2 \otimes N_{12} \\
&+ \zeta_4^2 \otimes N_1^8 + \zeta_4^2 \otimes N_4^2 + \zeta_4^2 \otimes N_2^4 + \zeta_4^2 \otimes N_2^2N_4 + \zeta_4^2 \otimes N_8 \\
&+ \zeta_5^2 \otimes 1]
\end{aligned}$$

References

[1] J. F. Adams, *Stable Homotopy and Generalised Homology*, University of Chicago Press, 1974.

[2] V. M. Buhštaber, *Topological applications of the theory of two-valued formal groups*, Math. USSR Izvestija **12** (1978), 177.

[3] V. Gorbunov & N. Ray, *Orientations of Spin bundles and symplectic cobordism*, to appear in Publ. RIMS Kyoto University.

[4] H. R. Miller, D. C. Ravenel & W. S. Wilson, *Periodic phenomena in the Adams-Novikov spectral sequence*, Annals of Math. **106** (1977), 469–516.

[5] D. C. Ravenel, *Complex Cobordism and the Stable Homotopy Groups of spheres*, Academic Press, 1986.

[6] N. Ray, *Indecomposables in* Tors MSp$_*$, Topology **10** (1971), 261–70.

Manchester University, Manchester, M13 9PL, England.
Current address:
Glasgow University, Glasgow G12 8QW, Scotland.

E-mail: gama43@uk.ac.glasgow.cms

On a conjecture of Mahowald concerning bordism with singularities

V. V. Vershinin
USSR 630090 Novosibirsk
Institute of Mathematics
Universitetskii pr. 4

V. G. Gorbunov
Department of Mathematics
The University
Manchester M13 9PL
England

1 Introduction

Theories of symplectic cobordism with singularities were introduced in [V] and studied in the subsequent paper [V,G]. Consider the following subsets of the symplectic cobordism ring: $\Sigma_n = (\phi_0, \ldots, \phi_{2^n-2})$, where ϕ_0 represents a bordism class of a non-trivially framed circle and the ϕ_{2^n-2} are the bordism classes introduced in [R]. Recall that $2\phi_{2^i-2} = 0$. We consider also the infinite set $\Sigma = (\phi_0, \ldots, \phi_{2^n-2}, \ldots)$ and the empty set, which we denote by Σ_0. By MSp^{Σ_n} and MSp^{Σ} we denote the generalized bordism theory based on the set of weakly almost symplectic manifolds with singularities from Σ_n and Σ respectivly. Note that there are canonical maps $\gamma_n \colon MSp^{\Sigma_n} \longrightarrow MSp^{\Sigma_{n+1}}$, for $n \geqslant 0$. It was proved in [V] that there is an equivalence in the 2-local stable category $MSp^{\Sigma}_{(2)} \cong \bigvee_i S^i BP$. So the spectra MSp^{Σ_n} are situated "between" the MSp and BP spectra.

Doug Ravenel introduced in [Rav] the spectra $X(n)$ using the ideas of M. Barratt and M. Mahowold [D,H,S]. The spectrum $X(n)$ is a Thom spectrum induced by the composition

$$\Omega SU(n) \longrightarrow \Omega SU \longrightarrow BU,$$

where the first map is associated with the inclusion $SU(n) \longrightarrow SU$ and the second is a homotopy inverse to the Bott map (e.g see [Sw]). The spectrum $X(n)$ is a commutative and associative ring spectrum. The inclusion $SU(n) \longrightarrow SU(n+1)$ induces a map of ring spectra $X(n) \longrightarrow X(n+1)$. By construction we have $MU = \lim X(n)$. Note also that $X(1) = S^0$. So we have another sequence of spectra situated "between" the sphere spectrum and MU. Mark Mahowald cojectured that there must be a connection between the spectra MSp^{Σ_k} and $X(n)$. Namely, he conjectured that the spectrum $MSp \wedge X(2^n)$ splits into an infinite number of copies of suspensions of spectra MSp^{Σ_k} in the 2-local stable category. We are going to prove that conjecture in this paper. We are grateful to Mark Mahowald, Doug Ravenel and Mike Hopkins for discussions on the subject of the paper.

2 Preliminary information

First we recall some known facts about the spectra $X(n)$. Consider the map $\beta_{m-1}: CP(m-1) \longrightarrow \Omega SU(m)$ which is a restriction of the Bott map. The Thom spectrum induced by this map is equivalent to $S^{-2}CP(m)$. So we get the following commutative diagram for any $m \geqslant 1$:

$$\begin{array}{ccc} S^{-2}CP(m) & \xrightarrow{\omega_m} & X(m) \\ \downarrow & & \downarrow \\ S^{-2}CP(m+1) & \xrightarrow{\omega_{m+1}} & X(m+1). \end{array} \tag{2.1}$$

Each spectrum $X(m)$ itself has an interesting filtration by subspectra $F_k(m)$, introduced in [D,H,S]. Let us consider the following fibration

$$SU(m-1) \longrightarrow SU(m) \xrightarrow{p} S^{2m-1}.$$

Applying the Ω–functor we obtain another fibration

$$\Omega SU(m-1) \longrightarrow \Omega SU(m) \xrightarrow{\Omega p} \Omega S^{2m-1}.$$

By $J_k S^{2m}$ we denote the k–stage of the James construction for ΩS^{2m-1}. The canonical inclusion $j_k: J_k S^{2m} \longrightarrow \Omega S^{2m-1}$ gives the following induced fibration

$$\Omega SU(m-1) \longrightarrow E_k(m) \longrightarrow J_k S^{2m-2},$$

which can be included into the diagram

$$\begin{array}{ccccc} \Omega SU(m-1) & \longrightarrow & E_k(m) & \xrightarrow{\widetilde{\Omega p}} & J_k(S^{2m-1}) \\ \| & & \downarrow \overline{j_k} & & \downarrow j_k \\ \Omega SU(m-1) & \longrightarrow & \Omega SU(m) & \xrightarrow{\Omega p} & \Omega S^{2m-1} \end{array} \tag{2.2}$$

The Thom spectrum over $E_k(m)$ induced by the map $\overline{j_k}$ we denote by $F_k(m)$. These spectra give the filtration of $X(m)$ we mentioned above.

Since the manifold $CP(m-1)$ has dimention $2m-2$, the composition $\Omega p \circ \beta_{m-1}$ can be factored through a $(2m-2)$–dimentional skeleton of ΩS^{2m-1} or in other words through $J_1 S^{2m-2}$. This means that there exists a map $v_{m-1}: CP(m-1) \longrightarrow J_1 S^{2m-2}$ such that $\Omega p \circ \beta_{m-1} = j_1 \circ v_{m-1}$. The usual properties of fibrations imply that there is a map $f_{m-1}: CP(m-1) \longrightarrow E_1(m)$ for $m \geqslant 2$ which give rise to commutative triangles $\overline{j_k} \circ f_{m-1} = \beta_{m-1}$ and $\widetilde{\Omega p} \circ f_{m-1} = v_{m-1}$. Passing to Thom spectra we get a map $f_{m,1}: S^{-2}CP(m) \longrightarrow F_1(m)$. Since there are natural maps $F_k(m) \longrightarrow F_{k+1}(m)$ we can form their compositions with $f_{m,1}$, which we denote by $f_{m,k}$. By construction these maps are compatible and give in the limit a map $\omega_m: S^{-2}CP(m) \longrightarrow X(m)$.

Recall now the structure of the homology modules of the spectra involved:

$$H_*(S^{-2}CP(m); \mathbf{Z}) \simeq \mathbf{Z}\{\beta_0, \ldots, \beta_{m-1}\},\ \dim\beta_i = 2i;$$
$$H_*(X(m); \mathbf{Z}) \simeq \mathbf{Z}[b_0, \ldots, b_{m-1}]/(b_0 - 1).$$

Note that each $F_k(m)$ is a module spectrum over $X(m)$ and that $H_*(F_k(m); \mathbf{Z})$ is a free module over $H_*(X(m); \mathbf{Z})$ with basis $\{1, b_m, \ldots, b_m^k\}$. By construction the maps ω_m and $f_{m,k}$ induce the following homomorphisms in homology:

$$(\omega_m)_*(\beta_i) = b_i,\ (f_{m,k})_*(\beta_i) = b_i \cdot 1.$$

3 Proof of the main theorem

We proceed to a proof of Mahowald's conjecture.

Proposition 3.1 *The Atiyah-Hirzebruch spectral sequences for $MSp_*^{\Sigma_q}(F_k(m))$ and for $MSp_*^{\Sigma_q}(X(m))$ collapse, if $k \geqslant 1$ and $2^{q+1} > m$. In the Atiyah-Hirzebruch spectral sequence for $MSp_*^{\Sigma_q}(F_k(2^{q+1}))$, $k \geqslant 1$ the formula for the first nontrivial differential is $d_{2^{q+2}-2}(b_{2^{q+1}-1}) = \phi_{2^{q-1}}$.*

Proof The map $\omega_m: CP(m) \longrightarrow X(m)$ induces a map of the Atiyah-Hirzebruch spectral sequences

$$\begin{array}{ccc} H_*(CP(m), MSp_*^{\Sigma_q}) & \Longrightarrow & MSp_*^{\Sigma_q}(CP(m)) \\ \big\downarrow & & \big\downarrow \\ H_*(X(m), MSp_*^{\Sigma_q}) & \Longrightarrow & MSp_*^{\Sigma_q}(X(m)). \end{array}$$

The first spectral sequence is trivial if $m < 2^{q+1}$ because the ring $MSp_*^{\Sigma_q}$ has no torsion in dimensions less than $2^{q+3} - 3$ [V]. Using the multiplicative structure of the spectrum $X(m)$ and the multiplicative properties of the Atiyah-Hirzebruch spectral sequence we obtain that the second spectral sequence is trivial for $m < 2^{q+1}$. The first nontrivial differential in the

Atiyah-Hirzebruch spectral sequence for $MSp_*^{\Sigma_q}(CP(m))$, $m = 2^{q+1}$ was computed in [VG], namely, $d_{2^{q+2}-2}(b_{2^{q+1}-1}) = \phi_{2^q-1}$. The above description of the map $(\omega_m)_*: H_*(CP(m)) \longrightarrow H_*(X(m))$ and the naturality of the Atiyah-Hirzebruch spectral sequence finish the proof. □

Now let us consider the diagram (2.1) for $k = 1$ and $m = 2^n$. Following the arguments from [D,H,S] we obtain that the inclusion of the fibre in the upper fibration generates the cofibre sequence:

$$\Omega SU(2^n-1) \longrightarrow E_1(2^n) \longrightarrow S^{2^{n+1}-2}\wedge(\Omega SU(2^n-1)_+) \xrightarrow{g} S\wedge(\Omega SU(2^n-1)_+),$$

where the map g is the composite:

$$S(S^{2^{n+1}-3}) \wedge (\Omega SU(2^n-1)_+) \xrightarrow{Sf\wedge 1}$$

$$S(\Omega SU(2^n-1)_+) \wedge (\Omega SU(2^n-1)_+) \xrightarrow{Sm} S(\Omega SU(2^n-1)).$$

Here f is the characteristic map of the top fibration in (2.2) and m is the multiplication map induced by loop sum. Passing now to Thom spectra we get the cofibre sequence:

$$X(2^n-1) \longrightarrow F_1(2^n) \longrightarrow S^{2^{n+1}-2}X(2^n-1) \xrightarrow{\psi_1} SX(2^n-1), \qquad (3.2)$$

where the map ψ_1 is the composition:

$$S(S^{2^{n+1}-3}) \wedge X(2^n-1) \xrightarrow{S\varphi\wedge 1} SX(2^n-1) \wedge X(2^n-1) \xrightarrow{S\mu} SX(2^n-1),$$

the map μ is a multiplication and $\varphi: S^{2^{n+1}-3} \longrightarrow X(2^n-1)$ is induced by f.

Proposition 3.3 *The map $\varphi: S^{2^{n+1}-3} \longrightarrow X(2^n-1)$ is not homotopic to zero, and the Hurewitz image of this map in $MSp_*(X(2^n-1))$ is not equal to zero.*
Proof (3.2) implies that $F_1(2^n)$ is obtained from $X(2^n-1)$ by attaching $X(2^n-1)$ with the help of ψ_1. If φ is trivial the same is true for ψ_1 by construction and it implies that $F_1(2^n)$ splits into two copies of $X(2^n-1)$; this means that Atiyah-Hirzebruch spectral sequence

$$H_*(F_1(2^n); MSp_*^{\Sigma_{n-1}}) \Longrightarrow MSp_*^{\Sigma_{n-1}}(F_1(2^n))$$

degenerates. This contradicts **Proposition 3.1** □

Note that by construction the map ψ_1 can be desuspended. Thus $\psi_1 = S\psi$ where ψ is the composition:

$$S^{2^{n+1}-3} \wedge X(2^n-1) \xrightarrow{\varphi\wedge 1} X(2^n-1) \wedge X(2^n-1) \xrightarrow{\mu} X(2^n-1).$$

Now let us state our main result

Theorem 3.4 *There exist four MSp−module maps, which are all homotopy equivalences, namely:*

$$X(m) \wedge MSp \simeq \bigvee_{\omega} S^{|\omega|} MSp^{\Sigma_n}, \tag{3.5}$$

where $2^n \leqslant m < 2^{n+1}$, $\omega = (i_1, \ldots, i_q)$, $m-1 \geqslant i_j \geqslant 1$, $|\omega| = \Sigma_{j=1}^{q} i_j$ *and the number of i such that* $i_j = 2^s - 1$ *for fixed s is even; and*

$$F_{2k-1}(m) \wedge MSp \simeq \bigvee_{\omega} S^{|\omega|} MSp^{\Sigma_n}, \tag{3.6}$$

where the numbers ω, n, m, i_j, j *are the same as in* (3.5) *with one more condition: the number of j such that* $i_j = m-1$ *is less or equal than* $2k-1$*; and*

$$F_{2k}(m) \wedge MSp \simeq \bigvee_{\omega} S^{|\omega|} MSp^{\Sigma_n} \bigvee (\bigvee_{\alpha} S^{|\alpha|} MSp^{\Sigma_{n-1}}), \tag{3.7}$$

where $2^n < m < 2^{n+1}$, *and the number of j such that* $i_j = m-1$ *is less or equal than* $2k$*; and*

$$F_{2k}(2^n) \wedge MSp \simeq \bigvee_{\omega} S^{|\omega|} MSp^{\Sigma_n} \bigvee (\bigvee_{\alpha} S^{|\alpha|} MSp^{\Sigma_{n-1}}), \tag{3.8}$$

where the ω *are the same as in* (3.6) *and*

$$\alpha = (\underbrace{m, \ldots, m}_{2k}, i_{2k+1}, \ldots, i_q),$$

where $i_j, j \geqslant 2k+1$ *are the same as in* (3.5) *for* $m-1$.

Proof We will prove the theorem by induction on m.

Let $m = 1$. In this case $X(1) = S^0$ and $F_1(2) = S^{-2}CP(2)$ so $S^{-2}CP(2) \wedge MSp = MSp^{\theta_1}$ as required.

Assume now that the statement of the theorem is proved for $X(2^n - 1)$. Smashing (3.2) with MSp we obtain the following cofibering

$$S^{2^{n+1}-3} \wedge X(2^n-1) \wedge MSp \longrightarrow X(2^n-1) \wedge MSp \longrightarrow$$

$$F_1 \wedge MSp \longrightarrow S^{2^{n+1}-2} \wedge X(2^n-1) \wedge MSp.$$

We claim that the map

$$S^{2^{n+1}-3} \wedge X(2^n-1) \wedge MSp \xrightarrow{\phi \wedge 1} X(2^n-1) \wedge MSp$$

is homotopic to the following composition:

$$S^{2^{n+1}-3} \wedge X(2^n-1) \wedge MSp \xrightarrow{\phi_{2^n-2}\wedge 1 \wedge 1} MSp \wedge X(2^n-1) \wedge MSp \xrightarrow{T\wedge 1}$$
$$X(2^n-1) \wedge MSp \wedge MSp \xrightarrow{1\wedge \mu} X(2^n-1) \wedge MSp.$$

This is because there exists only one nontrivial element in the group $\pi_{2^{n+1}-3}(X(2^n-1) \wedge MSp)$. It is the image of the nontrivial element φ in $\pi_{2^{n+1}-3}X(2^n-1)$ under the map

$$(1 \wedge e): X(2^n-1) \wedge S^0 \longrightarrow X(2^n-1) \wedge MSp,$$

where $e: S^0 \longrightarrow MSp$ is a unit, which is not equal to zero by **Proposition 3.2** and it is also the image of ϕ_{2^n-2} in $\pi_{2^{n+1}-3}(MSp)$ under the map

$$(e \wedge 1): S^0 \wedge MSp \longrightarrow X(2^n-1) \wedge MSp.$$

The first assertion follows from consideration of the Atiyah-Hirzebruch spectral sequence for $X(2^n-1)_* MSp$, and the second can be proved by comparing the Atiyah-Hirzebruch spectral sequences

$$\begin{array}{ccc} H_*(X(2^n-1), MSp_*) & \Longrightarrow & MSp_*(X(2^n-1)) \\ \downarrow & & \downarrow \\ H_*(X(2^n-1), MSp_*^{\Sigma_{n-1}}) & \Longrightarrow & MSp_*^{\Sigma_{n-1}}(X(2^n-1)). \end{array}$$

By the induction hypothesis we have an MSp-module map

$$X(2^n-1) \wedge MSp \longrightarrow \bigvee_\omega S^{|\omega|} MSp^{\Sigma_{n-1}},$$

which is a homotopy equivalence. This implies that we have a commutative diagram

$$\begin{array}{ccc} S^{2^{n+1}-3} \wedge X(2^n-1) \wedge MSp & \xrightarrow{\psi \wedge 1} & X(2^n-1) \wedge MSp \\ \downarrow & & \downarrow \\ S^{2^{n+1}-3} \wedge (\bigvee_\omega S^{|\omega|} MSp^{\Sigma_{n-1}}) & \longrightarrow & \bigvee_\omega S^{|\omega|} MSp^{\Sigma_{n-1}}, \end{array}$$

where the bottom map has the form $\bigvee_\omega f_\omega$ and each map

$$f_\omega: S^{2^{n+1}-3} \wedge S^{|\omega|} MSp^{\Sigma_{n-1}} \longrightarrow S^{|\omega|} MSp^{\Sigma_{n-1}}$$

is induced by multiplication by ϕ_{2^n-2}. So we obtain an MSp-module homotopy equivalence:

$$F_1(2^n) \wedge MSp \simeq \bigvee_\omega S^{|\omega|} MSp^{\Sigma_n}.$$

Consider now $F_2(2^n)$. We have the cofiber sequence

$$F_1(2^n) \longrightarrow F_2(2^n) \longrightarrow S^{4(2^n-1)} \wedge X(2^n-1) \xrightarrow{S\psi^0} SF_1, \tag{3.9}$$

in which the map $\psi^0\colon S^{4(2^n-1)-1} \wedge X(2^n-1) \longrightarrow F_1(2^n)$ is the following composition

$$S^{4(2^n-1)-1} \wedge X(2^n-1) \xrightarrow{\phi^0\wedge 1} F_1(2^n) \wedge X(2^n-1) \xrightarrow{\mu^0} F_1(2^n).$$

Here μ^0 defines the module structure of $F_1(2^n)$ over $X(2^n-1)$, and the attaching map is $\phi^0\colon S^{4(2^n-1)-1} \longrightarrow F_1(2^n)$. Now smashing (3.9) with MSp we get the cofibering

$$F_1(2^n) \wedge MSp \longrightarrow F_2(2^n) \wedge MSp \longrightarrow S^{4(2^n-1)} \wedge X(2^n-1) \wedge MSp$$

$$\xrightarrow{S\psi^0\wedge 1} SF_1(2^n) \wedge MSp. \tag{3.10}$$

Note that we can write the map $S\psi^0 \wedge 1$ in the form

$$S^{4(2^n-1)} \wedge X(2^n-1) \wedge MSp \xrightarrow{\chi\wedge 1\wedge 1} S(F_1(2^n) \wedge MSp) \wedge X(2^n-1) \wedge MSp \xrightarrow{1\wedge\tau\wedge 1}$$

$$S \wedge F_1(2^n) \wedge X(2^n-1) \wedge MSp \wedge MSp \xrightarrow{1\wedge\mu^0\wedge\mu} S \wedge F_1(2^n) \wedge MSp.$$

Here χ is the map $\phi^0 \wedge e$, where $e\colon S^0 \longrightarrow F_1(2)^n$ is the unit of $F_1(2^n)$. According to the induction hypothesis, $F_1(2^n) \wedge MSp$ is equivalent to a wedge of suspensions over MSp^{Σ_n}. As mentioned above the first nontrivial torsion element in $\pi_*(MSp^{\Sigma_n})$ lies in dimention $2^{n+2}-3$, so the same is true for $\pi_*(F_1(2^n) \wedge MSp)$, and hence $\pi_{4(2^n-1)}(F_1(2^n) \wedge MSp) = 0$. This means that the cofiber sequence (3.10) splits and therefore

$$F_2(2^n) \wedge MSp \simeq F_1(2^n) \wedge MSp \bigvee S^{4(2^n-1)} X(2^n-1) \wedge MSp \simeq$$

$$(\bigvee SMSp^{\Sigma_n}) \bigvee S^{4(2^n-1)}(\bigvee SMSp^{\Sigma_{n-1}}).$$

Consider now the case $X(2^n) \wedge MSp$. We have the canonical map $F_2(2^n) \wedge MSp \longrightarrow X(2^n) \wedge MSp$, which induce a map of Atiyah-Hirzebruch spectral sequences

$$\begin{array}{ccc} H_*(F_2(2^n); MSp_*) & \Longrightarrow & MSp_*(F_2(2^n)) \\ \downarrow & & \downarrow \\ H_*(X(2^n); MSp_*) & \Longrightarrow & MSp_*(X(2^n)). \end{array}$$

By the above remarks about the homology groups of the spectra involved, this map induces a monomorphism of the E_2–terms. Recall that $H_*(F_2(2^n); \mathbf{Z})$ is generated as an $H_*(X(2^n-1)); \mathbf{Z})$– module by the elements b_{2^n-1} and $b_{2^n-1}^2$. Since (3.10) splits the element $b_{2^n-1}^2$ must be a permanent cycle in the Atiyah-Hirzebruch spectral sequence for $MSp_*(F_2(2^n))$, and hence the image of this element is a permanent cycle in the Atiyah-Hirzebruch spectral sequence for $MSp_*(X(2^n))$. Since all spectra under consideration are multiplicative and the Atiyah-Hirzebruch spectral sequence has multiplicative properties we conclude that all the elements $b_{2^n-1}^{2l}$ are permanent cycles. These elements determine the elements σ_l in $\pi_{4l(2^n-1)}(X(2^n) \wedge MSp)$. Let us now consider the following composition:

$$\bigvee_{l=0}^{\infty} S^{4l(2^n-1)} \wedge F_1(2^n) \wedge MSp \xrightarrow{\sigma_l \wedge i \wedge 1} X(2^n) \wedge MSp \wedge X(2^n) \wedge MSp \xrightarrow{1 \wedge T \wedge 1}$$

$$X(2^n) \wedge X(2^n) \wedge MSp \wedge MSp \xrightarrow{\mu \wedge \mu} X(2^n) \wedge MSp.$$

A direct verification shows that this map induces an isomorphism of homotopy groups and hence is a homotopy equivalence. Combining it with the splitting for $F_1(2^n) \wedge MSp$ we obtain the splitting

$$X(2^n) \wedge MSp \simeq \bigvee_{\omega} S^{|\omega|} MSp^{\Sigma_n}. \qquad (3.11)$$

As for the statement of the theorem for $F_{2k-1}(2^n) \wedge MSp$ we can repeat the same arguments for $F_{2k-1}(2^n) \wedge MSp$ as we used for $X(2^n) \wedge MSp$. In this case we will obtain the desired splitting of $F_{2k-1}(2^n) \wedge MSp$ as a subsplitting of (3.11). Because of the same reasons there is an analogous splitting of $F_{2k}(2^n) \wedge MSp$:

$$F_{2k}(2^n) \wedge MSp \simeq \bigvee_{l \leqslant k-1} S^{4l(2^n-1)} \wedge F_1(2^n) \wedge MSp \bigvee S^{4l(2^n-1)} \wedge X(2^n-1) \wedge MSp.$$

This proves (3.7) so we have proved the statement of our theorem for $m = 2^n$.

Now let m be such that $2^n < m < 2^{n+1}$. In this case $F_1(m)$ is obtained by attaching $D^{2(m+1)} \wedge X(m-1)$ to $X(m-1)$ with the help of a map $S^{2m-3} \longrightarrow X(m-1)$. Because of dimensional arguments and the induction hypothesis this map is trivial. This implies that the element b_{m-1} and all its powers are permanent cycles in the Atiyah-Hirzebruch spectral sequence for $X(m)_* MSp$. So we can obtain the splittings (3.5), (3.6), (3.7) by the same means as before. □

Corollary 3.12 *The spectrum* $\bigvee_\omega S^{|\omega|} MSp^{\Sigma_n} = F_1(2^n) \wedge MSp$ *is a Thom spectrum.*

Remark 3.13 *It would be interesting to find a wedge of suspensions of* MSp^{Σ_n} *with less summands then in the above corollary which is still a Thom spectrum.*

Corollary 3.14 *After smashing the filtration*

$$F_1 \subset \ldots \subset F_{2k-1}(m) \subset \ldots \subset X(m)$$

with MSp we obtain a filtration of the wedge

$$\bigvee_{\omega} S^{|\omega|} MSp^{\Sigma_n}$$

by subwedges with an increasing number of summands.

References

[V] V. V. Vershinin, *Symplectic cobordism with singularities,* Math. USSR Izvestiya **22** (1984).

[V,G] V. V. Vershinin and V. G. Gorbunov, *Ray's elements as obstructions to orientibility of symplectic cobordism,* Soviet Math. Dokl **32** (1985).

[R] N. Ray, *Indecomposable in* MSp_*, Topology **10** (1970), 261-270.

[D,H,S] E. S. Devinatz, M. J. Hopkins and J. H. Smith,*Nilpotence and stable homotopy thery I,*Ann. of Math. **128** (1988), 207-242.

[Rav] D. S. Ravenel, *Localization with respect to certain periodic homology theories,* Amer. J. Math. **106** (1984), 351-414.

[Sw] R. M. Switzer, *Algebraic topology, homotopy and homology,* Springer-Verlag, Berlin and New York, 1976.

Topological gravity and algebraic topology

Jack Morava

Introduction

This paper is an account, directed toward algebraic topologists, of Ed Witten's topological interpretation of some recent beakthroughs in what physicists call two-dimentional gravity; it is thus a paper about applications of algebraic topology. That algebraic topology indeed *has* applications was the thesis of Solomon Lefschetz's last book [20], which is an introduction to Feynman integrals. Today, the singularities which so concerned Lefschetz are avoided by string theory, which deals in a very clean and direct way with the space of moduli of Riemann surfaces; and I suspect that he would have been fascinated and delighted by the elegance of the problems which now interest mathematical physicists.

In a sense, the topic of this paper in the infinite loop-spectrum defined by Graeme Segal's category (with circles as objects, and bordered Riemann surfaces as morphisms, [2,33]); but by current standards, this is an immense object, related to the theory of Riemann surfaces as the algebraic K-spectrum is to linear algebra, and much current research is devoted to finding some aspect of this elephant susceptible to analysis by means of the investigator's favorite tool. In the present case, the subject will be the construction of topological field theories [3], using complex cobordism [1].

I first heard of these developments from a talk by Witten in the IAS topology seminar, and I learned more about them during subsequent conversations with him, with Segal, and with Peter Landweber. At a conference last May I benefitted from tutorials by Peter Bouwknecht, Greg Moore, Nathan Seiberg and Steve Shenker on one hand, and Yukihiko Namikawa, Yuji Shimizu, Akihiro Tsuchiya and Kenji Ueno on the other. I would like to acknowledge helpful conversations at this Symposium with Carl-Friedrich Bödigheimer, Jim McClure, Haynes Miller, and Nigel Ray, while at Johns Hopkins I owe debts to George Kempf, Bernard Shiffman and Steve Zucker for help with the algebraic geometry. I owe all these and many others thanks for sharing their insights and excitement with me, and (along with all the participants here) I want to thank the Manchester School of Topology for the opportunity to attend this meeting.

There are four further sections to this paper; the first sketches a program, following [37, §3], for the construction of a $(1+1$-dimensional) topological field theory defined on a class of closed strings flying around in a compact complex manifold. [For a recent account of $2+1$-dimensional field theories directed at topologists, cf. [7].] The second section summarizes standard facts about complex cobordism and about the moduli space of curves, and contains the main construction, while the third section contains the main result. The point of the story is that there is an interesting homomorphism, analogous to the Chern character of K-theory, which maps the stable rational homology of the moduli space of curves to complex bordism made rational. It presumably has an interpretation as a morphism from the spectrum defined by Segal's category to $MU \otimes \mathbf{Q}$, although I have not gone into that here. The final section is essentially a list of suggestions for further reading.

§1. A (conjectural) topological field theory

1.0 To begin, we consider a category $\mathcal{C}$ having ordered finite sets as objects; a morphism $C_0 : F' \longrightarrow F$ is to be a stable complex algebraic curve C_0, not necessarily connected, together with an embedding of $F' + F$ as a set of smooth points of C_0. If $C_1 : F \longrightarrow F''$ is another such morphism, then gluing C_0 and C_1 along F (cf. [18]) defines a morphism

$$C_0 \cup_F C_1 : F' \longrightarrow F''.$$

We obtain by this construction a monoidal category [39], with disjoint union as the composition operation. We will be interested in monoidal functors

$$V : \mathcal{C} \longrightarrow (\mathbf{C} - \text{vectorspaces})$$

which send disjoint union to tensor product. We might also endow $\mathcal{C}$ with extra data, such as a spin or level structure.

1.1 More generally, let X be a projective algebraic variety; we will eventually want it to be a rational homology manifold. We consider the category $\mathcal{C}/X$ having as objects, finite ordered sets mapped to X; a morphism from F' to F in $\mathcal{C}/X$ is to be a holomorphic map $\phi : C \longrightarrow X$, together with an embedding of $F' + F$ in C such that the diagram

$$\begin{array}{ccc} F' + F & \hookrightarrow & C \\ & \searrow \quad \swarrow \phi & \\ & X & \end{array}$$

commutes. As before, $\mathcal{C}/X$ is a monoidal category under disjoint union. It will be important in what follows, that there is a scheme $H(C)$ of holomorphic maps from C to X, with quasi-projective components.

1.2 We will need the following

Lemma: Let F, C_0, and C_1 be as above, and let $H_0(C_0 \cup_F C_1)$ be a component of the space of holomorphic maps from $C_0 \cup_F C_1$ to X. Then

$$\begin{array}{ccc} H_0(C_0 \cup_F C_1) & \longrightarrow & H_0(C_0) \\ \downarrow & & \downarrow \\ H_0(C_1) & \longrightarrow & X^F \end{array}$$

is a fiber product square; where $H_0(C_0)$, $H_0(C_1)$ are the components of $H(C_0)$, $H(C_1)$ respectively, containing the images of $H_0(C_0 \cup_F C_1)$ under restriction. ■

1.3 The fact that these spaces of morphisms are not compact in general is the source of technical difficulties. We propose a ruse to evade them, by a perhaps irrelevant compactification, as follows. If C is a stable curve, which we may as well now assume connected, let D be its set of double points, and let $\mathrm{Hil}(C \times X)$ denote the Hilbert scheme [26] of closed subvarieties of $C \times X$. If F is a finite set of smooth points on C, let

$$H(C) \xrightarrow{gr \times ev} \mathrm{Hil}(C \times X) \times X^{D+F}$$

denote the product of the graph embedding of $H(C)$ in $\mathrm{Hil}(C \times X)$, with the evaluation map from $H(C)$ to X^{D+F}. This is an immersion with locally closed image, and by $H(C)^c_F$ we denote its closure. The diagram

$$\begin{array}{ccc} H(C_0 \cup_F C_1) & \longrightarrow & \mathrm{Hil}(C_0 \cup_F C_1) \times X) \times X^{F'+D+F''} \\ \downarrow & & \downarrow \\ H(C_0) & \longrightarrow & \mathrm{Hil}(C_0 \times X) \times X^{F+D_0+F'} \end{array}$$

(where D is the set of double points of $C_0 \cup_F C_1$, D_0 is the double points of C_0, etc.) defines a morphism

$$H(C_0 \cup_F C_1)^c_{F'+F''} \longrightarrow H(C_0)^c_{F+F'},$$

and thus a canonical map

$$H(C_0 \cup_F C_1)^c_{F'+F''} \longrightarrow H(C_0)^c_{F'+F} \times_{X^F} H(C_1)^c_{F+F''}.$$

On the other hand, the diagram

$$\begin{array}{ccc} H(C_0)^c_{F'+F} \times_{X^F} H(C_1)^c_{F+F''} & \longrightarrow & H(C_0 \cup_F C_1)^c_{F'+F''} \\ \downarrow & & \downarrow \\ (*) & \longrightarrow & \mathrm{Hil}((C_0 \cup_F C_1) \times X) \times X^{F'+D+F''} \end{array}$$

(where $(*) = \mathrm{Hil}(C_0 \times X) \times X^{F'+D_0+F} \times_{\mathrm{Hil}(F \times X)} \mathrm{Hil}(C \times X) \times X^{F+D_1+F''}$), defines a natural candidate for an inverse map.

1.4 *Definition* If $\phi : C \longrightarrow X$ is a holomorphic map as in 1.1, let

$$v(\phi, F) = \mathrm{pr}_*[H_\phi(C)^c_F] \in H^*(X^F)$$

denote the image of the class of the component $H_\phi(C)^c_F$ containing ϕ of $H(C)^c_F$, under the projection $\mathrm{Hil}(C \times X) \times X^F \longrightarrow X^F$. [A physicist would probably sum over these components, weighting them by a coupling constant u analogous to an inverse temperature, according to their codimension.]

In view of the preceding discussion, the

Assertion: Let $F' + F \subset C_0 \xrightarrow{\phi_0} X$, $F + F'' \subset C_1 \xrightarrow{\phi_1} X$ be morphisms in $\mathcal{C}/X$; then

$$v(\phi_0 \cup_F \phi_1, F' + F'') = \mathrm{pr}^F_*(v(\phi_0, F' + F) \otimes_{H^*(X^F)} v(\phi_1, F + F'')),$$

where $\mathrm{pr}^F : X^{F'+F+F''} \longrightarrow X^{F'+F''}$ is the canonical projection,

seems very likely. ∎

1.5 Poincaré duality defines an isomorphism

$$\mathcal{D}^{F'}_F : H^*(X^{F'+F}) \xrightarrow{\sim} \mathrm{Hom}_{\mathbf{Q}}(H^*(X^{F'}), H^*(X^F)).$$

In this language, the preceding assertion (true when the mapping spaces are compact) becomes the

Conjecture: The correspondence which assigns to the morphism $F' + F \hookrightarrow C \xrightarrow{\phi} X$ of $\mathcal{C}/X$, the homomorphism

$$\mathcal{D}^{F'}_F v(\phi, F' + F) : H^*(X^{F'}) \longrightarrow H^*(X^F),$$

defines a monoidal functor from the category $\mathcal{C}/X$ to the category of $\mathbf{Z}/2\mathbf{Z}$-graded $\mathbf{C}$-vector spaces.

§2 Jacobian homomorphisms

2.0 We recall some constructions from the theory of Riemann surfaces [6]:

$$M_g = \mathcal{J}_g/\Gamma_g$$

is the (complex analytic) space of moduli of smooth Riemann surfaces, of genus $g > 1$. Here $\mathcal{J}_g$ is the Teichmüller space, isomorphic to an open cell in $\mathbf{C}^{3(g-1)}$, upon which the mapping-class group

$$\Gamma_g = \pi_0(\mathrm{Diff}^+ S_g)$$

acts properly discontinuously; where $\mathrm{Diff}^+ S_g$ is the group of orientation-preserving diffeomorphisms of a surface of genus g. [It is a group with contractible components.] The action of Γ_g on $\mathcal{J}_g$ has fixed points, with finite isotropy groups, so M_g is a rational homology manifold.

Let $\mu_g : C_g \to M_g$ denote the family of complex curves parameterized by M_g; it is topologically almost a bundle, with fiber S_g. Some closely related spaces have been studied by Deligne, Mumford, and Knudsen; the first of these is a compactification $\mathbf{M}_g$ of M_g, the points at the boundary being stable curves (i.e., with only ordinary double point singularities, and such that each irreducible component of genus zero meets the rest of the curve in at least three points); we write

$$\boldsymbol{\mu}_{g,1} : \mathbf{M}_{g,1} \longrightarrow \mathbf{M}_g$$

for the corresponding universal family. There is also the n-fold fiber product

$$\mu_g^n : C_g^n := C_g \times_{M_g} \cdots \times_{M_g} C_g \longrightarrow M_g$$

of copies of the universal smooth curve over itself, which can be interpreted as the moduli object for data consisting of a smooth curve, together with n ordered (but not necessarily distinct) points upon it; and finally we write

$$\boldsymbol{\mu}_{g,n} : \mathbf{M}_{g,n} \longrightarrow \mathbf{M}_g$$

for the space of stable curves, with n ordered, distinct, smooth points marked upon them.

2.1 The action of the mapping-class group Γ_g on $H^*(S_g, \mathbf{Z})$ defines a surjection

$$\Gamma_g \longrightarrow Sp(2g, \mathbf{Z});$$

since M_g has the rational homotopy type of $B\Gamma_g$, this defines a homomorphism

$$H^*(B\,Sp(2g, \mathbf{Z}), \mathbf{Q}) \longrightarrow H^*(M_g, \mathbf{Q})$$

which is known by results of Harer and Miller [21] to be injective as $g \to \infty$. More precisely, the unitary group $U(g)$ is the maximal compact subgroup of $Sp(2g, \mathbf{R})$, and the resulting homomorphism

$$\mathbf{Q}[c_1, \ldots, c_g] \cong H^*(BU(g)) \longrightarrow H^*(B\,Sp(2g, \mathbf{R})) \longrightarrow H^*(M_g)$$

maps the classes c_{2i-1} injectively, when $2i - 1 < \frac{1}{3}g$. The class c_{2i} is the Pontrjagin class of a flat (i.e. $Sp(2g, \mathbf{Z})$) bundle and is thus rationally trivial, but there is a larger polynomial subalgebra of $H^*(M_g, \mathbf{Q})$, of classes κ_i defined by Mumford, which will be constructed below, cf. [25].

2.2 We write $MU^*(-)$ for the complex cobordism functor, but otherwise we follow the conventions of Quillen [30]: an element of $MU^{2q}(X)$ is to be interpreted as the cobordism class of a *proper* complex-oriented smooth map $Y \to X$, of codimension $2q$, between smooth manifolds. Similarly, we will interpret

$$MU^*_{\mathbf{Q}}(-) := MU^*(-) \otimes_{\mathbf{Z}} \mathbf{Q}$$

as cobordism of rational homology manifolds. This construction is contravariantly functorial with respect to smooth maps, and covariantly functorial with respect to proper complex-oriented smooth maps. The Thom homomorphism

$$\rho : MU^*(X) \longrightarrow H^*(X, \mathbf{Z}),$$

assigns to such a cobordism class, the (Poincaré dual of) its underlying cycle.

2.3 We also recall, very briefly, the basic facts about the Landweber-Novikov Hopf algebra $S_* = \mathbf{Z}[\,t_i \mid i \geq 1\,]$ of stable MU_*-cooperations, cf. [19,23]. If $f : Y \to X$ is a class as above, define

$$s_t[f] = \sum s_A[f]t^A = \sum f_* cf_A(\nu(f))t^A \in MU^{2q}(X)[t_i \mid i \geq 1],$$

where

$$\nu(f) = f^*T_X - T_Y$$

is the formal normal bundle of f,

$$cf_t(E) = \sum cf_A(E)t^A$$

denotes the total Conner-Floyd Chern class of the complex vector bundle E, in which

$$t^A = t_1^{a_1} \dots t_n^{a_n}$$

is a monomial in indeterminates t_i (of cohomological degree $-2i$) associated to the multiindex $A = (a_1, \dots, a_n)$, $a_i \geq 0$, and f_* is the covariant (Gysin) homomorphism defined by the complex-oriented proper map f. Composition of the Landweber-Novikov operations is given by

$$s_u \circ s_t = s_{t \circ u},$$

where

$$t(X) = \sum_{i \geq 0} t_i X^{i+1},$$

$$u(X) = \sum_{i \geq 0} u_i X^{i+1},$$

and

$$(t \circ u)(X) = t(u(X)),$$

with the convention that $t_0 = u_0 = 1$ (so $(t \circ u)_0 = 1$). However, there is a relevant enlargement

$$\tilde{S}_* = \mathbf{Z}[t_i \mid i \geq 1][t_0, t_0^{-1}]$$

of the Landweber-Novikov algebra to a family of mildly unstable operations, defined analogously by series $t(X)$ under composition, with t_0 required only to be a unit. [We take the total Conner-Floyd Chern class of a line bundle to be

$$cf_t(L) = \sum_{i \geq 0} t_i cf_1(L)^i,$$

and proceed as before.]

We will need one further remark. In the geometric constructions of §3, we will need operations

$$\check{s}_t[f] = \sum \check{s}_A[f] t^A = \sum f_* cf_A(-\nu(f)^*) t^A,$$

in which the normal bundle to f is replaced by the cotangent bundle. These operations can be related to the usual Landweber-Novikov operations by universal formulae in $MU^*(BU)$, in which the Chern classes of the universal bundle are expressed in terms of the Chern classes of its inverse dual. These upside-down and backwards operations behave very much like their usual counterparts.

2.4 *Definition:* If u is an indeterminate of cohomological degree +2, then the functor $MU^*_{\mathbf{Q}}(-)[u]$ on finite complexes extends to a cohomology functor, for which we will use the same symbol. Let

$$jh(u) = \sum_{k \geq 0} \prod_{g \geq 0} \mu_g^k \frac{u^k}{k!} \in MU^0_{\mathbf{Q}}(\sqcup_{g \geq 0} M_g)[u],$$

[where M_g for $g = 0, 1$ are as in [29;37 §2]]; we will call this element the *Jacobi character.*

[The kth symmetric power [40]

$$\sigma^k(f) = (\overbrace{Y \times_X \cdots \times_X Y}^{k\text{times}})/\Sigma_k \longrightarrow X$$

defines an interesting operation in rational cobordism, and if we write

$$\sigma_u(f) = \sum \sigma^k(f) u^k,$$

then

$$\mathrm{jh}(u) = \prod_{g \geq 0} \sigma_u(\mu_g)$$

is essentially the cobordism class defined by the classical Jacobian of the modular curve, cf. [17].]

In another direction, let

$$\mathbf{jh}(u) = \sum_{n \geq 0} \prod_{g \geq 0} \mu_{g,n} \frac{u^n}{n!} \in MU^0_{\mathbf{Q}}(\sqcup \mathbf{M}_g)[u],$$

where

$$\mu_{g,n} : \mathbf{M}_{g,n} \longrightarrow \mathbf{M}_g$$

is defined using the compactification of Mumford, Deligne, and Knudsen, as in §2.0. The open embedding

$$i : M_g \longrightarrow \mathbf{M}_g$$

defines a homomorphism

$$i^* : MU^*(\mathbf{M}_g) \longrightarrow MU^*(M_g),$$

but the relations between $i^*\mathbf{jh}(u)$, $jh(u)$, and $\mathrm{jh}(u)$ are complicated enough without bringing up the Artin-Hasse exponential.

§3 The main result

3.0 On cohomology, our generalized Jacobian character represents a homomorphism

$$\mathbf{jh}^* : H^*(\mathbf{MU}, \mathbf{Q}) \longrightarrow H^*(\sqcup \mathbf{M}_g, \mathbf{Q}[u]).$$

It will be convenient to use the symbol

$$x \longmapsto \langle 0 \mid x \mid 0 \rangle \; : \; H^*(\mathbf{M}_g, \mathbf{Q}) \longrightarrow \mathbf{Q}[u]$$

for the quantity

$$\langle 0 \mid x \mid 0 \rangle = x[\mathbf{M}_g] u^{3(g-1)}$$

defined by evaluating x on the fundamental class of $\mathbf{M}_g$. We also recall that the Thom isomorphism

$$\Phi : H^*(BU, \mathbf{Q}) \longrightarrow H^*(\mathbf{MU}, \mathbf{Q})$$

makes $H^*(\mathbf{MU}, \mathbf{Q})$ into a free module over the algebra of Chern classes.

3.1 With this notation, we can state the following

Theorem:

1) The homomorphism induced by $\mathbf{jh}$ on cohomology is injective, and its $\mathbf{Q}[u]$-linear extension to a homomorphism

$$\mathbf{jh}^* : H^*(\mathbf{MU}, \mathbf{Q}[u, u^{-1}]) \longrightarrow H^*(\sqcup_{g \geq 0} \mathbf{M}_g, \mathbf{Q}[u, u^{-1}])$$

has the Mumford-Miller-Morita algebra $\mathbf{Q}[\kappa_i \mid i \geq 0][u, u^{-1}]$ as image.

2) The formal power series

$$F = \langle 0 \mid \mathbf{jh}^*(\check{s}_t \Phi(1)) \mid 0 \rangle \in \mathbf{Q}[[u^i t_i \mid i \geq 0]]$$

is the topological free energy of Witten.

Proof: In both cases, the behavior of the Landweber-Novikov operations on the Jacobian homomorphism is central. We begin by examining

$$\check{s}_t(\exp(u\mu_g)) \in MU^*_{\mathbf{Q}}(M_g)[u][t_i \mid i \geq 0];$$

since the Landweber-Novikov operations are multiplicative, this is just exp $(u\, \check{s}_t(\mu_g))$. Now

$$\check{s}_t(\mu_g) = \sum_{i \geq 0} t_i \mu_{g*} cf_1(T^*)^i,$$

where T^* is the cotangent bundle along the fiber of μ_g, so

$$\check{s}_t(\exp(u\mu_g)) = \sum_{m \geq 0} \frac{u^m}{m!} \mu^m_{g*} \left(\sum_{i \geq 0} t_i cf_1(T^*)^i \right)^m,$$

with μ_g^m being the map of the m-fold fiber product of the universal curve with itself, to M_g.

Let

$$i_1 + \cdots + i_m = \sum_{k \geq 0} k r_k,$$

where r_k is the number of occurrences of the integer k in the sequence $i_1, \ldots, i_m$. Then

$$\check{s}_t(\exp(u\mu_g)) = \sum_{m \geq 0} \frac{u^m}{m!} \mu_{g*}^m \left(cf_1(T_1^*)^{i_1} \cdots cf_1(T_m^*)^{i_m} \right) \prod t_k{}^{r_k},$$

with the understanding that T_m^* is the cotangent space at the mth marked point on the modular curve. Now the classes

$$k_R^{(g)} := \mu_{g*}^m \left(cf_1(T_1^*)^{i_1} \cdots cf_1(T_m^*)^{i_m} \right), \quad R = (r_0, r_1, \ldots,),$$

are independent of the order of the marked points, and we conclude that

$$\begin{aligned} \check{s}_t(\exp(u\mu_g)) &= \sum_{m \geq 0} \frac{u^m}{m!} \begin{bmatrix} m \\ r_0, \ldots, r_h \end{bmatrix} k_R^{(g)} \prod_{k \geq 0} t_k^{r_k} \\ &= \sum k_R^{(g)} \gamma_R(ut), \end{aligned}$$

where

$$\gamma_R(ut) = \prod_{k \geq 0} \frac{(ut_k)^{r_k}}{r_k!}$$

is a multi-divided power.

It follows that

$$jh^*(\Phi c_R) = jh^*(s_R \Phi(1)) = \rho\left(s_R(jh(u))\right) = \kappa_R \frac{u^{|R|}}{R!}$$

where $|R| = \sum r_i$, $R! = \prod r_i!$, $\|R\| = \sum (i-1) r_i$, and

$$\kappa_R := \rho\left(\prod_{g \geq 0} k_R^{(g)} \right) \in H^{2\|R\|}\left(\sqcup_{g \geq 0} M_g, \mathbf{Q} \right).$$

Since

$$\rho \mu_{g*}^m \left(cf_1(T^*)^{i_1} \ldots cf_1(T^*)^{i_m} \right) = \prod_{k \geq 0} \mu_{g*} c_1(T^*)^{i_k},$$

we have

$$\kappa_R = \prod_{i \geq 0} \kappa_{i-1}{}^{r_i}$$

in Mumford's notation.

The second assertion follows similarly, for the same combinatorics show that

$$\langle 0 \mid \mathbf{jh}^*(\check{s}_t\Phi(1)) \mid 0\rangle = \sum \tau_R[\mathbf{M}_g]\gamma_R(\tilde{t}),$$

where $\tilde{t}_i = u^i t_i$ and $\tau_R = \mu_{g,m*}\left(\prod_{k\geq 0} c_1(T_k^*)^{i_k}\right)$ is as in Witten. ∎

3.2 The (co)algebra of Landweber-Novikov operations defined in §2.3 can be identified, over the rationals, with the enveloping algebra of the Lie algebra generated by the operations

$$v_k := \partial_k s_t \mid_{t_i=0},$$

where $\partial_k = \partial/\partial t_k$, $k \geq 0$; since the Landweber-Novikov algebra is the affine algebra of functions on the group scheme of formal automorphisms of the line, preserving the origin, we have the Lie bracket

$$[v_k, v_l] = (k-l)v_{k+l};$$

thus a comodule over the Landweber-Novikov operations is in a natural way a module over half of the Virasoro algebra. For example, on the natural geometric generators $\mathrm{CP}(n)$ of $MU^*_{\mathbf{Q}}(\mathrm{pt})$ we have

$$v_k\, \mathrm{CP}(n) = (n+1)\mathrm{CP}(n-k).$$

[Note that this is an action by derivations, unlike the standard charge one represention of the Virasoro algebra, cf. [15,32].] Similary, the v_k act on the t_i's by

$$v_k = \sum_{i\geq 0}(i+k+1)t_i\partial_{i+k}.$$

With this notation, we can state the

Corollary: If $\sum(i_k - 1) = 3(g-1)$, then

$$\langle 0 \mid v_{i_k} \dots v_{i_1}\Phi(1) \mid 0\rangle = (\tau_{i_n-1} \dots \tau_{i_1-1})[\mathbf{M}_g].$$

This provides one mathematical interpretation for the contact algebra studied by E. and H. Verlinde [35] and Dijkgraaf [10].

§4. Recent and future developments

4.0 Motivated by work on the Hermitian matrix model of two-dimensional gravity [4,5,8,12,13], Witten has shown that the free energy F of Proposition 3.1 is a formal solution to the Korteweg-deVries hierarchy of partial differential equations (cf. [9,11,27,34]); in particular he proves that this conjecture follows from the related conjecture, that the formal partition function $Z = \exp(F)$ is a highest weight vector for a representation of the full Virasoro algebra (cf. [16,38]). I don't have much to say about these ideas here, but it is tempting to note their obvious analogy with Quillen's Theorem (that the complex cobordism of $\mathbf{CP}(\infty)$ is a universal one-dimensional formal group): a complicated topological object (the rational cohomology of the moduli space of stable curves) is characterized by a surprising isomorphism with an apparently unrelated algebraically defined object.

4.1 Now in [37, §3c], Witten generalizes some of these notions to a context encompassing both two-dimensional gravity and nonlinear sigma-models. In terms of the constructions sketched in this paper we can do something similar: if, as in §1, X is a projective variety which is a rational homology manifold, and if $F \subset C \xrightarrow{\phi} X$ is a holomorphic map, then it becomes natural to ask if the evaluation morphism

$$H_\phi(C)_F^c \to X^F$$

defines a rational complex cobordism class, satisfying the analogue of Proposition 1.5. (Since MacPherson [22] has constructed Chern classes for general singular varieties, this seems not unreasonable.) This then suggests that the topological field theory sketched in §1 has an extraordinary analogue, taking values in the category of graded Landweber-Novikov comodules.

4.2 It seems that the Landweber-Novikov algebra, the Virasoro algebra, and (as I have learned from Nigel Ray and Bill Schmitt at this conference) the Faa di Bruno algebra of combinatorics, are all deeply related, in ways which we are just beginning to comprehend [24]. I have no idea what Frank would have said about this, and it is a continuing source of grief not to be able to have a quiet talk with him about it all. However, I take comfort in the idea that somewhere he and Lefschetz are running a lively seminar on related topics.

The Johns Hopkins University

1. J.F. Adams, *Stable homotopy and generalized homology*, University of Chicago Press (1974).
2. J.F. Adams, *Infinite loopspaces*, Annals of Math. Studies 90, Princeton University Press (1978).
3. M.F. Atiyah, Topological quantum field theories, Publ. Math. IHES 68 (1988) 175-186.
4. T. Banks, M. Douglas, N. Seiberg, S. Shenker, Microscopic and macroscopic loops in nonperturbative two-dimensional quantum gravity, Rutgers University preprint RU-89-50.
5. D. Bessis, C. Itzykson, J.B. Zuber, Quantum field theory techniques in graphical enumeration, Adv. App. Math. 1(1980) 109-157.
6. C.F. Bödigheimer, *On the topology of moduli spaces*, Habilitationschrift, Göttingen 1990.
7. S. Cappell, R. Lee, E. Miller, Invariants of 3-manifolds from conformal field theory, preprint, Courant Institute (1990).
8. C. Crnković, R. Ginsparg, G. Moore, The Ising model, the Yang-Lee edge singularity, and $2D$ quantum gravity, Yale preprint YCTP-P20-89 (1989).
9. E. Date, M. Kashiwara, M. Jimbo, T. Miwa, Transformation groups for soliton equations, in *Proceedings of RIMS Symposium on nonlinear integrable systems*, ed. M. Jimbo, T. Miwa, World Science Publishing, Singapore 1983.
10. R. Dijkgraaf, E. Verlinde, H. Verlinde, Loop equations and Virasoro constraints in non-perturbative $2D$ quantum gravity, IAS preprint HEP-90-48.
11. I. Frenkel, Representation of affine Lie algebras, Hecke modular forms and KdV-type equations, in Lecture Notes in Math 933 (1982) 71-110.
12. M. Fukuma, H. Kawai, R. Nakayama, Continuum Schwinger-Dyson equations and universal structures in two-dimensional quantum gravity, Univ. of Tokyo, preprint UT-562 (1990).
13. D. J. Gross, A.A. Migdal, A nonperturbative treatment of two-dimensional quantum gravity, Phys. Rev. Lett. 64(1990) 127.
14. S.A. Joni, G.C. Rota, Coalgebras and bialgebras in combinatorics, Studies in Applied Math. 61(1979) 93-139.
15. V.Kac, *Infinite dimensional Lie algebras* (2nd ed., 1985), Cambridge University Press.
16. V. Kac, A. Schwarz, Geometric interpretation of partition function of $2D$ gravity, MIT preprint, 1990.
17. G. Kempf, Weighted partitions and patterns, J. of Algebra 85(1983) 535-564.
18. F. F. Knudsen, The projectivity of the moduli space of stable curves II, III, Math. Scand. 52 (1983) 161-199; 200-212.

19. P. Landweber, Cobordism operations and Hopf algebras, Trans. AMS 129(1967), 94-110.
20. S. Lefschetz, *Applications of algebraic topology*, Applied Mathematical Sciences 16, Springer (1975).
21. E. Miller, The homology of the mapping-class group, J. Diff. Geo. 24(1986).
22. R.D. MacPherson, Chern classes for singular algebraic varieties, Ann. Math. 100(1974) 423-432.
23. J. Morava, The cobordism ring as a Fock representation, in *Homotopy theory and related topics*, ed. M. Mimura, Lecture Notes in Math. 1416, Springer.
24. J. Morava, Landweber-Novikov cooperations, the Faa di Bruno coalgebra, Segal Γ-spaces, Dyer-Lashof operations, moduli of curves, the Virasoro algebra, quantum groups, and Tannakian categories, preprint, Johns Hopkins, 1990.
25. S. Morita, Characteristic classes of surface bundles, Invent. Math. 90 (1987) 551-570.
26. D. Mumford, *Lectures on curves on an algebrac surface*, Annals of Math. Studies 59, Princeton (1966).
27. D. Mumford, *Tata Lectures on Theta* II, Birkhäuser, Progress in Math. 43 (1984).
28. D. Mumford, Towards an enumerative geometry of the moduli space of curves, in *Arithmetic and Geometry*, ed. M. Artin, J. Tate; Birkhäuser (1983).
29. Takayuki Oda, Etale homotopy type of the moduli spaces of algebraic curves, preprint, Johns Hopkins 1989.
30. D. Quillen, Elementary proofs of some properties of cobordism using Steenrod operations, Adv. in Math. 7(1971)29-56.
31. N. Ray, W. Schmitt, to appear.
32. G. Segal, Unitary representations of some infinite-dimensional groups, Comm. Math. Phys. 80(1981) 301-342.
33. G. Segal, *The definition of conformal field theory*, preprint, Oxford (1989).
34. G. Segal, G. Wilson, Loop groups and equations of KdV-type, Publ. Math. IHES/no. 61(1985) 5-65.
35. E. Verlinde, H. Verlinde, A solution of topological two-dimensional gravitry, IAS preprint HEP-90-40.
36. E. Witten, On the structure of the topological phase of two-dimensional gravity, IAS preprint (1989), Nuc. Phys B., to appear.
37. E. Witten, Two-dimensional gravity and intersection theory on moduli space, IAS preprint HEP 90/45 (1990).
38. H. Yamada, The Virasoro algebra and the KP hierachy, in *Infinite dimensional groups with applications*, ed. V. Kac, MSRI publications

4(1985), pp. 371-380.

39. D. Yetter, Quantum groups and representations of monoidal categories, Math. Proc. Camb. Phil. Soc. 108(1990) 261-290.
40. D. Zagier, *Equivariant Pontryagin classes and applications to orbit spaces*, Lecture Notes in Math. 190, Springer 1972.